Cashew Production and Processing Technology

Cashew Production and Processing Technology (Recent Advances)

R. C. Gajbhiye
B. V. Padhiar

NEW INDIA PUBLISHING AGENCY
New Delhi – 110 034

NEW INDIA PUBLISHING AGENCY
101, Vikas Surya Plaza, CU Block, LSC Market
Pitam Pura, New Delhi 110 034, India
Phone: + 91 (11) 27 34 17 17 Fax: + 91 (11) 27 34 16 16
Email: info@nipabooks.com
Web: www.nipabooks.com

Feedback at feedbacks@nipabooks.com

ISBN: 978-93-83305-82-7

Composed, Designed and Printed in India

Preface

The field of horticulture is very wide and is divided into Fruit Science (Pomology); Vegetable Science (Olericulture); Floriculture and Landscape architecture; Post Harvest Technology; Plantation, Spices, Medicinal and Aromatic Crops. The cashew nut (*Anacardium occidentale* L.) belongs to the Anacardiaceae family. It has a long history as a useful tree. Cashew is grown for its kernels along with cashew shell nut liquid (CSNL) and cashew apple, and today assumed a status of important dollar earning horticultural crop in India. It is native to Eastern Brazil and introduced into India by the Portuguese during the 16th Century which is one of the important gifts of the 'New World' to the 'Old World'. Cashew is also called as poor man's crop but rich man's food.

Small-scale local exploitation of cashew for its nuts and cashew apples is the pattern more than 300 years in Asia and Africa. It was not until early years of 20th Century most International trade in cashew kernels began with first export became an important commercial plantation crop in the last couple of years. In India, use of cashew apples and nuts was started by local people making cashew wine/liquor possessing a common practice in both Asia and Africa.

India is the country that nourished this crop and made it a commodity of International trade and acclaimed. Initially cashew plants were seen in and around Malabar region of Kerala and Goa in West Coast. Cochin served as dispersal point for the cashew in India perhaps SE Asia. Spread of this plant was mostly by natural dispersal and introduced in various parts of country for forestation and soil conservation. Presently, cashew plants in semi-wild state and in well managed orchards are seen in Maharashtra, Goa, Karnataka and Kerala along the West Coast, Tamil Nadu, Andhra Pradesh, Orissa and West Bengal on the East coast. For the increase in area and cashew production, the attempts were made for the collection of locally available seed material by the AICRP centres of the respective states viz. Kerala, Karnataka, Tamil Nadu, Andhra Pradesh, Maharashtra and Gujarat. As cashew is primarily a cross pollinated crop, it is heterozygous. Considerable segregation has resulted in large variation in the populations. Later on, only vegetatively propagated material is being collected and conserved in the National Cashew Gene Bank (NCGB). So far, all together 527 cashew germplasm accessions were made and planted in cashew field gene bank at NRCC, Puttur for evaluation. Apart from indigenous collection, 23 are exotic collections of different countries.

Seedling vis-à-vis vegetatively propagated materials were made considering the economic characters such as high yield, bold nuts, cluster bearing, apple qualities, canopy types and dwarfness of the plant. However, more comprehensive exploration is required for targeted collection of germplasm with specific traits such as resistance to TMB, C.S.R.B., high yield dwarf and compact, bold nut with cluster bearing, tolerance to drought, frost, salt and to problematic soils. Directorate of Cashew Research, Puttur has initiated this work on this time.

Presently, Vietnam is the single largest producer of cashew nuts. Nigeria ranks second and India is the third largest producer. Production of raw cashew nuts has been fourfold increased in Vietnam and other countries. In the last couple of years, India was importing raw cashew nuts from these countries and export was made of processed kernels to the countries of Europe and U.S. The largest consumers of cashews are countries with a high capita income such as France, Germany, Italy, the United States, Canada and Spain. Considering the facts with cashew import and export, Government of India had established cashew Export Promotion Council of India in 1955, with the active cooperation of the cashew industry with the object of promoting exports of cashew kernels and CSNL from India. In India, Maharashtra state leads for highest collection (302). In all cashew growing states, though area of Maharashtra is low (1,81,000 ha), the productivity is recording 1186 kg/ha and ranked first.

We take an opportunity to express our sincere thanks to New India Publishing Agency, New Delhi for their interest to bring out this publication. We also thank our family members for their kind patience during the course of compilation and preparation of manuscript. We are also thankful to scientists and friends for their help and encourage.

R.C. Gajbhiye
B.V. Padhiar

Contents

Preface v

1. **Introduction** 1
   - Initiation of Cashew Research in India 3
   - Importance of Cashew Tree 3
   - Medicinal Uses of Cashew Apple 5

2. **History: Origin and Geographical Distribution** 7
   - Origin and Distribution 7
   - Spread of Cashew in India 8
   - Dispersion of Cashew to SE Asia and Australia 9
   - Related Species of Cashew 10
   - Genetic Diversity 11
   - Exploration and Collection 13
   - Nomenclature 15
   - Botanic Description 15

3. **Cashew Area, Production, Productivity, Export and Import** 17
   - Global Scenario 17
   - National Scenario 20
   - Export and Import 26

4. **Soil and Climate** 29
   - Soil 29
   - Climate 30
   - Effect of Abiotic and Biotic Factors on Cashew Flowering 35
   - Influence of Altitude on Cashew Flowering 36

5. Biodiversity and Crop Improvement 39
Related Species of Cashew 40
Genetic Diversity 41
Exploration and Collection 43
Germplasm Conservation 45
Germplasm Registration 46
Germplasm Augmentation Through Introduction From Other Countries 46
The Biological Diversity Act, 2002 47
Biotechnology in Cashew 48
Rootstocks 49
Scions 50
Cashew Gene Bank 57
Genetic Improvement of Cashew 58
Application of Tissue Culture to Cashew Breeding 58

6. Cashew Varieties 59
Dwarf Cashew 83
Purple Leaved Cashew 83
Drought Tolerant Cashew Varieties 84
Performance of Cashew Variety 84

7. Propagation and Production of Planting Material 85
Vegetative Propagation 86
Softwood Grafting and Nursery Management in Cashew 87
Effect of Gamma Radiation Treatments to Raw Nut on Germination, Seedling Growth and Establishment 92

8. Cashew Morphology and Tree Characteristics 95
Pattern of Growth Flushes in a Bearing Tree 95
Branching Habit 95
Leaf Structure 96
Canopy 97
Roots 98
Vegetative Flush 99
Inflorescence 99
Floral Biology 100
Structure of Flower 101
Pollination, Fertilization and Fruit Development 105
Fruit Setting 105
Structure of the Fruit 106
Factor Affecting the Flowering in Cashew 110

9. Pollination Technique in Cashew **115**

The Procedure for Pollination Technique Involves the Following Steps 117

10. Establishment of Cashew Orchards **121**

Selection of Site and Land Preparation 121
Planting Season 121
Spacing 121
High Density Planting 121
Digging of Pits and Planting of Grafts 122
After Care 123
Critical Management Practices 123

11. Integrated Nutrient Management in Cashew **125**

Mineral Nutrition of Cashew Plants Under Stress 125
Gas Exchanges in Cashew Plants Under Stress 127
Nutrient Management 128
Response to Nutrients 128
Application of Manures and Fertilizers 130
Response to Major Nutrients 132
Response to Micronutrients 134
Nutrient Removal 135
Nutrient Concentration in Leaf 138
Factors Deciding Fertilizer Requirements 138
Nutrient Recommendation 139
Nutrition in relation to age 139
Time of fertilizer application 140
Method of Fertilizer Application 140
Biofertilizers 142
Foliar Application of Growth Regulators, Pollination Attractants and Nutrients for Improving Productivity in Cashew 143
Mechanisms of Penetration of Nutrients into the Plant 156

12. Canopy Management in Cashew **167**

Principles of Canopy Management 168
Canopy Management in Cashew 171
Canopy Management in Productive Orchards 173
High Density Planting in Cashew 173
Ultra High Density Planting in Cashew 174
Pruning System for Ultra High Density 175
Potential Benefits of Ultra Density Planting 176
Photosynthetic Efficiency 176
Dried Branches and Weed Biomass 177

Average Root Density ... 177
Soil Moisture ... 177
Management of High Density Plantations ... 178
Nut Yield ... 178

13. Rejuvenation of Old Cashew Orchard ... 179
Rejuvenation of Old Trees by Pruning ... 181
Rejuvenation by Top Working ... 183
Conditions and Points to be Considered Before Top Working ... 183
Pest Management in the Rejuvenated Trees ... 184

14. Irrigation and Water Management in Cashew ... 185
Effect of Irrigation on Photosynthetic Rate of Cashew ... 186
Irrigation and Water Management for Cashew ... 187

15. Intercropping and Covercropping ... 191
Intercropping ... 191
Covercropping ... 193

16. Interculture Operation ... 195
Weed Management ... 195
Mulching ... 196
Protection of Cashew from Fire ... 196

17. Maturity, Harvesting, Yield and Post Harvest Handling ... 197
Maturity of Cashewnut ... 197
Harvesting of Nuts ... 198
Harvesting and Yield ... 198

18. Post Harvest Storage ... 203
Introduction ... 203
Precaution During Post Harvest Storage of Raw Nuts ... 204

19. Marketing of Raw Nuts ... 207
Marketing of Raw Cashewnuts ... 207
Role of BAIF under 'WADI' Yojna ... 208

20. Nutritive Value of Cashew Kernel and Apple ... 209
Cashew Kernels ... 209
Comparison of Cashew with other Tree Nuts ... 212
Value Addition in Cashew ... 216
Cashew Apple ... 217

21. Cashew Nut Processing 221
Brief History of Cashew Nut Processing Industry in India 221
Cashew Processing 225
Drying Raw Nuts 228
Storage of Raw Nuts 228
Cashewnut Processing 229
Stages in Cashew Processing 229
Roasting/Steaming 229
Drying and Cooling of Shelled Kernels 232
Processing at Orchard Level 242
Cashew Nut Shell Liquid (CNSL) 243
Market Potential 249
Cashew Kernel Products 251

22. Cashew Apple Processing 255
Composition of Cashew Apple 256
Uses of Cashew Apple 257
Fresh Consumption 257
Beverages 259
Use in Nutraceutical Extractions 263
Agricultural Uses 264
Animal Feeds 265
Pest Management 266
Energy Production and Industrial Uses 266
Cashew Apple Processing Issues 266
Suitability of Cashew Apples for Processing 266
Pre-requirement for Processing 267
Commercial Cashew Apple Products 267
Economics of Cashew Apple Processing 277

23. Pests and Diseases of Cashew 281
Tea Mosquito Bug (*Helopeltis antonii*) 281
Cashew Stem and Root Borer (*Placaederus ferrugineus* L.) 284
Leaf and Blossom Webber (*Lamida moncusalis* W.) 285
Apple and Nut Borer (*Thylocoptila panrosema* M.) 286
Flower Thrips (*Rhynchothrips raoensis, Scirtothrips dorsalis*) 286
Leaf Miner (*Acrocerops syngramma* M.) 286
Shoot and Inflorescence Tip Borer (*Hileria haligramma* M.) 287
Mealy Bugs 287
Aphids 287
Bark Eating Caterpillars 287
Nut Crinkler (*Paradasynus rostratus* Distant) 288
Mites 288

Diseases of Cashew 288
Damping off Seedlings 288
Incidence of Inflorescence Blight 291
Anthracnose (*Colletotrichum gloeosporiodes* Penzig.) 291
Die-back or Pink Disease 292
Inflorescence Blight 292
Leaf Blight 293
Gummosis 293
Powdery Mildew 293
Sooty Mould 293
Leaf Spots (*Pestalotia paconiae Servazzi, P. dichatae*) 293
Yellow Leaf Spot 294

24. Economics for Cashew Cultivation 297

25. Constraints, Technologies Developed, Future and Development Strategies and Innovation in Horticulture 299
Technologies Developed for High Production 299
Transfer of Technology 302
Future Strategy 302
Crop Improvement 302
Crop Management 303
Post-harvest Technology 303
Crop Protection 303
Developmental Strategies 304

References 307

CHAPTER – 1

Introduction

The cashew nut (*Anacardium occidentale* L.) belongs to Anacardiaceae family which also includes the mango, pistachio and poison ivy (Nakasone and Paull, 1998). *Anacardium* contains 8 species all of which are native to the coastal parts of north eastern Brazil (Azam-Ali and Judge, 2001). Cashew nut tree provides food, employment, income and the wood is used for carpentry mainly in the construction of boats and ferries while other products derived from it include firewood and charcoal (Akinwale, 2000). Cashew resins are also used in the manufacture of insect repellents and natural insecticides, while different tree parts have medicinal value as they are used to cure various diseases (Cavalcante *et al.*, 2003). The cashew tree is a sprawling broad-leafed evergreen, well adapted to poor soils and dry sandy locations, drought resistant, but grows best on well drained sandy soils with pH of 4.5 to 6.5 (Aliyu, 2004). It favours temperatures between 15 to 35°C and rainfall over 400 mm with elevation of up to 1000 m above sea level.

Flowering for cashew is similar to the close relative mango in that both male and perfect flowers are borne on the same inflorescence. Trees with more of the perfect flowers bear more fruits and the number of inflorescence produced per plant is one of the major factors determining productivity (Azevedo *et al.*, 1998).

To improve the productivity of the nuts there is need to select desirable genotypes from the existing gene pool and use superior materials for tree improvement programme. Genetic improvement is limited by the lack of knowledge of genetic diversity of the existing germplasm. Breeding of cashew is mostly based on selection of useful phenotypic and agronomic traits such as nut size, nut weight, colour of apple, size of the fruits, tree canopy, length of panicle and overall yield production (Mneney *et al.,* 2001). Traditionally, genetic diversity evaluated in crop species is based on differences in morphological characters and qualitative traits. In many cases morphological characters have been used as a powerful tool in the classification of cultivars and as such morphological traits continue to be the first step in the studies of genetic relationships in most breeding programmes. Therefore, it becomes imperative to quantify the extent of variation among the cashew germplasm in India by using morphological traits.

Cashew (*Anacardium occidentale* L.) is commercially grown for its kernels; although cashew nut shell liquid (CNSL) and apple are also valuable by products. It is considered to be the most important edible nut in the world trade (Ascenso,1986) cashew crop which has been considered only useful for soil conservation, afforestation and for development of waste lands, has today assumed a status of important dollar earning horticultural crop in India (Chadha, 1991)

Cashew is one of the important gift of the "New world" to the "Old world". It is a native of tropical America; however, it became commercially important in its adopted home i.e. in India, Tanzania and Mozambique. It was originally introduced into India by the Portuguese during the 16th Century.

Cashew is also called as poor man's crop but rich man's food. It is hardy and drought resistant and grown in a wide variety of soils under marginal conditions of management.

The first introduction of cashew in India was made in Goa from where it spread to other parts of the country. In the beginning it was mainly considered as a crop for afforestation and soil binding to check erosions. The raw cashewnut contains 20-30 percent kernel and 2-3 percent papery seed coat (testa) of the raw nut weight; the remaining 70-75 per cent is the shell. The cashew kernels are used in confectionery and dessert. The shells contain high quality oil known as cashewnut shell liquid (CNSL) which has got wide industrial uses. Cashew apple is eaten fresh or mixed in fruit salads and a drink is prepared from the juice. Cashew apple can be distilled to produce alcoholic drinks (Fenny). Seed coats (testa) are used as poultry feed. The residue of the shell after the extraction of CNSL is used as fuel in the cashew processing units and CNSL extraction plants. Cashew wood is used as fuel or a low quality timber. Wounded trees exude a gum which is used as an adhesive (for wood work panels, plywood, book binding etc.) partly because it has insecticidal properties. The raw nuts on an average weigh 4-8 g, but some may be of 18 g even and cashew apples weigh 40-150 g. The kernel contains 21 per cent protein. The juice of cashew apple is rich in ascorbic acid (Vitamin C), riboflavin (Vitamin B_2) and calcium.

India is the first country to exploit international trade of cashew kernels in the early part of the 20th century and also the first one to initiate research in the early 1950s. However, in 1980's importance of cashew as horticultural crop was realized.

Initiation of Cashew Research in India

Cashew research in India dates back to 1950's when Indian Council of Agricultural Research sanctioned ad hoc schemes in the then composite Madras, Travancore, Cochin and Bombay States. Thus Kottarakkara (Kerala) in 1952; Ullal (Karnataka in 1953 and Vengurla (Maharashtra) in 1957 came into being as pioneering cashew centres in the country. More research schemes were started with ICAR's assistance in 1955 at Bapatla (Andhra Pradesh) and in 1956 at Deragaon (now shifted to Kothiali) in Assam. During the period Central Food Technological Research Institute, Mysore and National Chemical Laboratory, Pune were also engaged in developing technologies for economic utilisation of cashew apple and Cashew Nut Shell Liquid (CNSL). In the second phase of strengthening cashew research in India, the ICAR started All India Spices and Cashewnut Improvement Project in 1971 linking different research centres in the country as a network. Collection, conservation and evaluation of germplasm, hybridization and selection, vegetative propagation, and nutritional studies were the primary areas under this programme.

Importance of Cashew Tree

Besides the nutritive importance of cashew kernel and cashew apple as discussed above, the different parts of cashew plant are also useful. Different uses of cashew plant parts mentioned by Mandal (2007) are as follows.

Leaves

- Leaves contain about 23 % tannin which is used in tanning for dyeing fishing net. Water from the bark and leaves boiled together is astringent and is used as a lotion and mouth wash to relieve toothache and sore gums. It is also used for treating dysentery.
- Young leaves are used for diarrhoea and piles, also as tooth powder. The acid in the juice prevent the decay of the teeth and keep the gum in good condition. The young tender leaves are used for fish dishes and can be eaten as salad.
- Older and dry leaves form good mulch, when burnt or decomposed, serve as good manure.

Bark

- Bark contain high percentage of tannin (9-12%), which may be used for tanning or colouring purpose.

- The juice which comes from the incisions in the bark can be used in the manufacture of indelible marking ink.
- A yellow cashew gum is found on the bark in the tree, this resinous gum is similar to gum Arabic, which is obnoxious to insects.

Stem

- Fleshy stem is the source of refreshing beverages.
- The natives make intoxicating liquor from the juice of stem.
- A type varnish can be produced from the fleshy stem.

Wood

- The reddish-brown wood is light, which can be used for packing cases, house poles, even building boats. The wood can be used as charcoal.

Root

- A decoction of root is used as purgative.

Cashew shell

- The residue of the shell after the extraction of CNSL is used as fuel in the cashew processing units and CNSL extraction plants.

Cashew testa

- Cashew testa (peel of kernel) is a valuable by product of cashew industry which contains about 40 % tannin. Large quantities of testa may become an auxiliary material for the manufacture of commercial tannin extract. The composition of the husk shows high tannin content. The testa can be used as poultry feed. The chemical composition of testa is as follows.

Chemical composition of testa

Constituent	Percentage
Moisture	9-11
Tannins	24-26
Non-tannins	10-11
Insoluble (fibre)	52-57
Oil	3-4

Cashew kernel oil

- By cold pressure extraction from the kernel residue, about 40% oil can be obtained which is known as cashew kernel oil. This is light-yellow in colour, sweetish, odourless, non-desiccative and rich in unsaturated fatty acid (60% oleic acid and 20% linoleic acid). It is edible oil.
- The kernel oil can be used in the treatment of leprosy, warts, worms; ulcer, etc.
- It can be substituted for iodine.

Kernel cake

- The residue from the peeled kernel is used for producing a butter called 'cashew kernel butter'.

Medicinal Uses of Cashew Apple

- Cashew apple liquor is used for medicinal purposes, for which the ailments listed are: worm, sickness, cold, body ache, fever or flue, tooth ache, fresh wounds and cuts, cramps due to chilly weather, muscular pain, irregular movement of bowel, low blood pressure, loss of sleep for aged people, diarrhoea, cholera and fresh juice keep eye sight in good condition.

□□□□

CHAPTER – 2

History: Origin and Geographical Distribution

Cashew has a long history as a useful tree but only in the current century, it became an important tropical horticultural crop. Small-scale local exploitation of cashew for its nuts and cashew apples is the pattern for more than 300 years in Asia and Africa. It was not until early years of 20th century that international trade in cashew kernels began with first exports from India. Although with slow beginning, cashew became an important commercial crop in the last couple of years. In India, use of cashew apples and nuts was started by local people, and accounts from Africa are similar; making cashew wine/liquor possessing a common practice in both Asia and Africa (Johnson, 1973).

Cashew, an evergreen tree, produces both nuts and apples, has become a crop of importance owing to the highly valued kernel and wide adaptability to varying agro-climatic conditions. India is one of the largest producer, exporter and consumer of cashew in the world. Today, cashew has become a crop of high economy and has moved from a crop of waste lands to commercial crop, even in organized sector (Singh, 1998).

Origin and Distribution

Brazil is the original home of cashew. The earliest report of cashew was from Brazil coming from French, Portuguese and Dutch observers (Johnson, 1973). The French naturalist and monk, A. Thevet was the first to describe, in 1558 a wild plant extremely common in Brazil, cashew tree and its fruits. He recounted that cashew apple and its juice were consumed and that the nuts were roasted in fire and the kernels eaten. He provided the first drawing of the cashew showing the local people harvesting cashew fruits and squeezing juice from cashew apples into a large Jar (Johnson, 1973; Delogu and Haeuster, 1994).

There are indications that local Tupi Indians used cashew fruits for centuries. They probably played a major role in the dispersion in their temporary migrations towards the coast of North-Eastern Brazil where a considerable interspecific variation was recorded (Ascenso, 1986). Entire cashew fruit with pseudo-apple when matures usually floats on surface of water. This could account in Brazil for coastward dispersal of the species by river draining North and East. Fruit bats might have also been involved in seed movement. Within the Amazon forest, fruit bats were the most important agents of seed dispersal of tree species (Johnson, 1973). Thus, from its origin in North-Eastern Brazil, cashew spread into South and Central America.

Presence of cashew in other continents is to be attributed to man's intervention (Johnson, 1973). The Portuguese discovered cashew in Brazil and spread first to Mozambique (Africa) and later into India between sixteenth and seventeenth centuries (De Castro, 1994). According to Agnoloni and Giuliani (1977), it arrived in Africa during second half of 16th century, first on East coast and then on the West and lastly in the Islands. In Africa, although it can be guessed that cashew was introduced at an early period by the Portuguese, there are no records which provide specific period. Dispersal of cashew in East Africa may in part be due to the elephant, whose fondness for fruits is well known (Johnson, 1973). Attracted by colour of the false fruit, they swallowed this together with the nut which was too hard to be digested. This was then expelled with their droppings, natural manure, and trodden far enough into the ground by the animals leaving behind to root and grow, into a seedling first and then a tree. This is how cashew spread along the east coast of Africa facing the Indian Ocean (Massari, 1994). The spreading of the cashew within South American continent was gradual and spontaneous (Delogu and Haeuster, 1994).

Spread of Cashew in India

Though cashew is indigenous to Brazil, India is the country that nourished this crop and made it a commodity of international trade and acclaimed. It was brought to India as seed material by Portuguese vellers in 16th Century between 1563 and 1578. Initially cashew plants were seen in and around Malabar region of Kerala and Goa in West Coast and later the cashew spread to remaining areas of East Coast. It was first grown in the gardens of Cochin and in the Malabar Coast, and Cochin served as dispersal point for the cashew in India and perhaps SE Asia as well (Johnson, 1973). Spread of cashew was mostly by natural dispersal means through birds, animals and human carriers. Gradually, cashew was introduced to Eastern India and North-Eastern States under afforestation and soil conservation programmes (Balasubramaniam and Singh, 2000). Presently, cashew plants in semi-wild state and in well managed orchards are seen in Maharashtra, Goa, Karnataka and Kerala along the West Coast, Tamil Nadu, Andhra Pradesh, Orissa and West Bengal on the East coast. To a limited extent the crop is also seen growing in Chhattisgarh, Gujarat, Assam,

Arunachal Pradesh, Meghalaya, Tripura, Manipur, Nagaland and Andaman and Nicobar Islands (Singh, 1998).

Dispersion of Cashew to SE Asia and Australia

After India, it was introduced into SE Asia (Delogu and Haeuster, 1994). Dispersal in SE Asia was aided by monkeys. Whether cashew reached the Philippines via India is uncertain. It may have come directly from the New World on the Manila Galleons (Johnson, 1973). The cashew was later spread to Australia and some parts of North-American continent, such as Florida. Finally, its present diffusion can be geographically located between 31° North latitude and 31° South latitude, both as wild species and under cultivation (Delogu and Haeuster, 1994). At presently cashew is cultivated in many tropical countries, mainly in coastal areas (Ascenso, 1986; Van Eijnatten, 1991). In 19th century, proper plantations were raised and then spread to a number of other countries in Africa, Asia and Latin America (Massari, 1994). Traditionally, cashew has been cultivated on commercial scale in Brazil, India, Tanzania, Mozambique, Kenya and Madagascar, while in the recent years plantations are also raised in SE Asian countries.

Cashew (*Anacardium occidentale* L.), belong to family Anacardiaceae with the natural order Sapinadales, is an evergreen tree. It is a native of tropical Central and South America (Paul, 1936); but is now distributed all over the tropics and part of warm sub-tropics (Conrad, 1944). However, colder sub-tropical areas are not suitable for cashew due to cold and frost; although the tree may grow and even bear a few flowers. However, very little efforts have been made to collect historical evidence of the cashew cultivation, except the first illustrative description of cashew was given by French naturalist, Andrew Thevet in 1558 AD. It has been extensively spread in the coastal areas of East Africa, Tanganyika, Kenya, Mozambique, Uganda, Gold Coast, Nigeria and Angola in Africa; Madagascar and the Malay Peninsula; Florida, Peru, Hawaii, Tahiti, Mauritius and Seychelles. It has obtained the greatest extension in Sri Lanka and along the coastal tracts of India.

In the 16th century, the introduction of the spices by the Portuguese adventures was first to Goa and to the Malabar Coast in Kerala by the Spaniards and to other centres in India (De Costa, 1578). It was planted particularly along the shore and has become naturalized throughout the West Coast (Kerala, Karnataka, Goa, Sindhudurg and Ratnagiri tract of Maharashtra); East Coast (Tamil Nadu, Andhra Pradesh, Orissa, West Bengal) and Andaman Islands. It is reported that as many as 20 species of *Anacardium* Linn. are known to exist within Central and South America; among which *Anacardium occidentale* (Cashew) is the only species which has been introduced outside the new world and commercially grown in the tropical countries (Nayak and Bhat, 2005).

The other species include:

A. amilcarianum

A. brasiliense

A. ciratellaefolium

A. corymlosum

A. encardium

A. excelsum

A. giganteum

A. humile

A. kuhlanannianum

A. mediterraneum

A. microcarpum

A. microsepalum

A. nanum

A. negrense

A. orthonianum.

A. parvifolium

A. pumilum

A. rhinocarpus

A. rondonianum

A. spruceanum

A. tenuifolium

Related Species of Cashew

Genus *Anacardium* contains eight tropical American species (Baily, 1949; 1958) and ten different species reported by Parente (1972), but Peixoto (1960) named 20 different species, several of which have edible peduncles such as *A. nanum.* A very early bearing small shrub, *A. subterranium*, a small tree from with its trunk completely underground containing water resources, *A. microcarpum* a small tree from sandy savannas, *A. spruceanum* a large tree, and the largest species of the genus, *A. gigantism* which grows in the Amazon forest with good timber value. Valeriano (1972) mentioned five different species namely *A. occidentale, A. pumilum, A. gigantium, A. rhinocarpus* and *A. spruceanun.* He also suggested the recognition of only two species namely, *A. nanum* and

A. gigantium which can be further subdivided based on colour (yellow or red) and shape (round, pear-shaped or elongated) of the pseudo-fruit. Description provided by Peixoto (1960) indicating separate recognition of more than two species. It may be possible that some species originated due to intercrossing between species in nature (Nayak and Bhat, 2005). It appears from published literature that *A. occidentale* is the only species that has been introduced outside the new world. Within Central and South America, as many as 20 species of *Anacardium* exist (Table 1).

Table 1: Species of *Anacardium**

Botanical Name	Country
Anacardium amilcarianum Machado	Brazil
A. brasiliense Barb.	Brazil
A corymbosum Barb.	Brazil
A. curateilaetolium St. Hil *(= A. subcordatum Presl.)*	Brazil
A encardium Noronha	Malaya
A excelsum Skeels *(= Rtiinocarpus excelsa)*	Brazil
A. giganteum Hancock ex Engl.	Brazil
A. humile St. Hil *(= A subterraneum Liais)*	Brazil
A. kuhlmannianum Machado	Brazil
A. lenuifolium Ducke	Brazil
A mediterraneum Veil.	Brazil
A. microcarpum Ducke	Amazon region
A microsepalum Loes	Amazon region
A negrense Pires and Fro'es	Brazil
A. nanum St. Hil *(= A. humile Engl., A. pumilum Walp)*	Brazil
A occidentale L.	Brazil
A. parvifolium Ducke	Amazon region
A. rhinocarpus DC.	Brazil
A. rondonianum Machado	Brazil
A spruceanum Benth ex. Engl.	Brazil

**Source*: Nayak and Bhat (2005)

Genetic Diversity

Cashew is botanically known as *A. occidentale*. According to Bailey (1958) *Anacardium* is a small genus of eight species indigenous to South America. Mitchell and Mori (1987) recognized ten species in the genus *Anacardium*, one of which, *A. fruticosum*, is described as new (Table 2). The genus has a primary centre of diversity in Amazonia and a secondary centre in the Planalto of Brazil. Ascenso (1986) reported that cashew (*Anacardium occidentale*) is the only species in the genus which attained economic importance. This morphologically polymorphic species also exhibits chromosome polymorphism (Mitchell and Mori, 1987). Chromosome numbers reported in literature range from 2n=24 (Khosla *el al.*, 1973; Goldblott, 1984), 2n=30 (Machado, 1944), 2n=40 (Simmonds, 1954; Goldblott, 1984) to 2n=42 (Darlington and Janaki Ammal, 1945; Khosla *et al.*, 1973; Goldblott, 1984; Purseglove, 1988). Such chromosome polymorphism is well known in many domesticated plants (Khosla *et al.*, 1973). Wide variation within the accessions due to cross pollinated nature of crop was reported for 292 germplasm accessions of seed origin which were evaluated at Central Plantation Crop Research Institute, Kasargod, Regional Station, Vittal/ Shantigodu (Nayak and Bhat, 2005). Considerable variability was observed for some of the economic characters like flowering season, flowering duration, harvesting duration, nuts/panicle, apple weight, nut weight, apple to nut ratio, shelling percentage, kernel weight, kernel count/lb, shell thickness and mean yield/plant reported by Rao and Swamy (1994) (Table 3). Variability in shape, size and colour of nut and apple are enormous and much use in cashew improvement programme.

Table 2: Species of *Anacardium* and Synonyms

Species	Synonym (s)
A. excelsum	*Rhinocarpus excelsa Anacardium rhinocarpus Anacardia rhinocarpa Anacardium caracoii*
A. spruceanum	*Anacardium brasiliense*
A. occidentale	*Cassuvium pomiferum Acajuba occidentale Anacardium occidentale var. americanum A. occidentale var. indicum A. mediterraneum A. curatellifolium Cassuvium reniforme A. occidentale var. longifolium A. occidentale var. gardneri A. subcardatum A. microcarpum A. rondonianum A. amilcarianum A. Kuhlmannianum A. othonianum*
A.humile	*Monodynamus humilus A. subterraneum A. pumilum A. pumilum var. petiolata A. humile var. subacutum*
A. nanum	*A. pumila*
A. corymbosum	-
A. parvifolium	*A. tenuifolium*
A. fruticosum	*A. parvifolium var. fruticosa*
A. giganteum	-
A. microsepalum	*A. negrense*

Source: Mitchell and Mori (1987)

Table 3: Variability recorded for some of the characters in the germplasm accessions (seedling progenies)

Character	Range
Flowering season	October – January
Flowering duration	40-127 days
Harvesting duration	30-105 days
Number of fruits per panicle	1-8
Apple weight	30-150 g
Nut weight	2.4-18.0 g
Apple: nut ratio	4:1-12:1
Shelling percentage	19.0-35.0
Kernel weight	0.5-4.5 g
Kernel count/lb	100-900
Shell thickness	1.5-5.0 mm
Mean yield/plant/year (10 annual harvests)	0.50-11.75 kg

Source: Rao and Swamy (1994)

Exploration and Collection

The initially cashew plantations was of seedling origin. As cashew is primarily a cross pollinated crop, it is highly heterozygous. Considerable segregation has resulted in large variation in the populations (Rao and Bhat, 1996). The early attempts for germplasm collections were made during the early fifties with the sanctioning of ad-hoc schemes in the then composite states of Madras, Travancore, Cochin and Bombay. Various research stations started under these ad-hoc schemes in Kerala (Kottarakkara), Karnataka (Ullal), Tamil Nadu (Vridhachalam), Andhra Pradesh (Bapatla) and Maharashtra (Vengurla) took up the programme of collection of locally available elite plants for evaluation and further selection. These were the first attempts in collection of cashew germplasm in India. Many of the other research centres which were established subsequently have collected the seeds of germplasm from these centres, namely, Bapatla, Kottarakkara, Ullal, Vridhachalam and Vengurla. While making the initial collections of germplasm, these centres had confined their survey mainly to the respective States and hence they represent the local germplasm available in the States. Since the inception of All India Coordinated Spices and Cashewnut Improvement Project (AICS and CIP) in 1971, Central Plantation Crops Research Institute (CPCRI) Regional Station, Vittal, also took up the programme of cashew germplasm collection which mainly consisted of the seedling progenies of collections available at Bapatla, Vridhachalam, Vengurla, Anakkayam and a few collections made locally from Karnataka. Subsequent to the establishment of National Research Centre for Cashew (NRCC) at Puttur (Karnataka) in 1986, the germplasm collection through seeds has been discontinued and only the vegetatively propagated material is being collected and conserved in the National Cashew Gene Bank (NCGB). A coordinated approach was brought in the cashew gemplasm collection by organising joint survey teams consisting of scientists of NRCC and the centres of All India Coordinated Research Project on Cashew (AICRP on Cashew) of the respective States. The main criteria adopted for collection of germplasm were considering the economic characters such as high yield, bold nut, cluster bearing, apple qualities, canopy types and dwarfness of the plant or any other unique characters observed during the survey. Since 1986, cashew germplasm collection surveys are being undertaken by NRCC in collaboration with centres of AICRP on Cashew and other centres of cashew growing states, namely, Karnataka, Kerala, Maharashtra, Goa, Tamil Nadu, Andhra Pradesh, Orissa, Chhattisgarh and West Bengal. So far, all together 494 cashew germplasm accessions were made and planted in cashew field gene bank at NRCC, Puttur for evaluation (Nayak and Bhat, 2004 b). Till the establishment of NRC for Cashew, over one thousand accessions of seedling origin were collected by different Cashew Research Stations in the country. Among the seedling accessions, 23 are exotic collections, of which nine were collected from Brazil, Nairobi, Mtwara, Lindi, Nacala, Mozambique, Tanganyka, Singapore and Australia and 14 from the Republic of Panama (Nayak and Bhat, 2005). The detail of germplasm holding in India as reported by Bhat *et al.* (2010) is given in Tables 4. The non-traditional areas such as Garo Hills (Meghalaya) and Bastar (Chhattisgarh) were also surveyed for germplasm collection. However, more comprehensive exploration is required for targeted collection of germplasm with specific traits such as resistance to TMB, CSRB, high yield, dwarf and compact, bold nut with cluster bearing, tolerance to drought, frost, salt and to problematic soils (Nayak and Bhat, 2005).

Table 4: Cashew germplasm-holding in India

Centre	Number
(A) DCR, Puttur	527
(B) AICRP- Cashew Centers	
Bapatla	132
Bhubaneswar	98
Jhargram	119
Vridhachalam	208
Madakkathara	130
Pilicode	43
Vengurla	302
Chintamani	128
Jagdalpur	65
Total	1752

Source: Bhat *et al.* (2010)

Nomenclature

Cashew is known by different countries by different names. The Brazilian name for the cashew nut was 'acajaiba' and the Tupi name 'acaju' which the Portuguese converted into 'caju' and also commonly known as *caju* or *kaju* in India. The French name is acajou which the English converted into cashew.

Arabic (habb al-biladhir); Brazilian (acajaiba); Burmese (thiho thayetsi); Chinese (yao kuo); Dutch (kasjoe, mereke); English (cashew, cashew nut); Filipino (kasoy, balogo, bulabad); French (anacardier, acajou, pommier cajou, pomme d' cajou, anacadier, noix d' cajou, anacarde, anacardes, pomme acajou); German (Acajubaum, Kaschubaum, cashwenuß, elefantenlaus); Indonesian (jambu monyet, jambu mede); Italian (acagia); Japanese (kashu nattsu); Javanese (jambu mede, jambu monyet); Khmer (svaay chanti); Malay (gajus, jambu monyet); Mandinka (kasuowo, kasuwu); Nepali (kaaju); Portuguese (caju).

In India, cashew is known differently in different languages *viz.* Assamese - Caju badam, Bengali - Hijli badam or caju badam; Kashmiri, Marathi and Gujarati - kaju; Hindi - bojan, kashu-mavu, kaju, hijuli; Malayalam - Paragi andi or kasu andi, Oriya - Lanka beeja and Tamil - Mundiri.

Botanic Description

Study of Family Anacardiaceae Particularly for cashewnut

Scientific classification

Division: Spermatophyta

Kingdom: Plantae

Sub-division: Angiosperms

Class: Dicotyledonae

(unranked): Rosids

Order: Sapindales

Family: Anacardiaceae

Genus: *Anacardium*

Species: *Occidentale*

Binomial name

Anacardium occidentale L.

□□□□

CHAPTER – 3

Cashew Area, Production, Productivity, Export and Import

Global Scenario

Globally, India's share in the cashew area and production is 20 per cent and 16 per cent, respectively. India occupies the largest area followed by Ivory Coast (19%), Brazil (16%), Indonesia (12%), Vietnam (7%), Nigeria (7%), Benin (5%), Guinea-Bissau (5%), Mozambique (2%) and Philippines (1%). The highest production was from Vietnam (30%) followed by Nigeria (19%), India (16%), Ivory Coast (11%), Brazil (5%), Indonesia (3%), Philippines (3%), Guinea- Bissau (3%), Benin (2%) and Mozambique (2%). Sometimes back, India was credited as largest producer, processor and exporter of cashew in the world but as other cashew producing countries also acquired the know-how of processing and entered in the international market, currently the India's share in the International market has shrank to 30 per cent only. On the other hand, the domestic consumption in India has increased considerably, owing to largest consumption of cashew kernel. Every year, over 7-8 lakh tonnes of raw cashew nut are imported mainly from African countries and exporting cashew kernel to USA and UK after processing (Saroj, 2013).

Cashew is grown in India, Vietnam, Nigeria, Brazil, Ivory Coast, Tanzania, Mozambique, Indonesia, Philippines, Guinea-Bissau and other tropical Asian and African countries. The largest producers of cashews are Vietnam, India, Nigeria, Ivory Coast, Brazil and Indonesia (Fig.1). However, the largest consumers of cashews are countries with a high per capita income, such as France, Germany, Italy, the United States, Canada and Spain (Vieira *et al.*, 2005).

Latin American countries have registered approximately 3 fold increase during the same period. In recent times, India is facing stiff competition from Vietnam and Brazil in international cashew trade (Bhat *et al.*, 2007).

Presently, Vietnam is the single largest producer of cashewnuts (1.272 million tonnes) followed by Nigeria (0.813 million tonnes) and India (0.674 million tonnes) during 2011-12. The other major producing countries of cashew are Ivory Coast (0. 452 million tonnes) Brazil (0. 230 million tonnes), Tanzania (0.075 million tonnes) and Mozambique (0.072 million tonnes). Area under cashew in the world is 4.701 million hectares and production is 4.201 million tonnes (Table 5).

Vietnam has registered over four fold increase of raw cashewnuts production over the last 12 years. The increase in the raw cashewnuts production in India was 36.2 per cent during this period. Steady growth rate in production of raw cashewnuts in South-East Asian countries, particularly Vietnam is of serious concern to India. Ivory Coast, Brazil and Mozambique have also registered considerable increase in the raw cashewnuts production of 230.5, 51.6 and 15.5 per cent, respectively during 2009-10 (Bhat *et al.*, 2010). The world production of raw cashewnuts for the last 10 years has increased from 1.73 million tonnes (1999-2000) to 4.20 million tonnes (2011-12) amounting to 142.7 per cent.

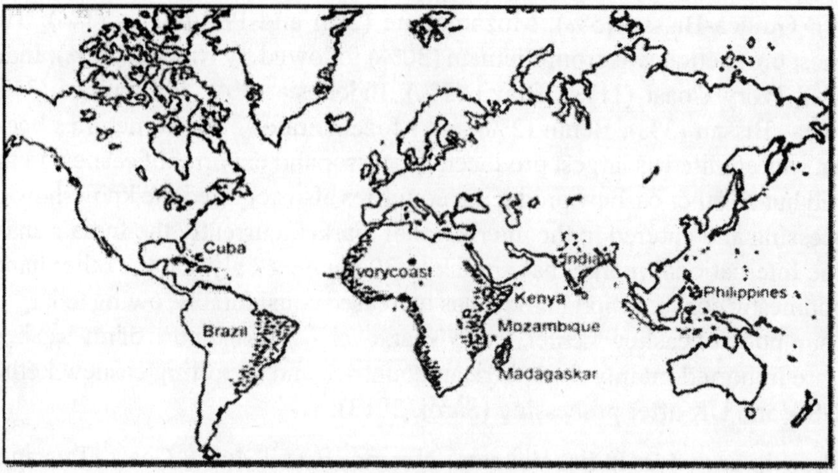

Fig. 1: Cashewnut producing area of the world

Table 5: World cashew area, production and productivity

Countries	Area (Lakh)	Production (MT)	Productivity (kg/ha)
Benin	250000	70000	285.8
Brazil	764472	230785	135.6
Ivoire Coast	877813	452656	4302
Ghana	60223	35736	481.4
Guinea-Bissau	222517	128684	378.8
India	953200	674600	700.0
Indonesia	570600	122100	562.4
Mozambique	77402	72263	872.7
Nigeria	330000	813023	1800.0
Philippines	28134	133388	478.7
Tanzania	80000	75000	1000.0
Viet Nam	331300	1272000	3407.0
Others	264456	120775	-
Total	4701983	4201010	-

Source: FAO Statistical Database, 2011-12

Global area harvested under cashew nuts

Cashew is mainly cultivated in Asia, Africa and Latin America. The global area harvested under cashew increased by 665 per cent, from 5,16,550 ha in 1961 to 39,53,175 ha in 2007 (FAO). The annual growth rate during this period was 4.55 per cent with variations ranging from 8.22 per cent to 20.08 per cent. Trends in global cashew area harvested (1961-2007) is shown in Table 6.

Table 6: Growth and variability in area harvested under cashew nuts

Years	Mean (ha)	CAGR (%)	C.V.(%)
1961-71	666496	5.44	17.08
1971-81	919633	2.04	8.22
1981-91	1115644	3.99	19.75
1991-2001	2499141	6.62	20.08
2001-07	2598030	4.17	13.41
1961-2007	1659350	4.55	67.91

Computed on the basis of data downloaded from FAO Statistical Database

National Scenario

Cashew Research in India

India is the first country to exploit international trade of cashew kernels in the early part of the 20th century, also the first one to initiate research in the early 1950s. However, in 1980s importance of cashew as horticultural crop was realized.

The research stations started during the early 1950s under the ad-hoc schemes in Maharashtra (Vengurla), Kerala (Kottarakkara), Karnataka (Ullal), Tamil Nadu (Vridhachalam) and Andhra Pradesh (Bapatla) and took up the programme of collection of locally available elite plants for evaluation and further selection. These were the first attempts in collection of cashew germplasm in India.

The All India Coordinated Spices and Cashewnut Improvement Project (AICS & CIP) was started in 1971 with its headquarters located at the CPCRI, Kasaragod. The ongoing project (AICS & CIP) was bifurcated into two separate projects, one on Cashew and another on Spices in 1986. The headquarters of the independent All India Coordinated Research Project (AICRP) on Cashew was shifted to the newly established National Research Centre for Cashew, Puttur in 1986. NRCC was upgraded and renamed as Directorate of Cashew Research (DCR) by ICAR in 2009. Presently, the total number of centres under AICRP-Cashew is 11. Recently AICRP (Cashew) centre initiated at AES, Paria, NAU, Navsari, Gujarat, in the year 2009.

Mandate of DCR

- To conduct mission-oriented research on all aspects of cashew for improving productivity and quality with special reference to export.
- To serve as a national repository for cashew germplasm and a clearing house for research information on cashew.
- To act as centre for training in research methodologies and technology updating of cashew and to coordinate national research projects.
- To provide consultancy on cashew production technology.
- To generate quality planting material.
- To collaborate with national and international agencies for achieving the mandate.

Mandate of AICRP (Cashew) Project

- Evolving high yielding varieties with export grade kernels, tolerant/ resistant to pests and diseases.
- Standardizing agro-technique for the Cashew crop under different agro-climatic conditions.
- Evolving cost effective and efficient pest and disease management practices.

Directorate of Cashew nut and Cocoa Development (DCCD), Cochin, Kerala

The Directorate of Cashew nut Development (DCD) established in 1966 as a primary field functionary under the Union Ministry of Agriculture gave a greater impetus for the development of cashew. This marked the first step towards the integration and co-ordination of Cashew Development in association with developmental agencies of States and Research Institutes. The Directorate of Cashew nut Development was handling only cashew which lateron got the mandate for development of cocoa in 1997 and name was change to Directorate of Cashew nut and Cocoa Development.

Cashew Export Promotion Council of India (CEPCI)

The Cashew Export Promotion Council of India (CEPC) was established by the Government of India in the year 1955, with the active cooperation of the cashew industry with the object of promoting exports of cashew kernels and CNSL from India. The CEPC Laboratory and Technical Division was established by the Cashew Export Promotion Council of India (CEPC) in the year 1997, fully funded by the Government of India for ensuring the quality of cashew kernels and allied products exported from India.

Cultivation of cashew in India confines mainly to the peninsular region. Presently, cashew is grown in Maharashtra, Goa, Karnataka and Kerala along the West coast; Tamil Nadu, Andhra Pradesh, Orissa and West Bengal on the East coast. To a limited extent, the crop is also seen growing in Chhattisgarh, Gujarat, Assam, Arunachal Pradesh, Meghalaya, Tripura, Manipur, Nagaland and Andaman and Nicobar Islands. During 2012-13, with total production of cashew in the country was 7.28 lakh tonnes from 9.82 lakh ha of land with a productivity of 772 kg/ha. Although Andhra Pradesh has the largest area under cashew, Maharashtra ranks first in both production and productivity (Table 7) (Anon., 2012a). Research and development efforts on cashew in India have resulted in achieving 5.6 times increase in production and 3.1 times increase in area from 1971-72 (Saroj, 2013).

Table 7: Area, production and productivity of cashew in India (2011-12)

State	Area (ha)	Production (MT)	Productivity (kg/ha)
Kerala	78,000	71,000	975
Karnataka	1,19,000	57,000	461
Goa	56,000	24,000	473
Maharashtra	1,81,000	2,08,000	1,186
Tamil Nadu	1,35,000	65,000	472
Andhra Pradesh	1,83,000	1,07,000	544
Orissa	1,49,000	91,000	641
West Bengal	11,000	11,000	909
Gujarat	7,100	21,300	700
Others	33,900	18,700	680
Total	9,53,000	6,74,000	Average 700

Traditionally, cashew has been an important crop in the Coastal Region (Western and Eastern) of the country, but has been recently spreading to non-traditional areas as well. There is great scope to expand area (Table 8) under cashew by about 7.1 lakh ha in both traditional and non traditional areas including Gujarat, Jharkhand, North Eastern Hilly (NEH) Regions, Andaman and Nicobar Islands, Chhattisgarh and Plains region of Karnataka in addition to the present area (Bhat *et al.*, 2010).

Table 8: Potential area available for expansion of cashew

State	Area (ha)
Bihar	25,000
Goa	10,000
Jharkhand	25,000
Chhattisgarh	50,000
Kerala	25,000
West Bengal	25,000
Gujarat	25,000
Tamil Nadu	25,000
NEH Region	50,000
Andhra Pradesh	1,00,000
Karnataka	1,00,000
Orissa	1,00,000
Maharashtra	1,50,000
Total	7,10,000

Source: Bhat *et al.* (2010)

Indian cashews are exported to more than 60 countries in the world and 60 per cent of cashew imports in USA are from India. USA, UK, Japan, UAE, France, Canada, Saudi Arabia, Singapore, Italy, Germany, Austria, Israel and Spain are the major international buyers of Indian Cashews. There is an ever-increasing demand for cashew kernel both in international and domestic markets. During 2011-12, India earned total Rs. 4450.14 crore as foreign exchange through export of cashew kernel and Cashew Nut Shell Liquid.

The cashew trade has an important contribution in India's international trade. It plays a vital role in contributing foreign exchange to the government of India. Cashew is the important horticultural crop which secures the third rank in the international market of horticulture crops. The cashew is a poor men's crop, but the rich men's food. It is a source of income and livelihood for poor farmers. Cashew industry may play very important role in the development of hilly region or rural area. India has historical importance in this regard that it is the first country in the world that started the international trade of cashew.

No other sector can create as many jobs for people of limited literacy as processing can. Cashew provides employment to over 5 lakh workers in 3796 processing factories in India. In cashew processing factories, over 95% of the workers are women belonging to economically backward strata. Cashew industry has the potential to play a leading role in the social and financial upliftment of rural poor. India requires about 14-16 lakh tons of raw cashewnuts to feed the large number of cashew processing units (3796 processing units) engaging over 5 lakh workers, especially women (Anon., 2012b). However, India produces only about 7 lakh tons of raw nuts required by processing industries which comes to only 50% of the requirement and hence balance quantity is imported from African and other countries. In future raw nut availability for importing by India may gradually decline or may all together stop as those countries have started processing their own nuts. So, there is urgent need to enhance domestic cashew nut production and become self sufficient in raw nut production.

Globalization put forth some opportunities such as: (i) increase in cashew cultivation and cashew production, (ii) employment generation, (iii) formation of cashew processing unit, (iv) alcohol production from cashew apple, (v) shield oil production, (vi) transportation facility for cashew industry, etc. India has tremendous potential in this regard. The globalization has an impact on the industry as a result of the economic asymmetries in trade of cashew. Foreign trade of cashew has been contributing Indian economy by earning valuable foreign exchange. The steps to stimuli for cashew-nut industry are required in the form of arranging campaign for awareness of cottage industry, strategy to increase domestic consumption, strengthening the channel of market information and extra initiative from the Ministry of Commerce and Trade, Government of India. Strong efforts, therefore, are required to make sustainable the Indian cashew industry in the context of globalization to face challenges and grab the opportunities (Mahajan and Patil, 2012).

The productivity level is not increased over the years as expected against the new technologies available for adoption. There are two ways of increasing production of raw cashewnuts in the country, firstly by increasing area of high yielding cashew varieties and secondly, increasing productivity. Cashew being a snack food crop, country cannot divert more area for cashew as food crop gets priority in the government sponsored programmes. However, some more area which falls under neglected/degraded soil conditions where other crops cannot come up, is available for cashew in different states.

Low productivity is the main concern in cashew cultivation in India. The productivity is low particularly in the states of Karnataka, Goa, Tamil Nadu, Andhra Pradesh and Orissa. The major factors for low productivity- in these states are the large plantations under seedling origin and poor orchard management practices. Mostly, the older plantations of nondescript material, have poor genetic potentiality and gone senile. There is a wide gap between potential productivity and present productivity. The productivity can be doubled if all senile plantations of seedling origin are replaced by improved varieties and with advance management practices (Saroj and Nayak, 2013).

The productivity of cashew in the country is highest in Maharashtra. This is because of the fact that in Maharashtra plantations have been established with high yielding varieties. Area under cashew in Maharashtra expressed that major area for expansion is in Konkan region. The cashew is being grown in this area as well as in Goa since about 400 years ago. Considering the suitability of this crop in Konkan and looking to its scope for extend the area, the State Government has encouraged its plantation on maximum possible area by providing cent per cent subsidy for cultivation. As a result, the area of 40,000 hectare under cashew in the year 1990 in Maharashtra has been increased to 1.81 lakh hectare by the year 2011. The revolution is continued and requires to be supported by availability of genuine grafts of high yielding cultivars of cashew. This shows that cashew production and productivity could be improved by providing genuine planting materials of recommended high yielding varieties and by adopting good management practices.

Cashew production is the monopoly of the Konkan region of Maharashtra. The maximum cashew plantation of Maharashtra state is in Konkan region of Sindhudurg and Ratnagiri districts and still there is good scope for area expansion in the said districts. Besides this, there is also great scope for cashew area expansion in other non-traditional area like Kolhapur district (Gargoti, Chandgad, Aajra, Gadhinglaj and Bhudargad Talukas), Nashik (Igatpuri) and Gadchiroli district of Maharashtra.

Maharashtra leading in, production and productivity of cashew is due to adoption of high yielding cashew varieties and technologies developed by the Dr. Balasheb Sawant Konkan Krishi Veedhyapeeth, Dapoli, Dist. Ratnagiri (M.S.). Because of the quantum jump in area and production of cashew in the region and this has led to positive impact on export scenario of the country.

Although Maharashtra only accounts for about 20 per cent of the total area under cashew, yields in this state are relatively high when compared to those of other states. Estimated cashew productivity in Maharashtra is 1500 kg raw cashew nuts per hectare, which is almost, double that in the rest of India. Area, production and productivity of cashew across the West and East coasts of India reported by Rao and Gopikumar (2010) are as follows.

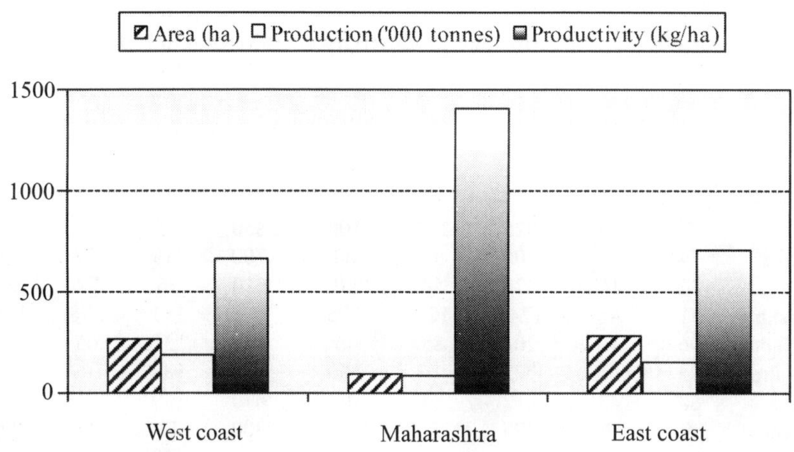

Fig. 2: Area, production and productivity of cashew across the West and East coasts of India

Source: Rao and Gopikumar (2010)

State-wise Trends in Area and Production of Raw Cashew nuts in India

Production of cashew nuts broadly depends on cashew area, yield rate, area under fruit bearing trees, age of plantations and breed of plantations. The Directorate of Cashew Research, Puttur, Directorate of Cashew and Cocoa, Cochin and State Agricultural Universities have been continuously intervening for the development of high yielding varieties, grafts and planting systems for augmenting the production of quality cashew nuts in our country. Area under cashew nut increased from 5.65 lakh hectare during 1993-94 to 9.53 lakh hectare during 2010-11. Cashew nuts production increased steadily from 3.48 lakh tonnes during 1993-94 to 6.74 lakh tonnes during 2010-11 (Table 9).

State-wise Trends in Productivity of Cashew nuts in India

Although the yield of Cashew nuts in India increased from 694 kg/ha during 1993-94 to 700 kg/ha during 2009-10, it was much lower when compared to the level of more than 2000 kg/ha reported to be realised by the latest varieties. As it can be seen from Table 9, it varied across the states with the maximum in Maharashtra with 1186 kg per hectare followed by Kerala (975 kg/ ha), West Bengal (909 kg/ha), Orissa (641 kg/ha) and Andhra Pradesh (544 kg/ha) for the triennium ending 2010-11. Improved varieties and management technology adopted by the farmers in Maharashtra has led to increase in productivity of over one tonne per ha.

Table 9: State-wise area, production and productivity of cashewnut in India (1993-2010). A- Area (000 ha), P- Production (000 MT) and Pr.- Productiity (kg/ha)

State	1993-94			1999-2000			2010-11		
	A	P	P	A	P	P	A	P	P
Kerala	155	140	925	122	100	850	78	71	975
Karnataka	74	31	500	91	60	700	119	57	461
Goa	46	16	370	54	30	610	56	24	473
Maharashtra	51	45	1246	121	125	1470	181	208	1186
Tamil Nadu	96	19	203	85	45	540	135	65	472
Andhra Pradesh	72	46	723	103	100	1100	183	107	544
Orissa	60	43	812	84	40	670	149	91	641
West Bengal	6	3	596	9	8	900	11	11	909
Others	14	2	299	17	12	800	339	18	680

Source: DCCD, Cochin

Export and Import

Sometimes back, India was credited as largest producer, processor and exporter of cashew in the world but as other cashew producing countries also acquired the know-how of processing and entered in the international market, currently the India's share in the International market has shrank to 30 per cent only. On the other hand, the domestic consumption in India has increased considerably, owing to largest consumption of cashew kernel. Every year, over 7-8 lakh tonnes of raw cashew nuts are imported mainly from African countries and exporting cashew kernel to USA and UK after processing. The country needs about 13.0-14.0 lakh tonnes of raw cashew nut per annum to feed large number of cashew processing units. Vietnam and Brazil are major competitors to India for cashew production and export (Saroj, 2013).

Now, India has the largest area under cashew and stands as the third largest producer of cashew in the world. Vietnam, Nigeria, Ivory Coast and Brazil are the main competitors to India for cashew production and export.

India's raw cashewnuts production is not sufficient to sustain the processing capacity established in the country. It is, therefore, urgent future need to increase our raw cashewnuts production and productivity to sustain itself in the international market. Countries such as Vietnam and Brazil are competing with India in the international market. Since, African countries have started processing raw cashewnuts themselves, availability of raw cashewnuts for importing by India may gradually decline or may altogether stop. This situation calls for increasing the domestic raw cashewnuts production and become self sufficient to provide adequate raw cashewnuts to the processing industries. Further, the raw cashewnuts have to be produced at cheaper rate in order to compete with other cashew growing countries of the world. It is also important to develop and expand domestic market for cashew kernels so that there will be continued good price for the nuts and thereby farmers are encouraged to grow cashew. Unless remunerative price is offered to the raw cashewnuts, farmers might switch over to other plantation crops such as rubber, which fetches very high price for its product. Efforts are being made for reviving cashew cultivation in Kerala through the efforts of Kerala State Agency for Expansion of Cashew Cultivation (KSACC). Value addition to cashew kernel is being attempted by Kerala State Cashew Development Corporation (KSCDC) (Anon., 2011).

In 2009-10, India accounted for over 26.8 per cent world exports of cashew kernels worth 2,906 crores. Between 2000 and 2010, raw cashew nuts production in India increased by 36.2 per cent. In order to keep place in the international market, India has to achieve the raw cashewnuts production of at least 2.50 million tonnes by 2030 (Anon., 2011).

India has been exporting cashew kernels since the early part of the 20th century. The country exports about 0.100-0.115 million tonnes of cashew kernels annually. Cashew kernels are exported to more than 65 countries of which the largest buyers of Indian cashew kernels are USA and Netherlands. Export of Cashew Kernel and CNSL from India and import of raw nut during 1990-91 to 2009-10 is presented in Table 10.

Table 10: Export of cashew kernel, cashewnut shell liquid (CNSL) and import of raw nuts (quantity in MT) and value in lakhs

Year	Cashew Kernel Export		CNSL Export		Raw nut import	
	Quantity (MT)	Value (₹ in lakhs)	Quantity (MT)	Value (₹ in lakhs)	Quantity (MT)	Value (₹ in lakhs)
1990-1991	49874	44224	5658	556	82639	13400
1991-1992	47738	66909	4542	402	106080	26668
1992-1993	53436	74549	4258	381	134985	37633
1993-1994	69884	104602	3625	290	191322	48270
1994-1995	77000	124602	3807	244	228109	69094
1995-1996	70334	124050	760	145	222819	76008
1996-1997	68663	128550	1735	277	192285	64060
1997-1998	76593	139610	4446	717	224968	74400
1998-1999	75026	160990	1572	326	181009	68024
1999-2000	92461	245145	764	184	200584	95352
2000-2001	89155	204975	2246	3894	249318	96080
2001-2002	97550	177680	1814	419	356566	96001
2002-2003	104137	193302	7215	925	400659	123657
2003-2004	100828	180442	6926	703	452898	140093
2004-2005	126667	270924	7474	791	578884	218324
2005-2006	114143	251486	6405	709	565400	216295
2006-2007	118540	245515	5589	920	592604	181162
2007-2008	114340	228890	7813	1197	605970	174680
2008-2009	109522	298840	9099	2606	605850	263241
2009-2010	108120	290582	9748	2412	752894	303735

Source: CEPCI, Kochi, Kerala; DCCD, Kochi

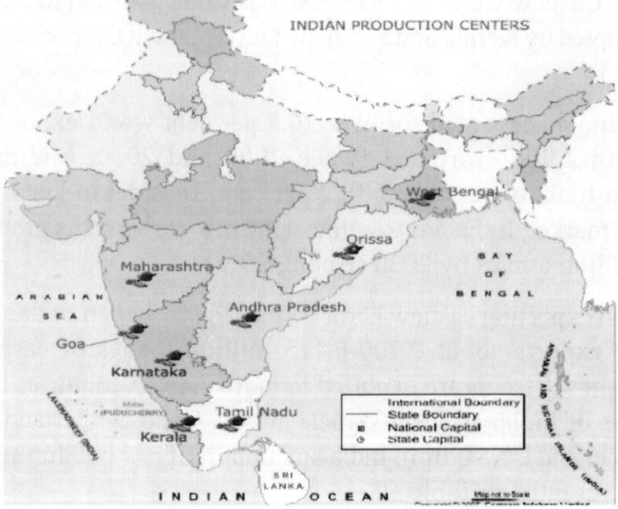

Fig. 3: Major cashew producing State in India

CHAPTER – 4

Soil and Climate

Soil

Cashew is grown on a wide variety of soils in India. It comes up well in all types of soils at an altitude up to 700 m. It cannot stand in water logged or saline soils. It can come up in very sandy soils also. Adoption of cashew has been found in wide variety of soils from coastal sands to degraded laterites. The worst soils, not suited to other crops, are normally allotted for cashew. In Sri Lanka and Philippines, it is grown in alluvial soils. Cashew comes up well even in volcanic soils of Indonesia, Fiji and some part of Philippines. Cashew is not suitable in heavy clays, water logged conditions and ill drained soils. It is grown abundantly in loose sandy soils particularly in South India, East Africa and Brazil. In Karnataka and Kerala, it grows well in poor laterite soils. Cashew prefers well-drained aerated sandy loam soils with adequate moisture. Coastal plains and plateau are very good for cashew cultivation.

In India, cashew is mostly grown on laterite and red soils and coastal sands. The tree has an extensive root system, which helps it to tolerate a wide range of moisture levels and soil types, but commercial production is only advisable in well-drained, sandy loam or red soils with slight acidic pH (4.5-6). The cashew tree can flourish in the sand of open beaches, but it grows poorly in heavy clay or limestone.

Though cashew can be grown on a variety of soils, a well-drained 1-1.5 m deep soil having a water table between 2 and 10 m depth and pH of 5.6 to 7.0 is preferred. Laterites, red and coastal soils rich with organic matter are best for cashew growing. Even the marginal soils can be used for cashew cultivation but the waterlogged condition should be avoided as the cashew is very susceptible to it (Farooki and Sreeramu, 1999).

Growth of cashew is affected in shallow soils also. In deep soils, growth and development is vigorous. The growth and performance of trees in the coastal soil is generally good suggesting tolerance of the crop to soil salinity. However, laboratory trials indicated that cashew has only a little tolerance for soil salinity and also that difference in tolerance exists amongst cashew trees (Rocchetti, 1970). Cashew cannot grow and yield well in saline soils (Marlos *et al.*, 2007). Electrical conductivity of irrigation water at 1.48 dSm^{-1} is threshold tolerance for precocious cashew during its initial growth (Carneiro *et al.*, 2002). Higher soil-temperature and density resulted in reduction of plant growth and roots (Oliveira *et al.*, 2003). Flowering in cashew requires mild winters and availability of soil moisture plays a key role in kernel development (Rao *et al.*, 2001).

Cashew grows excellently well in the laterite soils of Sindhudurg and Ratnagiri districts of Maharashtra, where other crops fail to establish. Cashew grows excellently in rich laterites, in forest loams soils. The growth rate, flowering and productivity are quite high in such soils. It is also found growing in semi-hard lateritic rock to some extent.

Currently, cashew is widely grown in the wastelands of Maharashtra, Gujarat, Tamil Nadu, Karnataka and Orissa where other crops are not successful. Although these soils are less endowed with moisture and nutrients, cashew is capable of showing satisfactory performance particularly due to its capacity to tolerate drought. Large areas of wasteland, in Maharashtra, Goa, Gujarat, Andhra Pradesh, Tamil Nadu, Orissa, Bihar and Chhattisgarh are also being brought under cashew in the recent years.

In northern Kerala, there exist extensive areas of hard lateritic wastelands which are not suitable for other crops. These soils are highly eroded with hard crust of iron and aluminium. Vegetation is scanty and it is difficult to take even pits in these soils. Excellent growth and better yield are noticed in the cashew trees grown in these environments. The cashew trees of these areas remain as living examples to confirm its potential for the utilisation of hard lateritic waste lands that are unsuitable for other crops (Salam and Peter, 2010)

Climate

Cashew may require relatively dry atmosphere with mild winter for better flowering. The mild winter may defined as "low minimum surface air temperature ranging between 15 and 20°C coupled with more dew nights having moderate dew" in the humid climates.

Cashew requires high temperature; but, it will not tolerate prolonged periods of extremes of cold and frost. The tree is very sensitive to cold when young but become fairly hardy with age and can withstand light frost for short periods. Cashew is grown in areas with rainfall ranging from 600-4500 mm per annum. Distribution of rainfall is more important than the quantity. Heavy rain during flowering and fruit set affects adversely and ultimately on yield. Thus, dry weather during this period seems to be conducive. The Kokan region of Maharashtra is characterized by modest temperature conditions (17 ^{0}C to 35 ^{0}C), high humidity (more than 80%), high rainfall (3000 to 3500 mm/year) and well-drained lateritic and sandy soils along the coastal belt with undulating topography. Such kind of soil and climatic situations is considered as most favourable for cashew plantation.

Cashew is a hardy and drought-resistant tropical crop. It grows well from mean sea level to an elevation of 1,000 metre above it. It thrives well in the areas with an annual rainfall range from 600 to 4,000 mm and temperature between 15-45°C, but a relatively high humidity and temperature between 19° and 35°C, is the best condition. It is highly susceptible to frost, thus situations with frequent occurrence of frost should be avoided. There should be dry spell during flowering, as cloudy weather during flowering enhances tea mosquito infestation. Heavy rain during flowering will lead to poor fruit-set. High temperature during marble stage of fruit development causes fruit drop (Farooki and Sreeramu, 1999).

Relative humidity during pre-flowering is a key factor in explaining yield variation in cashew (Haldankar *et al.*, 2003). Cashew can tolerate mild to moderate levels of moisture stress and growth of seedlings is unaffected (Latha and Salam, 2003; Souza *et al.*, 2004). Strong and severe water stress resulted in 20 and 22 per cent reduction in the number of scions produced, respectively (Shingre *et al.*, 2003).

Rainfall and temperature are the two major climatic factors contributing to production in cashew. A dry weather with clear sky and bright sunshine during flowering and fruit setting stages ensures good harvest. A well distributed back rain of North East monsoon contributes to high soil moisture status during the flowering and fruiting periods and ensures uniform and profuse flowering, better fruit set and development. If the monsoon is erratic and scarce, the moisture content of the soil depletes which results in partial failure of the emergence of new flushes and flower panicles. Cloudy weather during flowering also results in drying up of flowers. Heavy rains during flowering or fruit development may result in the crop loss. On the contrary, light showers during flowering and fruit set may sometimes be beneficial in arid cashew tracks. Normally, dry spell during January to May with occasional light summer showers ensure good crop production in cashew. The early monsoon showers during March-April may be more beneficial especially to late cashew varieties. Cashew needs a climate with at least 4-5 months well defined dry season to achieve best results. This might be one of the reasons for better cashew quality and higher production in the Kokan region of Maharashtra under rainfed conditions.

Cashew tree produces more fruits on the peripheral branches, which receive full sunlight. Cashew is sun loving plant, hence shady conditions is not permissible to plant. Under extreme shade condition, branches grow contiguous vertical manner till it gets the sufficient light, resulting into intermingling of branches, disturbing the natural canopy of the tree, more deadwood production, low fruiting area and poor yield. The cashew responds well towards sunlight and produces more new flushes (laterals), flowering panicles and fruit set on exposed side than on the shaded branches. Flowering, fruit set and yield differ with canopy direction. Flowering and fruiting in cashew is always more on south-west direction of the canopy (60-80%), while, it is less (20-30%) in the northern side of the canopy.

Cashew is essentially a tropical crop, grows best in the warm, moist and typically tropical climate. The distribution of cashew extends up to 27°N and 28°S latitude and is restricted to altitudes below 700 m where the temperature does not fall below 15° C for prolonged periods, although it finds growing at an elevation up to 1000 m. Most favourable optimum temperature lies between 24° C to 28° C. The cashew is hardy and drought resistant, but it is damaged by frost. Cashew does not tolerate prolonged period of extremes of cold and frost. Sunshine of minimum 7-8 hrs is required for better growth and production. Because of its adaptive ability in wide range of agro-climatic conditions with quality and higher production, it has become economical crop and attained the status of an export oriented commodity bringing considerable foreign exchange to the country.

The climatic factors influencing the cashew growth and productions

- Dry spell during flowering and fruit setting ensures better harvest
- Cloudy weather during flowering enhances scorching of flowers due to tea mosquito infestation
- Heavy rains during flowering and fruit set damages production
- High temp. (39-42° C) during marble stage of fruit development causes fruit drop
- Cashew performs better when the period of drought is shorter.

Recent studies conducted on "Impact of climate change in cashew" at DCR, Puttur indicated that the rainfed cashew crop is highly sensitive to changes in climate particularly during reproductive phase. Cashew requires relatively dry atmosphere and mild winter (15-20^0 C min. temperature) coupled with moderate dew during night for profuse flowering. High temperature (>34.4^0 C) and low RH of <20 % during afternoon causes drying of flowers resulting in yield reduction (Table 11). Unseasonal rains and heavy dew during flowering and fruiting period aggravated the incidence of pests and diseases. Pollinating insect activities also reduced to the minimum resulting in poor setting of nuts during that period. All these situations resulted in reduced yield (Anon., 2012c).

Table 11: Impact of climate change in cashew

S.N.	Temperature	RH % (A/N)	Dew factor	Flowering Yield
1.	15-20^{0}C (Min)	>20	Moderate	Optimum Satisfactory
2.	21-22^{0}C (Min)	>20	Poor	Reduced flowering Low yield
3.	>34.4^{0}C (Max)	<20	High	Drying of flowers Low yield due to Tea Mosquito Bug (TMB) and Flower drying

Nair *et al.* (1979) reported that dry spell during flowering and fruit setting ensures better harvest, cloudy weather during flowering enhances scorching of flowers due to tea mosquito bug infestation. Climate change in the Western Ghats, coastal districts and NE region is projected to significantly affect crop production. However, the impacts are crop specific and simple adaptation strategies such as change in variety and altered agronomy, high input delivery and use efficiency can offset the negative impacts of climate change (Kumar *et al.*, 2011).

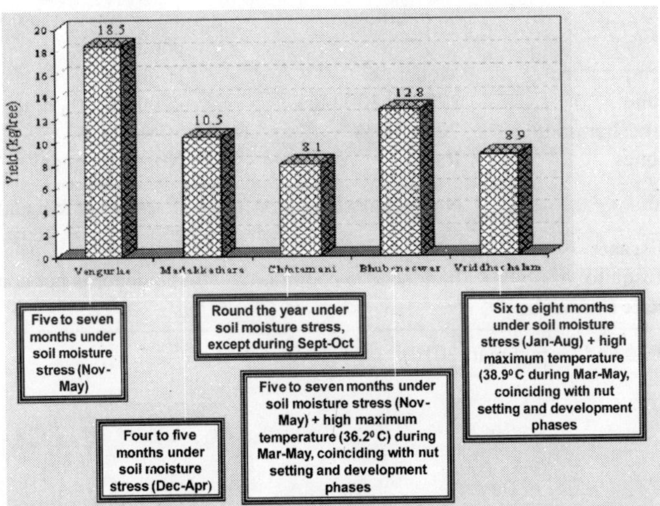

Fig. 4: Mean cashew yield of superior varieties at selected locations under agro-climatic conditions

Souce: Rao and Gopikumar (2010).

The effect on mean cashew yield of superior varieties under selected locations agro-climatic condition, Agro-climatological zones for cashew and Influence of altitude on cashew flowering reported by Rao and Gopikumar (2010) is presented in Figure 4, 5 and 6, respectively. Cashew and its environment across the West and East Coasts of India are presented in Table 12.

Table 12: Cashew and its environment across the West and East Coasts of India

Environment	West Coast	East Coast
Plant		
i. Planting Material	Pedigree & Senile	Relatively new plantations
ii. Crop management including crop protection measures	Poor	Relatively better
iii. Dormant phase	No moisture stress	Soil moisture stress
iv. Reproductive phase	Soil moisture stress	Soil moisture stress
v. Flowering	December/January	February/March
Soil		
i. Soil	Laterite	Littoral sand and sandy loam
ii. Topography	Hill Slopes	Plain
Climate		
i. Coastal climate	Same	Same (except Chinthamani & Jagadalpur)
ii. Annual rainfall	2000 to 3600 mm	600 to 2000 mm
iii. Max. Temperature	< 35°C (30-35°C) – may be optimum	> 35°C (occasionally up to 40°C) - may not be conducive
iv. Min. Temperature	15°C (15-20°C) – may be optimum	<15°C (occasionally goes below 10°C in north) - may not be conducive
v. Soil temperature	Moderate	High
vi. Sunshine	2700 to 3000 hrs	2700 to 3000 hrs
viii. Relative humidity	70 to 90 %	50 to 70 %
viii. Cyclones	Rare	Not uncommon
Irrigation	No summer irrigation	Farmers practice irrigation
Pest & Diseases		
i. Tea mosquito Moderate to severe	Incidence is common	Incidence is not common Low

(*Source*: Rao and Gopikumar, 2010)

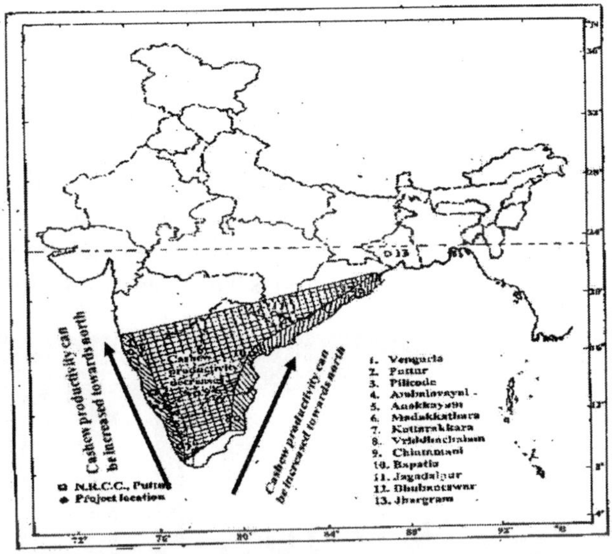

Fig. 5: Agro-climatological zones for cashew
(*Source*: Rao and Gopikumar, 2010)

Effect of Abiotic and Biotic Factors on Cashew Flowering

Flowering of cashew is highly erratic due to following reasons

- Crop management practices
- Exposure to sunlight
- Latitude
- Altitude
- Climate (Rainfall & its distribution, Surface air temperature, Sunshine, Humidity, Dewfall , Wind and Cold waves)
- Soils (Sandy to Red sandy loam and Lateritic gravel)
- Early, Medium and Late flowering types
- Variety/Cultivar
- Age of plantations
- Planting material; grafts-seedling? & Adaptability
- Crop disturbance due to pest and disease in the previous year

Influence of Altitude on Cashew Flowering

Variation in the flowering season of cashew is related to altitude and latitude. Cashew trees at higher altitudes, irrespective of latitude, show almost the same tendency for late flowering and fruiting (Fig. 6). Latitude also influences the flowering period in cashew. At Anakkayam (11°N) and Mangalore (13°N) the West coast the peak flowering is in the early January and the peak harvest is in the early April. At Bapatla (17°N) in the East coast, the peak flowering time is mid January to mid February and the crop is harvested by end of May (Fig. 7).

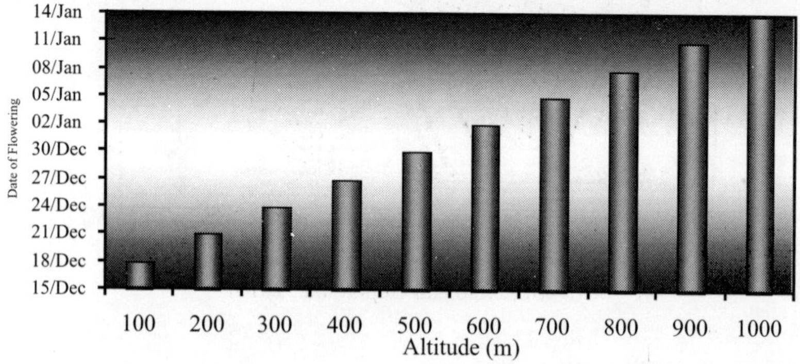

Fig. 6: Influence of altitude on cashew flowering

(*Source:* Rao and Gopikumar, 2010)

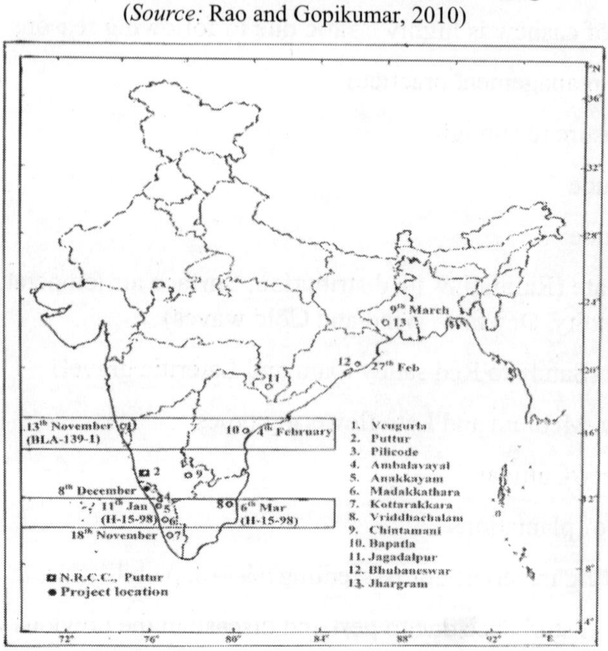

Fig.7: Latitudinal effect of cashew flowering across India

(*Source*: Rao and Gopikumar, 2010)

Mandal (1992) attempted to rate cashew growing environments as very good, good, fair and poor based on variation in altitude, rainfall, proximity to sea, maximum and minimum temperature, humidity and occurrence of frost. The ratings and the range of these parameters are indicated in Table 13.

Table 13: Environmental factors for growing cashew

Parameters	Very Good		Good	Fair	Poor
	Class I	Class II	Class III	Class IV	Class V
Altitude	20	20-120	120-450	450-750	
Rainfall (mm/yr.)	1500-2000	-	1300-1500	1100-1300	900-1100
Proximity to sea (km)	<80	80-160	160-240	240-320	-
Maximum temperature in summer (^{0}C)	28-32	32-33	33-34	34-35	-
Minimum temperature in winter (^{0}C)	19	18-19	17-18	-	15-17
Humidity (%)	70-80	65-70	60-65		50-60
Occurrence of frost	None	None	Very rare	Once in 5 yrs	-

Source: Mandal, R. C., 1992. Cashew Production and Processing Technology.

□□□□

CHAPTER – 5

Biodiversity and Crop Improvement

Anacardium occidentale is a medium-sized tree, spreading, and evergreen, much branched; grows to a height of 12 m. When grown on lateritic, gravelly, coastal sandy areas, it rarely exceeds 6 m and develops a spreading habit and globose shape with crown diameter to 12 m but, if grown inland on loams; it reaches 15 m tall and is much branched, with a smaller (4-6 m) crown diameter. The root system of a mature *A. occidentale*, when grown from the seed, consists of a very prominent taproot and a well developed and extensive network of lateral and sinker roots. Leaves simple, alternate, coriaceous, glabrous, obovate, rounded at ends, 10-18 x 8-15 cm, with short petiole, pale green or reddish when young and dark green at mature. Rai and Vatsala (1985) studied the dermal morphology and stomatal ontogeny of *Spondias pinnata* Kurz., *Lannea coromandelica* (Houtt.) Merr., *Anacardium occidentale* L., *Holigarna arnottiana* Hook. F. and *Mangifera indica* L. are studied. The cuticle is thick in *H. arnottiana, A. occidentale* and *M. indica*, but is thin in *S. pinnata* and *L. coromandelica*. Based on the shape of epidermal cells, the five species are categorised into three: cells in both upper and lower epidermis sinuous (*A. occidentale, H. arnottiana, S. pinnata*); upper sinuous and lower polygonal (*M. indica*) and upper polygonal and lower sinuous (*L. coromandelica*). Stomatal ontogeny aperigenous (*L. coromandelica*), paramesoperigenous (*A. occidentale*) and anisomesogenous (*M. indica, H. arnottiana*). *S. pinnata* is peculiar in that it possesses three different types of stomata, aperigenous, hemiparameso-perigenous and paramesoperigenous; whereas, in all the other four species, only one type of stomatal ontogenesis exists. Except *S. pinnata*, all the other species possess epidermal appendages. Unicellular trichomes occur in *A. occidentale* and *H. arnottiana*, and multicellular glandular trichomes in *A. occidentale, M. indica* and *L. coromandelica*. The multicellular hairs are found in *A. occidentale* in specialized pockets in two rows on either side of the major veins - a unique feature which is reported for the first time.

The inflorescence is a terminal panicle-like cluster commonly bearing male and hermaphroditic flowers. The male flowers are the most numerous and usually bear 1 exerted stamen and 9 small inserted ones. *A. occidentale* normally comes into flowering in 1st year itself. However, production can be taken from 3rd year after planting in order to develop canopy. The nut, which is the true fruit, dries and does not split open. Inside the thick shell, there is a large curved seed, nearly 2.5 cm long, the edible cashew nut (kernel). As the nut matures, the stalk (receptacle) at the base enlarges rapidly within a few days into the fleshy fruit like structure, broadest at the apex, popularly known as the fruit (cashew apple). This thin-skinned edible cashew fruit has a light yellow spongy flesh, which is very juicy, pleasantly acidic and slightly astringent when eaten raw and highly astringent at green stage. The generic name was given by Linnaeus and refers to the vaguely heartshaped look of its false fruit.

The individual productive capacity of the common cashew tree varies greatly, with plants that produce from less than one to more than 50 kg of nuts per harvest. The weight of nut also varies greatly, ranging from 3 to 33 g, with the weight of peduncle ranging from 20 to 200 g. The minimum age for stable production is greater than 8 years, and normally between 12 to 14 years. The dwarf cashew tree is characterized by growth that begins earlier in the season, its smaller size and early productivity. Other characteristics that differentiate the dwarf cashew tree from the common types are: a compact and homogenous crown, with an average span reaching approximately twice the height; smaller, lighter green leaves; smaller stem diameter; initial branches closer to the ground; early seasonal flowering; a prolonged period of fructification; smaller nuts; and generally larger peduncles (Barros, 1988). Nut weight in natural populations varies from 3 to 10 g and the peduncle (cashew apple) from 20 to 160 g, thus showing a smaller variability in relation to the common type (Vieira *et al.*, 2005). The individual reproductive capacity of the dwarf type is also lower with maximum known productions not higher than 65 g of nuts per harvest (Barros, 1995).

In Brazil, there are two groups of cashew. The most prevalent is the common type (giant), which grows to heights ranging from 5 to 8 m, but able to reach 15 m. The diameter of the crown generally varies from 12 to 14 m and, in exceptional cases up to 20 m (Barros, 1995). The other group, the dwarf cashews grow up to 4 m, on an average, with a crown diameter of 6 to 8 m. This group of dwarf cashew flowers between 6 and 18 months (Barros *et al.,* 1998).

Related Species of Cashew

The family Anacardiaceae to which cashew belongs comprises of 60 genera and 400 species of tropical and subtropical trees and shrubs.

According to Baily (1958), genus Anacardium contains eight tropical American species. Parente (1972) reported ten different species, but Peixoto (1960) named 20 different species; several of which have edible peduncles such as *A. nanum*. A very early bearing type, *A. subterranium*, a small shrub with its trunk completely underground containing water resources, *A. microcarpum* a small tree from sandy savannas, *A. spruceanum* a large tree, and the largest species of the genus, *A. gigantium* which grows in the Amazon forest with good timber value. Valeriano (1972) named five different species namely, *A. occidentale*, *A. pumilum*, *A. gigantium*, *A. rhinocarpus* and *A. spruceanun*. He also suggested the recognition of only two species namely, *A. nanum* and *A. gigantium* and further subdivided based on colour (yellow or red) and shape (round, pear-shaped or elongated) of the pseudo-fruit. Description provided by Peixoto (1960) separated recognition of more than two species. There is possibility that some species originated due to intercrossing between species. Nambiar *et al.* (1990) reported 20 species of *Anacardium* (Table 14).

Table 14: Species of *Anacardium**

Species	Centre of diversity
A. brasiliense	Brazil
A. curatellaefolium	Brazil
A. encardium	Malaysia
A. giganteum	Brazil
A. humile	Brazil
A. mediterraneum	Brazil
A. nanum	Brazil
A. occidentale	Brazil
A. rhinocarpus	Columbia, Venezuela
A. spruceanum	Brazil
A. microsepalum	Brazil
A. corymbosum	Amazon region
A. excelsum	Brazil
A. parvifolium	Central America
A. amilcarianum	Amazon region
A. kuhlmannianum	Brazil
A. regrense	Brazil
A, rondonianum	Brazil
A. tenuefolium	Brazil
A. microcarpum	Amazon region

Source: *Nambiar *et al.* (1990).

Genetic Diversity

Cashew is botanically known as *A. occidentale*. According to Bailey (1958), Anacardium is a small genus of eight species indigenous to South America. It appears from the published accounts that *A. occidentale* is the only species, introduced outside the New World although as many as 20 species of Anacardium exist within Central and South America. Mitchell and Mori (1987) recognized ten species in the genus Anacardium (Table 15). The genus has a primary centre of diversity in Amazonia and a secondary centre in Brazil. Ascenso (1986) reported that cashew (*Anacardium occidentale*) is the only species in the genus which attained economic importance. This morphologically polymorphic species also exhibits chromosome polymorphism (Mitchell and Mori, 1987). Chromosome numbers reported in literature ranges from 2n=24 (Khosla *et al.*, 1973; Goldblott, 1984), 2n=30 (Machado, 1944), 2n=40 (Simmonds, 1954; Goldblott, 1984) to 2n=42 (Khosla *et al.*, 1973; Goldblott, 1984; Darlington and Janaki Ammal, 1945; Purseglove, 1988). Such chromosome polymorphism is well known in many domesticated plants (Khosla *et al.*, 1973). Wide variation within the accessions due to cross pollinating nature of crop was reported for 292 germplasm accessions of seed origin which were evaluated at CPCRI Regional Station, Vittal and Shantigodu. Considerable variability was observed for some of the economic characters like flowering season, flowering duration, harvesting duration, nuts/panicle, apple weight, nut weight, apple to nut ratio, shelling percentage, kernel weight, kernel count/lb, shell thickness and mean yield/plant (Table 16) (Rao and Swamy, 1994). Variability in nut and apple characters such as shape, size and colour are enormous and are of much use in crop improvement programme.

Table 15: Species of Anacardium and Synonyms

Species	Synonym (s)
A. excelsum	*Rhinoc Rhinocarpus excelsa Anacardium rhinocarpus Anacardia rhinocarpa, Anacardium caracoii*
A. spruceanum	*Anacardium brasiliense*
A.occidentale	*Cassuvium pomiferum Acajuba occidentaleAnacardium occidentale var. americanum A. occidentale var. indicum A. mediterraneum, A. curatellifoliumCassuvium reniforme A. occidentate var. longifolium A. occidentale var. gardneri A. subcardatum, A. microcarpum A. rondonianum, A. amilcarianum A. Kuhlmannianum, A. othonianum*
A.humile	*Monodynamus humilus A. subterraneum, A. pumilum A. pumilum var. petiolata A. humile var. subacutum*
A. nanum	*A. pumila*
A. corymbosum	-
A. parvifolium	*A. tenuifolium*
A. fruticosum	*A. parvifolium var. fruticosd*
A. giganteum	-
A. microsepalum	*A. negrense*

(*Source*: Mitchell and Mori, 1987).

Table 16: Variability recorded for some of the characters in the germplasm accessions (seedling progenies)

Character	Range
Flowering season	October – January
Flowering duration	40-127 days
Harvesting duration	30-105 days
Number of fruits per panicle	1-8
Apple weight	30-150 g
Nut weight	2.4-18.0 g
Apple: nut ratio	4:1-12:1
Shelling percentage	19.0-35.0
Kernel weight	0.5-4.5 g
Kernel count/lb	100-900
Shell thickness	1.5-5.0 mm
Mean yield/plant/year (10 annual harvests)	0.50-11.75 kg

(*Source:* Rao and Swamy, 1994)

Exploration and Collection

Even though no reliable record of the introductions are available, it is presumed that the initial productions in the Malabar coast of Kerala were from only few trees and due to the hardy nature of the crop, it had spread to all the coastal regions of India naturally. All these introductions were belonging to *A. occidentale.* The initial emphasis was only on the establishment of plantations of seedling origin. As cashew is primarily a cross pollinated crop, it is highly heterozygous. Considerable segregation has resulted in large variation in the populations (Rao and Bhat, 1996). Early attempts for germplasm collections were made during the early fifties with the sanctioning of *ad-hoc schemes* in the composite states of Madras, Travancore, Cochin and Bombay. The research stations started under these *ad-hoc schemes* in Kerala (Kottarakkara), Karnataka (Ullal), Tamil Nadu (Vridhachalam), Andhra Pradesh (Bapatla) and Maharashtra (Vengurla) took up the programme of collection of locally available elite plants for evaluation and further selection for commercial exploitation and for breeding programme as well. These were the first attempts in collection of cashew germplasm in India. Many of the other research centres which were established subsequently have collected the seeds of germplasm from these centers, namely, Bapatla, Kottarakkara, Ullal, Vridhachalam and Vengurla. While making the initial collections of germplasm, these centres had confined their survey mainly to the respective states and hence they represent the local germplasm available in the states. Since the inception of All India Coordinated Spices and Cashewnut Improvement Project (AICS and CIP) in 1971, Central Plantation Crops Research Institute (CPCRI) Regional Station, Vittal, also took up the programme of cashew germplasm accession which mainly consisted of the seedling progenies of collections available at Bapatla, Vridhachalam, Vengurla, Anakkayam and a few collections made locally from Karnataka. Subsequent to the establishment of National Research Centre for Cashew (NRCC) at Puttur (Karnataka) in 1986, the germplasm collection through seeds has been discontinued and only the vegetatively propagated material is being collected and conserved in the National Cashew Gene Bank (NCGB). A coordinated approach was brought in the cashew germplasm collection by organizing joint survey teams consisting of scientists of NRCC and the centres of All India Coordinated Research Project on Cashew (AICRP on Cashew) of the respective states. The main criteria adopted for collection of germplasm were the superior economic characters such as high yield, bold nut, cluster bearing, apple qualities, canopy types and dwarfness of the plant or any other unique characters observed during the survey. Since 1986, cashew germplasm collection surveys are being undertaken by NRCC in collaboration with centres of AICRP on Cashew and other cashew centres of cashew growing States, namely, Karnataka, Kerala, Maharashtra, Goa, Tamil Nadu, Andhra Pradesh, Orissa and West Bengal. So far, all together 494 cashew germplasm accessions were made and planted in cashew field gene bank at NRCC, Puttur for evaluation (Nayak and Bhat, 2004a). Till the establishment of NRC for Cashew, over one thousand accessions of seedling origin were collected by different Cashew Research Stations in the country. Among the seedling accessions, 23 are exotic types, of which nine were collected from Brazil, Nairobi, Mtwara, Lindi, Nacala, Mozambique, Tanganyika, Singapore and Australia and 14 from the Republic of Panama. The non-traditional areas such as Garo Hills (Meghalaya) and Bastar (Chhattisgarh) were also surveyed for germplasm collection. However, more comprehensive exploration is required for targeted collection of germplasm with specific traits such as resistance to TMB, CSRB, high yield, dwarf and compact canopy, bold nut with cluster bearing, tolerance to drought, frost, salt and to problematic soils (Nayak and Bhat, 2005).

Characterization of cashew germplasm has been taken up at NRCC using the "Cashew Descriptors" suggested by IBPGR (Presently IFGRI), Rome, Italy. The first set of 56 accessions planted in 1986 have been characterized and published as the "Catalogue of Minimum Descriptors of Cashew Germplasm Accessions-I" (Swamy et al., 1997). The second set of 97 accessions planted in 1987 and 1988 have been characterized and documented in the "Catalogue of Minimum Descriptors of Cashew Germplasm Accessions-II, (Swamy et al., 1998). The third set of 102 accessions planted in 1989 and 1990 have been characterized and documented in the "Catalogue of Minimum Descriptors of Cashew Germplasm Accessions-III" (Swamy et al., 2000). These are the first few efforts made in characterization of clonal accessions of cashew in the world and so far, 255 accessions have been characterized and catalogued.

Germplasm Conservation

In situ conservation of cashew germplasm is done only in the Amazon forests of Brazil (original home of cashew). However, ex situ conservation is also being followed in cashew as in all other tree crops. The cashew germplasm is maintained as an active collection in the field gene bank which locks up the land for a long time. Further, field maintenance also has the inherent danger of losing the valuable collections during the natural calamities as it often happens in the East Coast, especially at Bapatla, Andhra Pradesh and Bhubaneswar, Orissa centres due to cyclones. In order to overcome this danger as well as to effectively utilize the land for other field experiments, suitable *in vitro* preservation / cryopreservation methods with the aid of biotechnology are to be developed. An effort in this direction is being made by establishing a Biotechnology Laboratory at NRC for Cashew, Puttur. In the NCGB so far 433 clonal accessions which were collected from 1986 to 2000 have been planted @ 6 grafts per accession by adopting a spacing of 6m x 6m and are being maintained as active collections in the field gene bank. Similarly, Regional Cashew Gene Banks (RCGBs) has been established at AICRP on Cashew Centers which are maintaining a total of 1225 accessions. The germplasm accessions conserved in the NCGB at NRCC, Puttur include the diverse types-high yield, bold sized nut, semi-tall plant, cashew nut shell liquid (CNSL) free, purple plant, high shelling percentage, cluster bearing, big sized apple, early maturity, etc. and the related species, namely, *Anacardium pumilum, A. othonianum* and *A. microcarpum*. Recently, a germplasm conservation block has been established by-planting 4 grafts per evaluated accessions at a closer spacing of 4m x 4m so as to reduce the land utilized for maintenance of large number of germplasm collections.

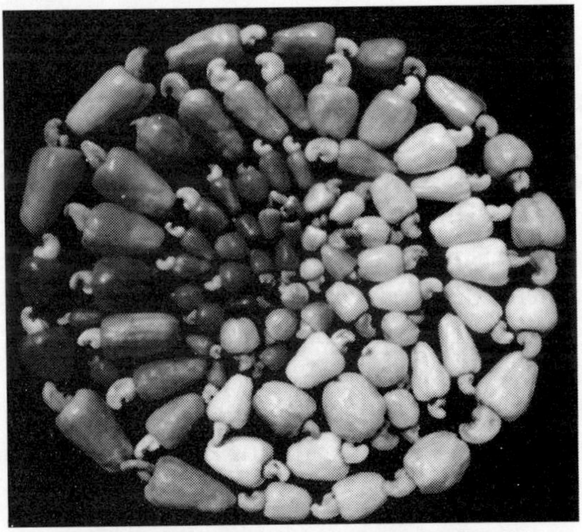

Fig. 8: Germplasm diversity in cashew
(*Source*: Bhat *et al*., 2010)

Germplasm Registration

Germplasm of agri-horticultural, commercial and other economic crop, which are unique and has verified attributes of scientific/commercial value are being registered by the Plant Germplasm Registration Committee of ICAR at NBPGR, New Delhi. Nine germplasm accessions from National Cashew Field Gene Bank at Puttur for unique characters viz., NRC-59 (VTH 196/18) for big apple, bold nut size and high shelling percentage, NRC-111 (Mardol-4) for mid season flowering, big apple and big nut size, NRC-116 (CNSL Free) for cashewnut shell liquid free type, NRC-120 (Nairobi) for early season flowering, bold apple and nut size, NRC-121 (Purple genotype) for purple stem and leaves and high shelling percentage, NRC-140 (VTH 155 L) for semi-tall, early season (Nov-Dec) and long flowering duration (120 days), NRC 142 (VTH 578/1) for wild relative with genetic diversity, NRC-152 (VTH 713/4) for wild relative with genetic diversity and NRC-201 (Pl.No.1254) for upright and compact habit, semi-tall type were registered with NBPGR, New Delhi (Nayak and Bhat, 2004b).

Germplasm Augmentation Through Introduction From Other Countries

One of the immediate priorities of the Indian cashew germplasm programme is to enhance the genetic variability in the collections, as all the plantations were raised from the limited initial introductions and genetic base available in the country is only moderate. Therefore, introduction from Central America and Brazil where considerable variability including the dwarf types are reported and as these are also the original home of cashew is essential. Introduction of *A. gigantium* from Surinam which is reported to have the biggest apple (200 g) will be advantageous, especially in Goa where the cashew apple utilization for processing industry contributes substantially to the economy of the state. Introduction of dwarf clones reported from Brazil (Ascenso, 1986) will be useful in developing high density orchards which will enhance productivity considerably. One of the main problems in cashew is that all the existing germplasm accessions are susceptible to cashew stem and root borer (CSRB) which kills the trees. There is a need to screen the allied species (for their suitability as root stocks) which have relatively hard wood and also possess smooth bark. It is suggested that introduction from Brazil of other species like *A. rhinocarpus* and *A. spruceanum* which are reported to possess hard wood seem to be useful for testing their suitability as root stocks (Rao and Swamy, 1994).

The Biological Diversity Act, 2002

The Indian Parliament enacted the Biological Diversity Act, 2002 for conservation, sustainable utilization and equitable sharing of the benefits arising out of utilization of genetic resources. The primary objectives of the Biological Diversity Act - 2002 are as follows:

- Conservation of biological diversity
- Sustainable use of the components of biodiversity
- Fair and equitable sharing of benefits arising out of the utilization genetic resources and bio resources.

The Biodiversity Act, 2002 covers all biological resources occurring and naturalized in India. Besides affirming that biological resources are sovereign property of the State, the Act also encompasses all indigenous and traditional knowledge associated with the biological resources as these forms are intangible component of biological resources. Basically the Act promotes conservation and sustainable use of biodiversity and protects and prevents the activities, which are detrimental to the conservation principles. The access to bio-resources is regulated and mutual transfer agreements on mutually agreed terms are entered into for such access, transfer of results, and IPR issues. The export of bio-resources should not take place without the permission of access granted by the National Biodiversity Authority. The access may be denied in case of threatened/ endangered species/strains and also in specified circumstances like threat to the environment and Bio-safety. The salient features of the Biological Diversity act, 2002 reported by Kannaiyan *et al.* (2010) are given below:

- To regulate access to biological resources of the country equitable share in benefits arising out of the use of biological resources.
- To conserve and sustainable use of biological diversity.
- Setting up of National Biodiversity Authority (NBA), State Biodiversity Board (SBB) and Biodiversity Management Committees (BMCs).
- NBA and SBB are required to consult BMCs in decisions relating to bio-resource/related knowledge within their Jurisdiction.
- To respect and protect knowledge of local communities traditional knowledge related to biodiversity.
- To secure sharing of benefits with local people as conservers of biological resources and holders of knowledge and information relating to the use of biological resources.

- All foreign nationals/organizations require prior approval of NBA for obtaining biological resources and/or associated knowledge for use.
- Indian scientists/individuals require approval of NBA for transferring results of research to foreign nationals/organizations.
- Conservation and development of areas of importance from the standpoint of biological diversity by declaring them as biological diversity heritage sites.
- Protection and rehabilitation of threatened species.
- Involvement of institutions of State Government in the broad scheme of the implementation of the BD Act through constitution of committees.
- Protect India's rich biodiversity and associated knowledge against their use by foreign individuals and organizations without sharing benefits arising out of such use and check Biopiracy.

Biotechnology in Cashew

Work on cashew tissue culture has been in progress on developing multiplication protocols from cashew explants. However, regeneration protocols from mature explants are yet to be developed, although, reports are available on multiplication and field establishment of cashew plant regenerated from young nodal cuttings (Philip, 1984; D'Silva and D'Souza, 1992; Lievens *et al.*, 1989; Keshavachandran and Khader, 1990; Leva and Falcone, 1990; Nair and Mohanakumar, 1993; Das *et al.*, 1996; Thimmappaiah and Shirly, 2000). Somatic embryogenesis was induced from nucellar tissue cultured from 2-3 week old immature nuts. So far, nucellar tissues from 14 varieties have been tested for embryogenesis. Embryogenesis and germination of somatic embryos was achieved in two varieties (Thimmappiah, 1997; Shirly and Thimmappaiah, 2005). Attempts have been made to fingerprint cashew using RAPD markers. Twenty Tanzanian accessions (Mneney *et al.,* 1998), 90 germplasm accessions at DCR, Puttur (Dhanaraj *et al.*, 2002) and 19 germplasm accessions (Archak *et al.*, 2002) have been fingerprinted. Based on this analysis, a moderate range of genetic variability among the accessions analysed in India. About 239 accessions at DCR have been fingerprinted by Thimmappaiah *et al.* (2009a). Low-level diversity has been observed in 40 elite varieties using RAPD, ISSR and SSR markers at DCR, Puttur. Moderate diversity has been reported in cashew population using protein isoenzyme electrophoretic analysis (Aliyu and Awopetu, 2007; Maranan and Mendiore, 2008; Thimmappaiah *et al.*, (2009a, b). Cashew microsatellites from non-enriched genomic library and sequences of 21 polymorphic SSR markers have been detected and used for multiplex analysis in cashew (Croxford *et al.*, 2005). Gene cloning studies have been reported (Wang *et al.*, 2002). Neto *et al.* (1995) showed utility of RAPD markers in distinguishing dwarf seedlings in cashew. Bulk Segregant Analysis (BSA) in germplasm bulks at DCR, Puttur, could identify four RAPD markers linked to economic characters like nut weight and plant stature.

Regeneration from explants of mature origin has met with difficulties due to a high rate of contamination and poor response (Thimmappaiah and Shirly, 2000; Thimmappaiah *et al.,* 2002a, c). Micrografting technique was standardised in cashew (Mantell *et al.,* 1997; Ramanayake and Kovoor, 1999, Thimmappiah *et al.*, 2002b). Embryogenesis was induced in immature cotyledonary segments (Hegde *et al.*, 1994; Nair and Mohanakumar, 1993; Sy *et al.,* 1991; Thimmappiah, 1997) and nucellar tissue (Nair and Mohanakumar, 1993; Ananthakrishnan *et al.*, 1999 a, b; Gogate and Nadgauda, 2000; Cardoza and D'Souza, 2002).

The analysis of genetic relationships in cashew using morphological traits and RAPD banding data can be useful for plant improvement, descriptions of new varieties and also for assessment of variety purity in plant certification programmes (Rejani Nadgauda *et al.*, 2005).

Relatively little genetic information is available about the cashew (2n = 2x = 42) because there was little breeding effort until relatively recently. In Brazil and possibly Venezuela, the cashew tree appears to be very heterogeneous, although elsewhere, e.g. Africa and Asia, the cashew tree is probably fairly homozygous. The generation period for cashew is only 3-5 years, and hence it should be possible to gather basic genetic information in a committed breeding programme. There are two groupings, which are based upon yellow and red fruit colour (Rejani Nadgauda *et al.*, 2005).

Rootstocks

Cashew seedlings have aggressive taproot and feeder root systems, which are thought to confer some level of tolerance to soil stress. Cashew trees can thrive on soils of low fertility and in littoral regions (Srurtz, 1985). Elite cashew selections have been propagated by grafting, air layering and rooting of cuttings, although most orchards consist of seedling trees. There are no soil-borne pests or stresses that cause serious production problems. Although vegetative propagation is used only rarely, there is no production problem that is associated with lack of uniformity of rootstock. Perhaps the most serious disorder associated with roots is damage caused by the root and stem borer *Plocaederus ferrugineus* L. (Coleoptera: Cerambycidae) (Rajapakse, 1998). This is also the most serious insect pest of cashew, and can cause significant crop losses

(Rajapakse, 1998). Heavy infestations can cause tree death (Pillai *et al.*, 1976). Rao (1998) suggested that *A. rhinocarpus and A. spruceanum*, having smooth bark, should be tested for their graft compatibility with the cashew tree to provide protection from this pest.

Scions

Major breeding objectives

The potential yield of cashew trees is thought to be >10 t raw nuts per hectare per annum (Ascenso, 1986a). Yields have rarely exceeded 3-4 t raw nuts per hectare when vegetatively propagated trees have been grown under careful management (Van Eijnatten, 1992). More typically, yields in orchards of seedling trees have rarely exceeded 0.4-0.71 ton raw nuts per hectare. Therefore there is considerable potential to improve the efficiency of nut production through planting of superior cultivars (Rejani Nadgauda *et al.*, 2005).

Diseases

According to Freire *et al.* (1995) the most important disease of cashew in Brazil is anthracnose, caused by *Colletotrichum gloeosporioides* Penz. The pathogen attacks leaves, young stems and shoots, floral parts and fruit. Blossom blight (Le. anthracnose) of cashew is vectored primarily by *Helopeltis antonii* Sign. (Heteroptera: Miridae) (Nambier *et al.*, 1973). As much as > 50% yield losses can be caused by anthracnose, and fruit and nut quality can be significantly reduced (Cardoso *et al.*, 1994). Anthracnose disease is most problematic in growing regions with high rainfall and relative humidity. According to Van Eijnatten (1992), control of the disease by periodic systemic fungicide application is possible but uneconomic. The application of elemental sulphur has been utilized to control anthracnose and powdery mildew on cashew, but this can be prohibitively expensive in many growing areas. Moreover, there is evidence that prolonged dusting of cashew with sulphur causes increased soil leaching, lowered pH and increased aluminium in the soil to toxic levels (Majule *et al.*, 1998). Currently, control of anthracnose is achieved by removal and burning of affected plant parts. Other pathogens implicated in blossom blight include *Pestalotiopsis* spp., *Gloeosporium mangiferae* and *Botryodiplodia* spp. (Rajapakse, 1998).

Powdery mildew disease, caused by *Oidium anacardii* Noack, is the leading production problem of cashew in East Africa (Boma *et al.*, 1998b). The disease affects leaves in new flushes, inflorescences, young fruit, nuts and cashew apples. If left uncontrolled, production is lost. Wettable sulphur and Dinocap fungicide are effective for control; however, long-term problems have been associated with the use of sulphur (see above) and Dinocap is too expensive for the main production areas in East Africa.

Dwarfness and precocity

Cashew tree size and shape are highly variable. According to Van Eijnatten (1992), canopy surface area determines production in a grove, because flowers and fruit are borne on the outer edge of the canopy. Therefore production can be maximized with small trees or with regular hedging to control tree size. 'Kanaka' and 'Priyanka', both of which resulted from controlled pollinations, are precocious bearers (Rao and Bhat, 1996).

Insect pests

The tea mosquito *H. antonii* sign causes heavy damage to young leaves and inflorescences of the cashew tree (Rajapakse, 1998). Losses that can exceed 30% have been reported in East Africa and India (Nambiar *et al.*, 1990). According to Boma *et al.* (1998a), damage caused by the tea mosquito until recently caused the greatest crop losses in East Africa. Blossom blight, caused by *C. gloeosporioides, Pestalotiopsis* spp., *G. mangiferae* and *Botryodiplodia* spp., is associated with damage caused by this sucking insect pest.

Fruit and nut quality

For juice processing a large cashew apple is desirable. The primary goal of existing breeding programmes has been increased nut yield, nut quality, kernel size and shelling percentage. Nuts with high protein and low sugar levels are preferred. CNSL, a by-product of cashew nut processing, is a mixture of phenol-based monomers, for which there is high industrial demand. CNSL is a resin with a high level of heat tolerance, and is used as a base in paint and varnish and for brake linings. CNSL is highly flammable, is a powerful skin irritant and can be very toxic. One of the constituents of CNSL is cardol, an allergen and irritant. Tree selections have not been made that would have industrial utility, i.e. high levels of CNSL. Conversely, there has been no attempt to select such trees that have nuts with low levels of CNSL. Contamination of cashew nut with CNSL is currently a major constraint of processing (Rejani Nadgauda *et al.*, 2005).

Breeding accomplishment

Although breeding programmes have existed two or more decades in many of the major cashew-growing countries, such as Brazil, India and some African countries, relatively little information is available in the public domain. Lack of movement of cashew germplasm from the centre of origin to cashew-growing countries has severely limited cashew breeding. Results obtained through classical breeding have not been encouraging because of the moderately long juvenile period and a narrow genetic base (Rejani Nadgauda *et al.*, 2005). Breeding objectives of ten lack clear definition, and crop improvement has been aimed to meet local needs. For example, high-density planting of improved selections with compact canopies is important in India, and replanting old seedling orchards with compact canopy varieties is preferred (Rao, 1999). Most orchards are still seed-propagated, and vegetatively propagated selections are considered to be uneconomic, as is the use of pesticides and chemical fertilizer inputs. Consequently, cashew production is more akin to agroforestry than horticulture (Rejani Nadgauda *et al.*, 2005).

Disease and pest resistance

Preliminary investigations have been reported in which sources of resistance to specific insects and diseases have been tentatively described. Millanzi (1998a) determined that damage caused by the tea mosquito is related to the flushing pattern of cashew trees. Early flowering is associated with severe damage. A correlation has been established that relates vigour of the inflorescence bud with resistance to powdery mildew disease (Millanzi, 1998b).

Fruit and Nut Quality

Breeding programmes to date have concentrated upon increased yield over other nut quality parameters. According to Rao (1998), the narrow genetic base of cashews in many countries limits the potential for cultivar development by conventional breeding. In addition, some of the related species have horticultural characters that could possibly be transferred to cashew. Cashew breeding in India at nine different research centres has resulted in 35 improved cultivars, out of which 24 are selections from openly pollinated seedlings and 11 are derived from controlled pollinations.

Molecular Genetics

Molecular studies would be very useful for characterizing the genetic diversity among different cashew cultivars and for identifying genes of commercial interest. This information could be used effectively for cultivar development (Rejani Nadgauda *et al.*, 2005). Neto *et al.* (1995) have evaluated the use of random amplified polymorphic DNAs (RAPDs) for genetic finger printing. Genomic DNA was extracted from seedlings of four dwarf clones, and amplified using six 10-mer arbitrary primers. Twenty-seven amplification products could be used for DNA finger printing. Only one primer generated bands sufficient for distinguishing each of the four clones. The results indicated that the RAPD technique is feasible for germplasm analysis and characterization in cashew.

Mneney *et al.* (1998) attempted to assess the level of genetic diversity in cashew genetic resources using RAPD markers. DNA from 25 cashew accessions representing various collections was amplified using random 10-mer primers; however, only limited DNA polymorphisms were observed. Dhanaraj *et al.* (2001) demonstrated that RAPD markers were effective for assessing diversity among 90 accessions from different parts of India. RAPD markers generated by seven Operon primers (10-base-long) were used. Their analysis indicated that the diversity in Indian cashew is not as narrow as reported earlier, but moderate (maximum dissimilarity 63%, minimum 12%). A core collection of 58 selected individuals revealed the same diversity as the entire population. This is an initial step towards efficient germplasm management of cashew in India, and will assist in selection of diverse parents in breeding programmes (Rejani Nadgauda *et al.*, 2005).

Micropropagation

All of the reports concerning the regeneration of cashew plants from shoot tip and nodal cultures have involved either *in vitro* germinated or greenhouse-grown seedlings (D'Silva and D'Souza, 1992; Thomas, 1995; Das *et al.*, 1996; Mantell *et al.*, 1998; Aranthakrishnan *et al.*, 1999b; Boggetti *et al.*, 1999; Thimmappaiah and Samuel, 1999). Successful regeneration has been reported from 6-15-month-old greenhouse-grown seedlings (Lievens *et al.*, 1989), 1-year-old greenhouse-grown seedlings (Boggetti *et al.*, 1999), 3-year-old greenhouse-grown seedlings (Leva and Falcone, 1990) and 5-year-old seedlings (Boggetti *et al.*, 1999). Proliferating shoot cultures can be established from highly juvenile material; however, there is progressive decline in shoot initiation with age (Das *et al.*, 1996; Mantell *et al.*, 1998; Boggetti *et al.*, 1999). Although bud break and some shoot elongation occurred with explants from 5-year-old seedlings, axillary bud proliferation and rooting have not been obtained. High phenolic exudation from field-grown shoot tips of cashew is considered to be the major hindrance for in vitro regeneration (Thomas, 1995; Das *et al.*, 1996).

Multiple shoot formation from cultured shoot tips and nodes was reported on Murashige and Skoog (MS) (1962) medium, although Boggetti *et al.* (1999) employed MS with half strength major salts and Lievens *et al.* (1989) used Lepoivre medium. Optimum growth regulator formulations for culture initiation have varied, and this may be a function of genotype or stage of development of the material. Relatively high concentrations of cytokinin(s) have been used, e.g. 5-20 μM benzyladenine (BA) (Boggetti *et al.*, 1999), 22.2 M-M BA (D'Silva and D'Souza, 1992) and 2.32 μM kinetin + 9.12 μM zeatin + 4.4 μM BA (Das *et al.*, 1996). Maltose significantly enhances shoot proliferation (D'Silva and D'Souza, 1992; Boggetti *et al.*, 1999). Shoots have been rooted in various ways: (i) inoculation with *Agrobacterium rhizogcnes* (Das *et al*., 1996); (ii) medium supplemented with 2.9 μM indoleacetic acid (IAA) and 4.9 μM indole-3-butyric acid (IBA) (D'Silva and D'Souza, 1992); and (iii) pulsing with 100 μM IBA for 5 days (Boggetti *et al*., 1999). Rooted micro-propagated shoots have survived transfer to soil.

Micrografting

Micrografting of microscions from mature trees on to *in vitro*-germinated seedling rootstocks has been described by Mantell *et al.* (1998) and Mneney and Mantell (2001). Nuts can be germinated in vitro on semi-solid or in liquid MS (half-strength) medium supplemented with 20 g/l sucrose. Seedlings are decapitated below the cotyledons, and shoot tip and side-grafts (< 0.5 mm to > 5 nun) have been equally successful. The grafted plantlets can be potted out after 10-12 weeks.

Micrografting of seedling shoot tips and axillary buds on *in vitro* germinated root-stock has also been described by Ramanayake and Kovoor (1999). Scions were treated with IAA, and cut surfaces were irrigated with citric acid solution to control browning. An exogenous auxin- NAA (Naphthalene Acetic Acid in culture medium) was required for graft fusion and development.

Somatic Cell Genetics

An important prerequisite for the application of somatic cell genetics to cashew improvement is regeneration from elite selections. Srinivasachar (1938), in a comparative study of the morphology of embryogeny within the Anacardiaceae, described the development of probable proembryos from the cashew nucellus. He compared these to the adventitious embryos that developed from the nucleus of polyembryonic mangoes.

Regeneration

Organogenesis

Philip (1984) induced direct shoot organogenesis from (non-elite) cotyledon explants of mature cashew seeds. Lin and Staba (1961) plant growth medium supplemented with 2.85 μM IAA and 2.32 μM kinetin was utilized. It is possible that the *in vitro* growth conditions in this study were suboptimal, because the author described Tripolar differentiation of organs within the meristematic mound. Hegde *et al.* (1990) reported organogenesis from embryonic ends of cotyledons from mature seeds. Sections of the cotyledons cultured on Linsmaier and Skoog (LS) medium with NAA and kinetin produced multiple shoots and roots in 4 weeks. After 2 months of culture, however further growth of these multiple shoots ceased, and the root and shoot desiccated.

Somatic embryogenesis

The first attempts to induce embryogenic cashew cultures involved seed explants (Jha, 1988; Lakshmi Sita, 1989; Hegde *et al.*, 1990; Sy *et al.*, 1991). Somatic embryos were obtained from cotyledon pieces of 6-8-week-old germinated seedlings (Lakshmi Sita, 1989), cotyledons from mature seeds (Sy *et al.*, 1991) and sections of immature cotyledons (Hegde *et al.*, 1994). Cotyledon explants on medium with NAA, 2, 4-dichlorophe-noxyacetic acid (2,4-D) and BA formed embryogenic cultures, and somatic embryos developed in the presence of NAA, BA or kinetin (Lakshmi Sita, 1989). In studies by Sy *et al.* (1991), the proximal portion of cotyledons produced somatic embryos when they were cultured initially on Schenk and Hildebrandt (SH) (1972) medium with NAA and BA, followed by culture on medium with casein hydrolysate and adenine sulphate. Hegde *et al.* (1994) observed that cotyledon pieces formed somatic embryos on LS medium with Ca pantothenate, IAA and BA under a 20 h photoperiod. Leafy shoots and roots developed on medium supplemented with activated charcoal; however, there was generally poor organization of shoot meristems.

Ananthakrishnan *et al.* (1999a) and Gogte and Nadgauda (2000) induced embryogenic cultures from (elite) nucellar cultures.

Induction

This protocol is based largely upon the procedures of Ananthakrishnan *et al.* (1999a) and Gogate and Nadgauda (2000). Immature cashew seeds are bisected longitudinally, and each half is placed into culture so that the nucellus is in contact with the plant growth medium. The induction medium of Ananthakrishnan *et al.* (1999a) consists of MS, modified as follows: 60 g/l sucrose, 400 mg/l glutamine, 100 mg/l ascorbic acid 100 mg/l casein hydrolysate, 8 g/l agar, 20% (v/v) coconut water and 6.78 µM 2, 4-D. Somatic embryogenesis from nucellar tissue has also been reported by Cardoza and D'Souza (2000) on MS medium with 0.5 mg/l picloram. Gogate and Nadgauda (2000) also utilized MS medium, but supplemented with 5.0 \M 2, 4-D + 15.0 µM gibberellic acid (GA_3) + 5.0 µM BA. Cultures were maintained at 25°C either with a 16 h photoperiod (40 µ mol/m2/s) (Ananthakrishnan *et al.*, 1999a) or in darkness (Gogate and Nadgauda, 2000). Embryogenic cultures were evident 7-21 days after explanting.

Maintenance

Gogate and Nadgauda (2000) maintained embryogenic cultures on MS supplemented with 10.0 µM 2, 4-D + 15.0 µM GA_3 + 10 % coconut water with 40 g/l sucrose and 0.5% activated charcoal. Ananthakrishnan *et al.* (1999a), on the other hand, transferred embryogenic cultures directly into liquid MS medium supplemented with 400 mg/l glutamine, 100 mg/l ascorbic acid, 200 mg/l casein hydrolysate and 4.52 µM 2,4-D for rapid proliferation. Suspension cultures were incubated at 100 rpm on a rotary shaker in darkness.

Maturation

Conditions for development and maturation of cashew somatic embryos have not been well defined. Ananthakrishnan *et al.* (1999 a, b) described the development of heart and torpedo stage somatic embryos from embryogenic suspension cultures; however, it is apparent that development beyond this stage was not observed. It would also appear from the figures that somatic embryo development was sporadic. Gogate and Nadgauda (2000) initiated somatic embryo development on MS medium supplemented with 5.0 µM 2, 4-D + 30 µM GA_3 + 10 % coconut water together with 40 g/l sucrose and 0.05% casein hydrolysate. Somatic embryos developed up to the cotyledonary stage and further germinated when transferred to MS + 0.5% activated charcoal + 3% sucrose + 0.2% Gelrite. The development of cotyledons was, however, negligible as compared to size of cotyledon in a germinating mature zygotic embryo. Complete growth and development of cotyledons of somatic embryos would help in better maturation and conversion rates (Rejani Nadgauda *et al.*, 2005).

In studies reported by Cardoza and D'Souza (2002) embryogenic callus was formed in presence of 0.5 mg/l picloram. On further transfer to medium with 2.07 µM picloram + 1.0 mg/l putrescine somatic embryos were formed. Somatic embryos germinated in the presence of 1.89 µM ABA, whereas, Ananthakrishnan *et al.* (1999a, b) obtained reversion of torpedo stage embryos to callus, in the presence of ABA.

Recently somatic embryogenesis has also been reported using the immature zygotic embryo as an explant (Cardoza and d'Souza 2000; Gogate and Nadgauda, 2003).

In both cases, embryogenesis was direct, where somatic embryos were formed directly from radicle tip/radicle end of the immature zygotic embryo. Somatic embryos were formed in the presence of picloram (2.07 x M) (Cardoza and D'Souza, 2000) or 5 µM 2,4-D + 5 µM BAP + M 3 (3 µM) (Gogate and Nadgauda, 2003). Medium with 20 (µM ABA and 3% maltose was used for maturation and hormone-free medium for germination of somatic embryos (Gogate and Nadgauda, 2003). Cardoza and D'Souza (2000) also reported germination on medium devoid of growth regulators.

Haploid recovery

Although there have been no reports of anther culture and haploid plant recovery, there is a report of in vitro germination of cashew pollen (Subbaiah, 1984). Boron was critical for pollen germination and tube growth, and pollen tubes grown in boron free medium were distorted in shape. Optimal medium contained 30% polyethylene glycol, 20% sucrose and 20 mg/l each of calcium and boron. These studies may be helpful for devising a medium for anther culture (Rejani Nadgauda *et al.*, 2005).

Despite the importance of the cashew as a fruit and nut crop, there have been few breeding and production advances. This is due to the narrow genetic base outside the centre of origin, barriers that have been erected to discourage the exchange of germplasm and the lack of resources, particularly in Africa. Basic in vitro strategies have been developed; however, these have not been refined to the point that, for example, genetic transformation and mutation induction and selection could be used to address specific breeding objectives. With little effort, after screening advanced cultivars for their embryogenic potential, some of the disease problems of cashew (*anthracnose, blossom blight, etc.*) could be addressed using one or more of these procedures. In the medium and long term, cashew is more likely to be improved by genetic engineering than by conventional breeding (Rejani Nadgauda *et al.*, 2005).

Cashew Gene Bank

National Cashew Field Gene Bank (NCFGB) has been established at the Directorate of Cashew Research with a mandate to collect and conserve the available diverse cashew germplasm. The Directorate has the largest germplasm collection of cashew in the country with 527 accessions. A total of 433 cashew accessions have been assigned with National Collection numbers (IC Nos.). A total of 285 accessions have been characterised as per IPGRI descriptors. Three germplasm catalogues for 255 accessions have been brought out (Swamy *et al.,* 1997, 1998 and 2000). In addition, over 1200 cashew accessions are conserved in Regional Cashew Field Gene Bank at the centres under AICRP on Cashew. They are being utilized in crop improvement programme for developing new high yielding varieties. A total of 40 varieties have been released so far, most of which are popular among farmers. Collection, conservation, evaluation and cataloguing of both exotic and indigenous cashew germplasm especially, dwarf and compact cashew types would be further emphasized.

Hybridization programme using germplasm having desirable traits such as higher number of panicles/m^2, better nut weight and high yield as parents would be focused.

Genetic Improvement of Cashew

Varieties with high yield, resistance to biotic and abiotic stresses, better flowering behaviour (synchronized and staggered) and better kernel quality for internal consumption and export would be evolved. Critical areas to be addressed include development of dwarf and compact high yielding cashew varieties suitable for high density planting. Systematic efforts would be made for releasing improved varieties in terms of yield, nut quality and processing parameters. Characterization of the released varieties or hybrids would be done for various physiological and other growth characters to suit the agro-ecological requirements. (Anon., 2011).

Application of Tissue Culture to Cashew Breeding

The recalcitrant nature of cashew has been attributed to the limited success recorded so far in the *in vitro* culture of the crop and abnormal development has been reported in the calli derived from its explants. Browning of explants in cashew was found to be due to the presence of high secondary metabolites and it has been reduced through frequent transfer of explants, addition of activated charcoal and dark treatment. Explants necrosis has also been traced to the effect of strong sterilization. Meanwhile, the use of explants from in vitro germinated seedlings or fungicidal treated young flush has been found to improve the success rate significantly. The use of MS base salt supplemented with two-step treatment of cytokinins enhances the response of cashew explants and development of derived plantlets (Aliyu, 2005).

□□□□

CHAPTER – 6

Cashew Varieties

Cashew research in India dates back to 1950's when Indian Council of Agricultural Research sanctioned ad-hoc schemes in the then composite Madras, Travancore, Cochin and Bombay States. Thus, Kottarakkara (Kerala) in 1952; Ullal (Karnataka in 1953 and Vengurla (Maharashtra) in 1957 came into being as pioneering cashew centres in the country. More research schemes were started with ICAR's assistance in 1955 at Bapatla (Andhra Pradesh) and in 1956 at Deragaon (now shifted to Kothiali) in Assam. During the period, Central Food Technological Research Institute, Mysore and National Chemical Laboratory, Pune were also engaged in developing technologies for economic utilisation of cashew apple and Cashew Nut Shell Liquid (CNSL). In the second phase of strengthening cashew research in India, the ICAR started All India Spices and Cashewnut Improvement Project in 1971 linking countrywide network among different research centres. Collection, conservation and evaluation of germplasm, hybridization and selection, vegetative propagation, and nutritional studies were the primary areas under this programme.

Maharashtra, the first state was to release cashew variety under the name Vengurla-1 through Dr. Balasaheb Sawant Konkan Krishi Vidyapeeth, Dapoli in the year 1974. So far, 8 varieties have been released from Regional Fruit Research Station, Vengurla (M.S.), of which, Vengurla-4, Vengurla-6, Vengurla-7 and Vengurla-8 are widely accepted for commercial plantation.

Among the pre released promising F_1 hybrids planted in 1983 and 1984, the performance of hybrid No. 255, 303, 320, and 367 was promising in respect of nut yield, size of nut, shelling percentage and whole kernel count (Table 17). Therefore, these hybrids have been included in multilocational trial since 1998.

In the year 1990-91 onwards new hybridization programme for the improvement of nut size of released varieties/type was launched. Cashew varieties, Vengurla-2 (4.3 g nut weight) and Vengurla-5 (4.5 g nut weight) crossed with bold nut type accessions, Kankadi (16 g), B.T.1, B.T.6, B.T.10, B.T.22, B.T.65 and other bold accessions resulting F_1 hybrids attaining bold nuts. So far, more than 3000 F_1 hybrid seedlings (include more than 350 cross combinations) have been planted at RFRS, Vengurla during 2000 to 2011 for further evaluation.

Manoj and George (1993) have reported that heterosis in cashew crop and calculated the selection index value in promising hybrids for ranking their superiority, it is expected that these outstanding crosses, if propagated clonally, will not only increase the indigenous production, but also will help to boost the export of good quality kernels in global market and earn sizable foreign exchange for the nation.

Table 17: Performance of promising hybrids planted in 1983 and 1984 at RFRS, BSKKV, Vengurla, Dist. Sindhudurg (M.S.)

S.N.	Characters	H-303	H- 320	H- 367
1.	Cross combinations	V-4 x VRI-I	VRI-II x Vetore-56	V-4 x VRI-I
2.	Cumulative yield (10 yrs)	67.760	72.750	104.67
3.	Av. Yield nut (kg)	6.77	7.275	10.467
4.	Maximum yield (kg/plant)	12.965	14.20	22.465
5.	Av.nut Wt.(g)	8.92	7.45	11.58
6.	Shelling (%)	28.00	32.00	28.00
7.	Kernel count (W)	210	240	180
8.	Apple weight (g)	75	80	100
9.	Tree canopy	Compact	Upright and Compact	Compact
10.	Special feature	Bold nuts	Bunch bearing and high shelling %	Big nut and apple

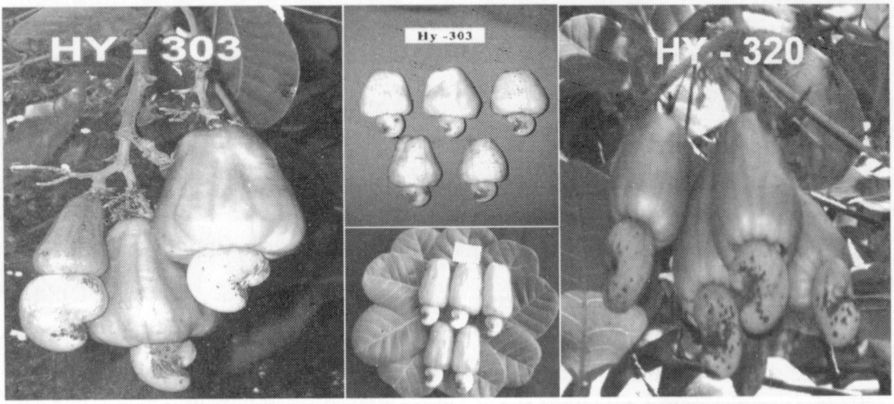

The first attempt in cashew improvement was made in India under the ad-hoc project of Indian Council of Agricultural Research by which five Research Stations were set up in various states. The Cashew Co-ordinated Research Project is now being carried out at 11 centres and a Directorate of Cashew Research, (formerly National Research Station for Cashew) Puttur, with the main objective of evolving high yielding varieties tolerant/resistant to pests and diseases and to develop agro-techniques including plant protection. From commercial point of view, the ideal cashew plant should have dwarf and compact canopy with intensive branching habit, short and flowering fruiting phase, more than 20 per cent perfect flowers, 8 to 10 nuts per panicle, medium to bold nut (8 to 10 g each), high shelling per cent (28 %) and high yield potential of more than 20 kg/tree/year and tolerant / resistant to major pests and diseases. In order to develop such an ideal type, efforts are being made at various research stations through introductions, selections, hybridization and mutation breeding.

In view of the need for stepping up the production and maintain our permanent position in the global market, it is felt necessary to develop such varieties either through selection or breeding which could produce high yield per unit tree canopy, bold nut size with high shelling conforming the requirement of internal market (Rao, 1989). The work on crop improvement was initiated in 1957 at Regional Fruit Research Station, Vengurle, Dist. Sindhudurg with the objective to evolve high yielding variety with medium to bold nut size and high shelling percentage to cope up with the farmers and market demand. Since 1970, efforts are being made in the direction to evolve high yielding hybrid varieties having almost desirable characters using promising selections from the existing accessions available at RFRS, Vengurla (Maharashtra).

In India more than 40 cashew varieties have been released so far, of which 14 are hybrids and 26 are selections. All these 40 cashew varieties are not presently recommended for commercial plantation because initially, achievement of high yield was only considered while releasing the variety; but due to high demand of bold export grade kernels both in international vis-á-vis in domestic market and there is more breakage of kernels while processing of small nuts. Therefore, presently only those varieties are recommended which are having bold size nuts, high shelling percentage and high yield.

Recommendation of high yielding varieties for the region shall be made for new planting and re-planting. By principle, preference is to be given for pest and disease resistant/tolerant varieties. Unfortunately, none of the improved varieties are found to be resistant to major pests of cashew. However, the variety Damodar is apparently tolerant to Tea Mosquito Bug. Likewise, Goa-11/6 (Bhaskara), released by National Research Centre for Cashew, Puttur is reported to perform well even under un-sprayed situations.

Cashew varieties nominated for different states of India is given in Table 18.

Table 18: Improved varieties of cashew recommended for different states of India are as follows.

S.N.	States	Nominated varieties of cashew
1.	Karnataka	Selection-1, Selection-2, Ullal-1, Ullal-2, Ullal-3, UN-50, VRI-1, Vengurla-1*,Vengurla-4*
2.	Kerala	Madakattara-1, Madakattara-2, K-22-1, Dhana, Priyanka, Damodar (H-1600)
3.	Maharashtra and Goa	Vengurla-1, Vengera-4, Vengurla-6, Vengurla-7, Vengurla-8
4.	Tamil Nadu	VRI-1, VRI-2,VRI-3
5.	Andhra Pradesh	BPP-4, BPP-6, BPP-8, VRI-2
6.	Orissa	VRI-2, Bhubaneshwar-1
7.	West Bengal	Jhargram-1, Jhargram-2 (recently released)
8.	Madhya Pradesh	No.40, Vengurla-4

* Recommended for North Karnataka

The Regional Fruit Research Station, Vengurla (M.S.) has given a very simple and popular technique for large scale clonal multiplication called as soft wood grafting technique in cashew. By this method about 60 to 70 lakh grafts are prepared every year in the country. In Maharashtra state, during the last decade, all the new plantations have been done by using clonal material of various improved varieties, which has resulted in increased yield *i.e.* about 1.5 t/ha. In some of the well managed plantations, surprisingly the variety Vengurla-4 yielded about 3 t/ha, *i.e.* on an average 15 kg/plant.

Flowering commences in cashew as per the following schedule given in Table 19.

Table 19: Flowering time in early, mid and late variety of cashew

Category of variety	Flowering Time
Early variety	November to December
Mid season variety	December to January
Late variety	January to February

The salient features of the high-yielding cashew varieties released in India are summarized here.

The Dr. B. S. Konkan Krishi Vidhyapeeth, Dapoli, so far released 8 high-yielding varieties of which 6 are hybrids and 2 are selections.

Vengurla-1

Year of release	1974
Centre	RCRS, Vengurla
Hybrid/Selection	Selection (Ansur-1)
Canopy type	Compact
Branching habit	Intensive
Flowering	Sept. to Nov.
Fruiting	December to March
Hermaphrodite Flower	8 %
Apple colour	Yellow
Apple weight	60 g
Juice percentage	65%
Nut weight	6.25 g
Number of nuts/kg	160
Kernel weight	1.39 g
Shelling percentage	31%
Export grade	W240
Mean nut yield/tree	15.74kg
Special character	Medium sized nut

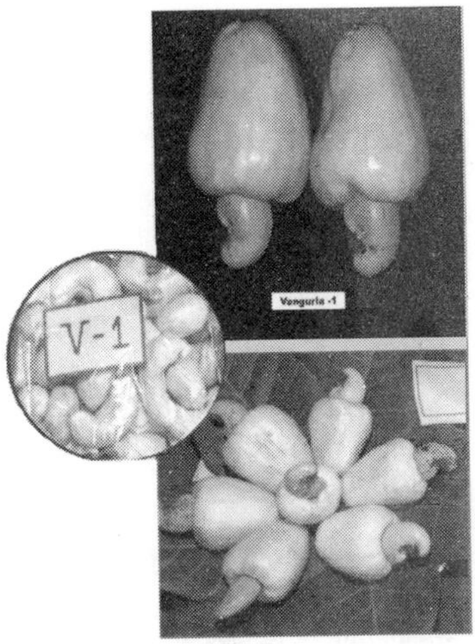

Vengurla-2

Year of release	1979
Centre	RCRS, Vengurla
Hybrid/Selection	Selection WBDC-VI
Canopy type	Compact
Branching habit	Intensive
Flowering	October to December
Fruiting	January to April
Hermaphrodite Flower	21 %
Apple colour	Red
Apple weight	37 g
Juice percentage	45%
Nut weight	4.35 g
Number of nuts/kg	230
Kernel weight	1g
Shelling percentage	32%
Export grade	W320
Mean nut yield/tree	23.10 kg
Special character	Short duration and high yield

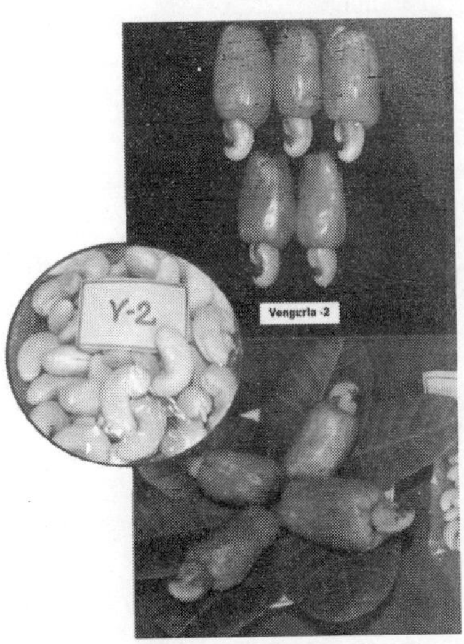

Vengurla-3

Year of release	:	1982
Centre	:	RCRS, Vengurla
Hybrid/Selection	:	Hybrid
Parentage		Vengurla-1xVetore 56
Canopy type	:	Compact
Branching habit	:	Intensive
Flowering	:	Nov. to December
Fruiting	:	February to May
Hermaphrodite Flower		25 %
Apple colour	:	Yellow
Apple weight		78 g
Juice percentage	:	77%
Nut weight	:	9.09 g
Number of nuts/kg	:	115
Kernel weight	:	2.09g
Shelling percentage	:	27%
Export grade	:	W210
Mean nut yield/tree	:	16.66 kg

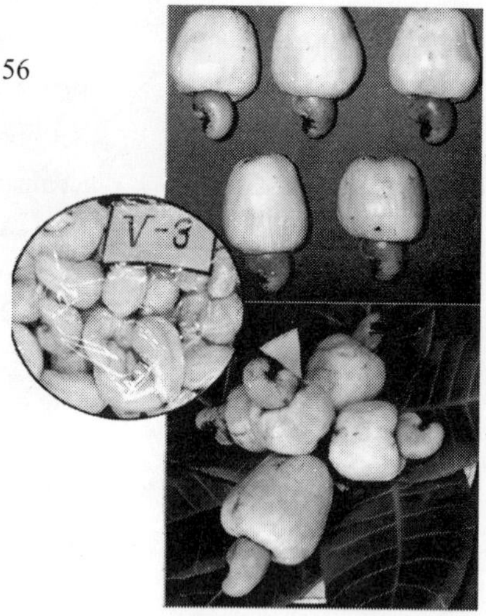

Vengurla-4

Year of release	:	1984
Centre	:	RCRS, Vengurla
Hybrid/Selection	:	Hybrid
Parentage		Midnapur Red X
Canopy type	:	Open
Branching habit	:	Extensive
Flowering	:	Nov. to Decembe
Fruiting	:	February to May
Hermaphrodite Flower		26 %
Apple colour	:	Red
Apple weight		46 g
Juice percentage	:	76%
Nut weight	:	7.69 g
Number of nuts/kg	:	140
Kernel weight	:	1.91g
Shelling percentage	:	31%
Export grade	:	W210
Mean nut yield/tree	:	19.08 kg
Special character	:	All purpose variety

Vengurla-5

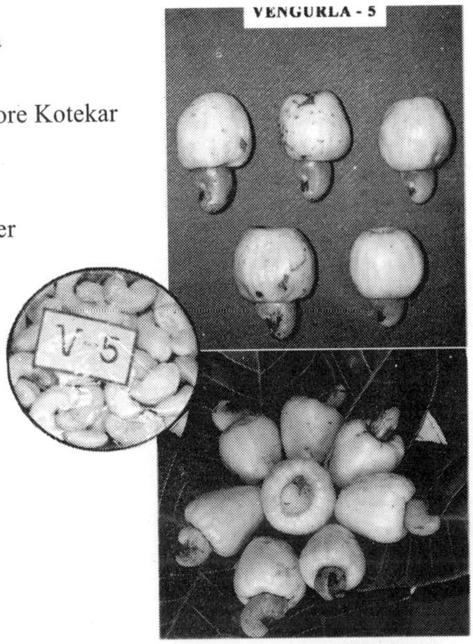

Year of release	1984
Centre	RCRS, Vengurla
Hybrid/Selection	Hybrid
Parentage	Ansur E. x Mysore Kotekar
Canopy type	Compact
Branching habit	Intensive
Flowering	Oct. to December
Fruiting	January to April
Hermaphroditc Flower	50 %
Apple colour	Yellow
Apple weight	30 g
Juice percentage	86%
Nut weight	4.54 g
Number of nuts/kg	220
Kernel weight	1.0 g
Shelling percentage	30%
Export grade	W320
Mean nut yield/tree	25.6 kg
Special character	Compact canopy with limited spread

Vengurla-6

Year of release	1992
Centre	RCRS, Vengurla
Hybrid/Selection	Hybrid
Parentage	Vetore 56 x V-1
Canopy type	Compact
Branching habit	Intensive
Flowering	Nov. to December
Fruiting	February to May
Hermaphrodite Flower	7.5 %
Apple colour	Yellow
Apple weight	70 g
Juice percentage	85 %
Nut weight	7.90 g
Number of nuts/kg	125
Kernel weight	1.95 g
Shelling percentage	28 %
Export grade	W320
Mean nut yield/tree	17.0 kg
Special character	Less spread & Bold-sized nut

Vengurla-7

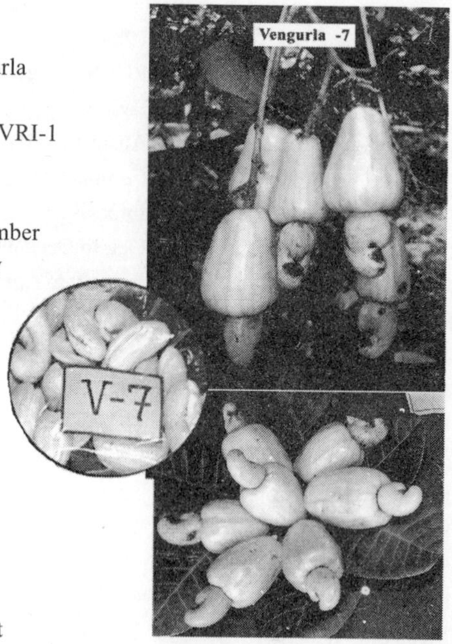

Year of release	:	1997
Centre	:	RCRS, Vengurla
Hybrid/Selection	:	Hybrid
Parentage		Vengurla-3 x VRI-1
Canopy type	:	Compact
Branching habit	:	Intensive
Flowering	:	Nov. to December
Fruiting	:	March to May
Hermaphrodite Flower		26.5 %
Apple colour	:	Yellow
Apple weight		65 g
Juice percentage	:	85%
Nut weight	:	10.0 g
Number of nuts/kg	:	100
Kernel weight	:	2.5 g
Shelling percentage	:	30.5 %
Export grade	:	W180
Mean nut yield/tree	:	14.94 kg
Special character	:	Bold-sized nut

Vengurla-8 (Gajbhiye *et al.*, 2005)

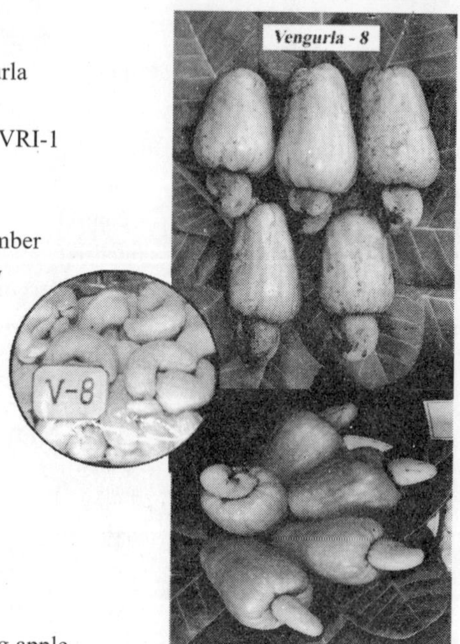

Year of release	:	2001
Centre	:	RCRS, Vengurla
Hybrid/Selection	:	Hybrid
Parentage		Vengurla-4 x VRI-1
Canopy type	:	Compact
Branching habit	:	Intensive
Flowering	:	Nov. to December
Fruiting	:	March to May
Hermaphrodite Flower		28.5 %
Apple colour	:	Red
Apple weight		100 g
Juice percentage	:	85%
Nut weight	:	11.5 g
Number of nuts/kg	:	90
Kernel weight	:	2.7 g
Shelling percentage	:	28.5 %
Export grade	:	W180
Mean nut yield/tree	:	16.5 kg
Special character	:	Bold-nut & big apple

The Kerala Agricultural University, so far, released 12 high-yielding varieties of which 7 are hybrids and 5 are selections. The characteristics of the released varieties are given below.

Anakkayam 1 (BLA-139-1)

Year of release	:	1982
Centre	:	CRS, Madakkattara
Hybrid/Selection	:	Selection
Canopy type	:	Compact
Branching habit	:	Intensive
Flowering	:	October to November
Fruiting	:	December to January
Special character	:	Early variety
Apple colour	:	Pinkish yellow
Juice percentage	:	71.0%
Nut weight	:	5.95g
Number of nuts/kg	:	168
Kernel weight	:	1.67 g
Shelling percentage	:	27.99%
Export grade	:	W280
Mean nut yield/tree	:	12 kg

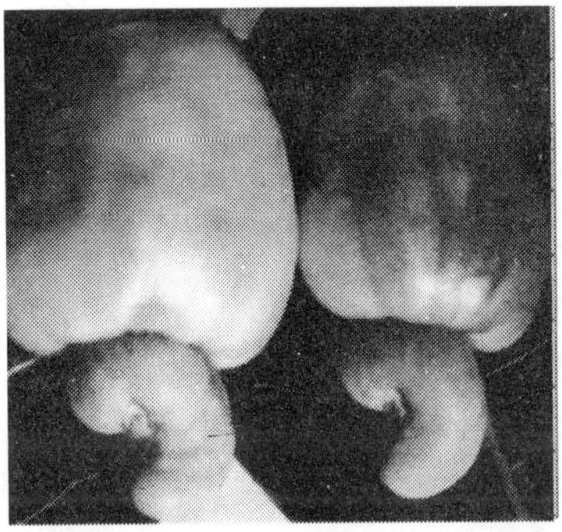

K-22-1

Year of release	:	1984
Centre	:	CRS, Madakkattara
Hybrid/Selection	:	Selection (Kottarakkara-22)
Canopy type	:	Compact
Branching habit	:	Intensive
Flowering	:	December to February
Fruiting	:	February to March
Special character	:	Mid-season variety
Apple colour	:	Red
Juice percentage	:	67.5%
Nut weight	:	6.2 g
Number of nuts/kg	:	161
Kernel weight	:	1.6 g
Shelling percentage	:	26.5
Export grade	:	W280
Mean nut yield/tree	:	13.2 kg

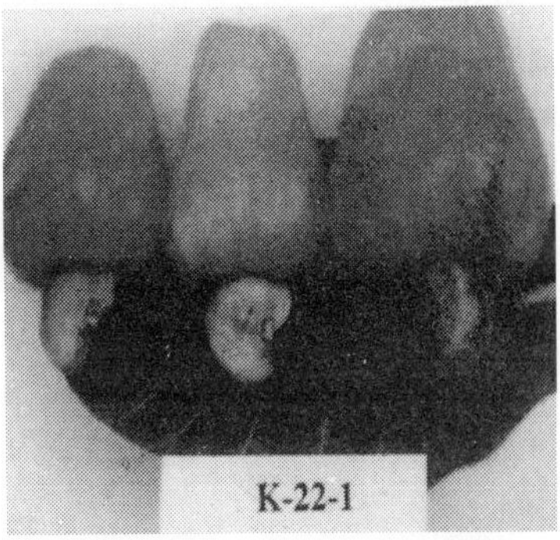

Madakkathara 1

Year of release	:	1990
Centre	:	CRS, Madakkattara
Hybrid/Selection	:	Selection
Canopy type	:	Compact
Branching habit	:	Intensive
Flowering	:	November
Fruiting	:	January to March
Special character	:	Early
Apple colour	:	Yellow
Juice percentage	:	72 %
Nut weight	:	6.2 g
Number of nuts/kg	:	61
Kernel weight	:	64 g
Shelling percentage	:	26.8%
Export grade	:	W280
Mean nut yield/tree	:	13.8 kg

Madakkathara 2 (NDR-2-1)

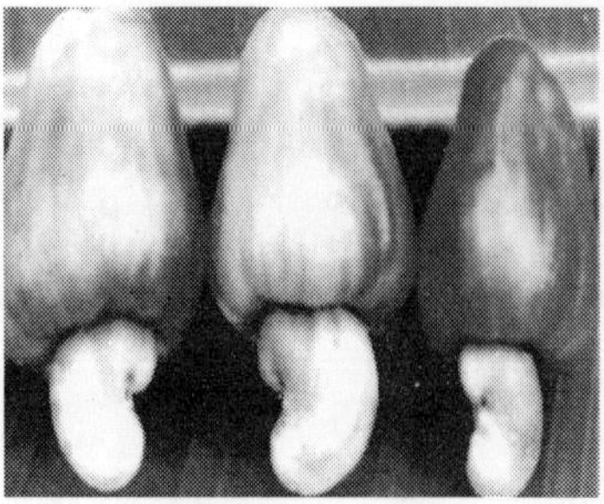

Year of release	:	1990
Centre	:	CRS, Madakkattara
Hybrid/Selection	:	Selection
Canopy type	:	Open
Branching habit	:	Intensive
Flowering	:	January to March
Fruiting	:	February to May
Special character	:	Late variety
Apple colour	:	Red
Juice percentage	:	68%
Nut weight	:	7.25 g
Number of nuts/kg	:	138
Kernel weight	:	1.88 g
Shelling percentage	:	26%
Export grade	:	W210
Mean nut yield/tree	:	17 kg

Kanaka (H1597)

Year of release	:	1993
Centre	:	CRS, Madakkattara
Hybrid/Selection	:	Hybrid
Parentage		BLA-139-1 x H3-13
Canopy type	:	Open
Branching habit	:	Intensive
Flowering	:	November to December
Fruiting	:	December to March
Special character	:	Mid-season variety
Apple colour	:	Yellow
Juice percentage	:	70 %
Nut weight	:	6.8 %
Number of nuts/kg	:	147
Kernel weight	:	2.08 g
Shelling percentage	:	30.58 %
Export grade	:	W280
Mean nut yield/tree	:	12.8 kg

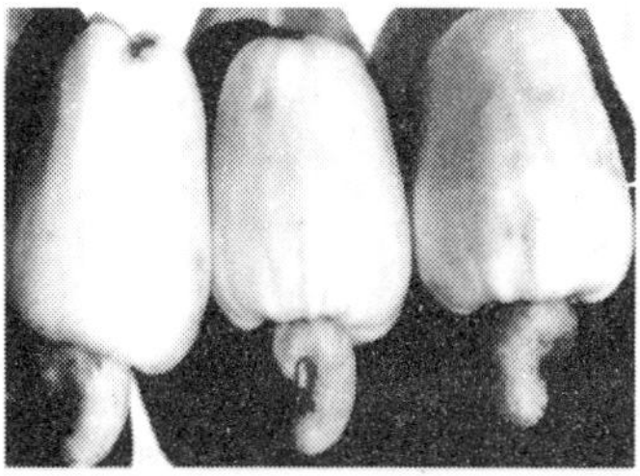

Dhana (H1608)

Year of release	:	1993
Centre	:	CRS, Madakkattara
Hybrid/Selection	:	Hybrid
Parentage		ALGD-1 x K-30-1
Canopy type	:	Compact
Branching habit	:	Intensive
Flowering	:	December to January
Fruiting	:	January to March
Special character	:	Mid-season variety
Apple colour	:	Yellow
Juice percentage	:	72 %
Nut weight	:	8.2 g
Number of nuts/kg	:	122
Kernel weight	:	2.44 g
Shelling percentage	:	29.8 %
Export grade	:	W210
Mean nut yield/tree	:	10.66 kg

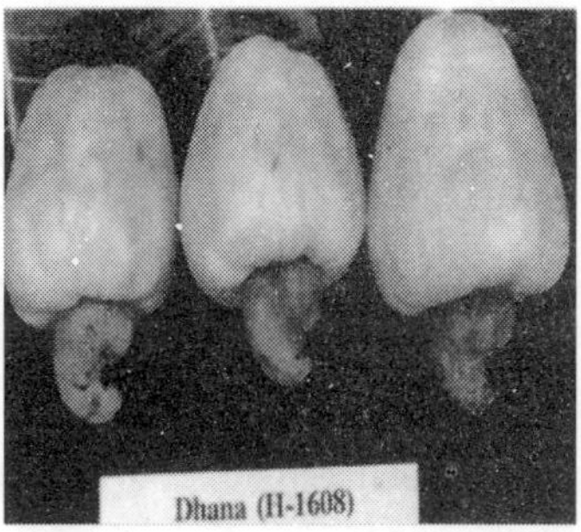

Priyanka (H1591)

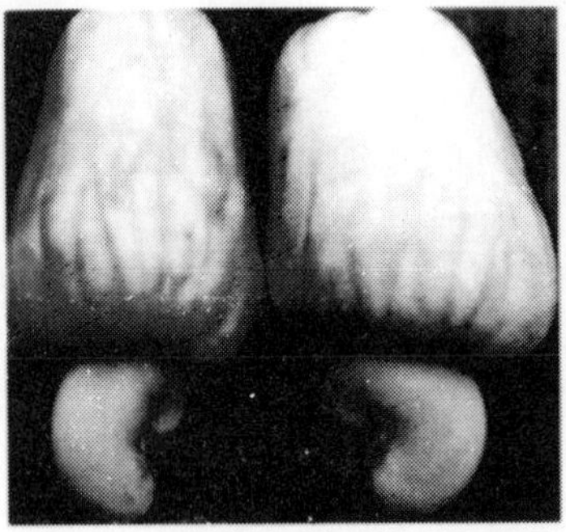

Year of release	:	1995
Centre	:	CRS, Madakkattara
Hybrid/Selection	:	Hybrid
Parentage		BLA-39-1 x K-30-1
Canopy type	:	Open
Branching habit	:	Intensive
Flowering	:	December to February
Fruiting	:	February to May
Special character	:	Bold nut variety, mid-season and drought tolerant
Apple colour	:	Yellowish Red
Juice percentage	:	67 %
Nut weight	:	8.2 g
Number of nuts/kg	:	122
Kernel weight	:	2.87 g
Shelling percentage	:	26.57%
Export grade	:	W180
Mean nut yield/tree	:	17.03 kg

Sulabha (K-10-2)

Year of release	:	1996
Centre	:	CRS, Madakkattara
Hybrid/Selection	:	Selection
Canopy type	:	Compact
Branching habit	:	Intensive
Flowering	:	January to February
Fruiting	:	February to April
Special character	:	Bold nut, late season
Apple colour	:	Light orange
Juice percentage	:	70 %
Nut weight	:	9.8 g
Number of nuts/kg	:	102
Kernel weight	:	2.5 g
Shelling percentage	:	25.51 %
Export grade	:	W210
Mean nut yield/tree	:	21.9 kg

Dhanashree (H-3-17)

Year of release	:	1996
Centre	:	CRS, Anakkayam
Hybrid/Selection	:	Hybrid
Parentage		T30 x Brazil-18
Canopy type	:	Compact
Branching habit	:	Intensive
Flowering	:	December to January
Fruiting	:	January to march
Special character	:	Mid-season
Apple colour	:	Yellowish pink
Juice percentage	:	67 %
Nut weight	:	7.8 g
Number of nuts/kg	:	128
Kernel weight	:	2.4 g
Shelling percentage	:	30.5 %
Export grade	:	W240
Mean nut yield/tree	:	15.02 kg

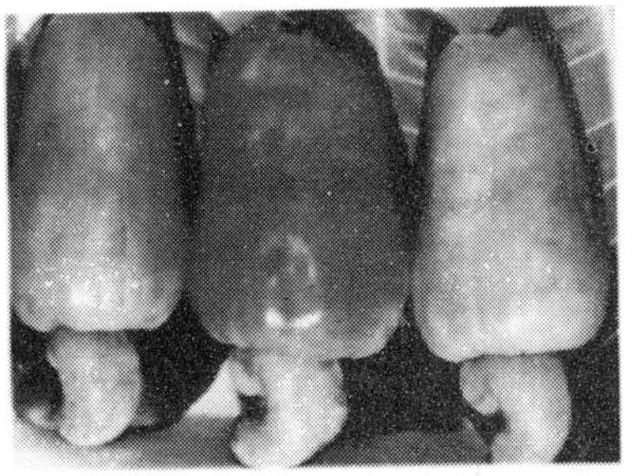

Amrutha (H-1597)

Year of release	:	1998
Centre	:	CRS, Madakkattara
Hybrid/Selection	:	Hybrid
Parentage		BLA-139 x H-3-13
Canopy type	:	Spreading
Branching habit	:	Extensive
Flowering	:	December to January
Fruiting	:	January to march
Special character	:	Mid-season variety
Apple colour	:	Yellow
Juice percentage	:	72.2 %
Nut weight	:	7.18 g
Number of nuts/kg	:	139
Kernel weight	:	2.24 g
Shelling percentage	:	31.58 %
Export grade	:	W210
Mean nut yield/tree	:	18.35 kg

Akshaya (H-7-6)

Year of release	:	1998
Centre	:	CRS, Anakkayam
Hybrid/Selection	:	Hybrid
Parentage		H-4-7 x K-30-1
Canopy type	:	Compact
Branching habit	:	Intensive
Flowering	:	December to January
Fruiting	:	January to March
Special character	:	Mid-season variety
Apple colour	:	Yellow
Juice percentage	:	68 %
Nut weight	:	11 g
Number of nuts/kg	:	91
Kernel weight	:	3.12 g
Shelling percentage	:	28.36 %
Export grade	:	W180
Mean nut yield/tree	:	11.78 kg

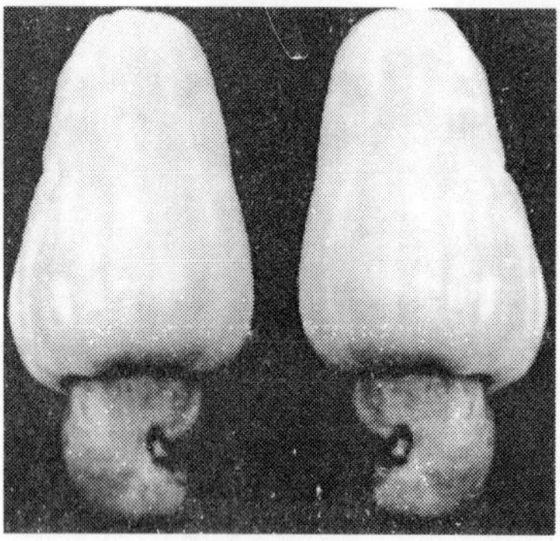

Anagha (H-8-1)

Year of release	:	1998
Centre	:	CRS, Anakkayam
Hybrid/Selection	:	Hybrid
Parentage		T20 x K-30-1
Canopy type	:	Compact
Branching habit	:	Intensive
Flowering	:	January to February
Fruiting	:	February to April
Special character	:	Mid-season variety
Apple colour	:	Orange Red
Juice percentage	:	66 %
Nut weight	:	10 g
Number of nuts/kg	:	100
Kernel weight	:	2.9 g
Shelling percentage	:	29 %
Export grade	:	W180
Mean nut yield/tree	:	13.73 kg

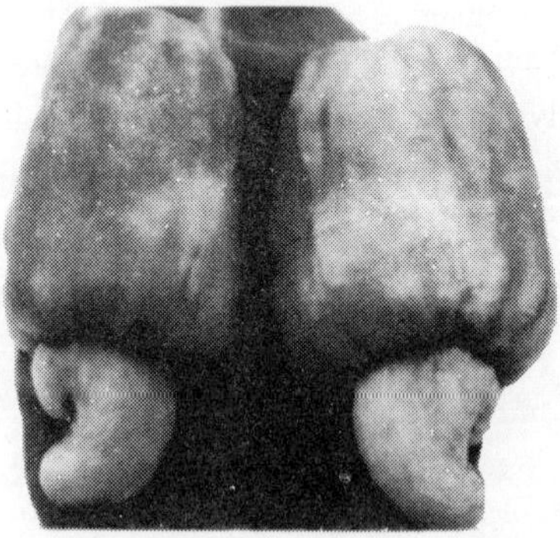

Acharya N. G. Ranga Agricultural University, Hyderabad has released 7 varieties of which 3 are hybrids and 4 are selections. The characteristics of the varieties are given below.

BPP-1

Year of release	:	1980
Centre	:	CRS, Bapatla
Hybrid/Selection	:	Hybrid
Parentage		T1 x T273
Canopy type	:	Open
Branching habit	:	Extensive
Flowering	:	February to April
Fruiting	:	April to May
Special character	:	Flush colour is pinkish
Apple colour	:	Yellow
Juice percentage	:	68 %
Nut weight	:	5 g
Number of nuts/kg	:	200-205
Kernel weight	:	1.375 g
Shelling percentage	:	27.5 %
Export grade	:	W400
Mean nut yield/tree	:	10 kg

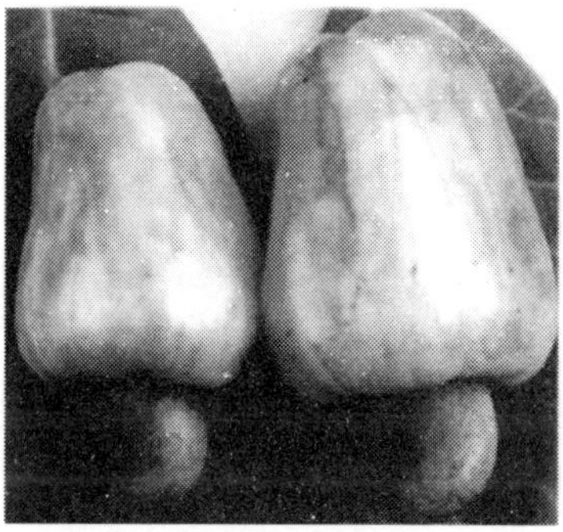

BPP-2

Year of release	:	1981
Centre	:	CRS, Bapatla
Hybrid/Selection	:	Hybrid
Parentage		T1 x T273
Canopy type	:	Open
Branching habit	:	Extensive
Flowering	:	February to April
Fruiting	:	April to May
Special character	:	Bunch bearer, small nut
Apple colour	:	Yellow
Juice percentage	:	67%
Nut weight	:	4 g
Number of nuts/kg	:	250
Kernel weight	:	1.04 g
Shelling percentage	:	25.7%
Export grade	:	W450
Mean nut yield/tree	:	11 kg

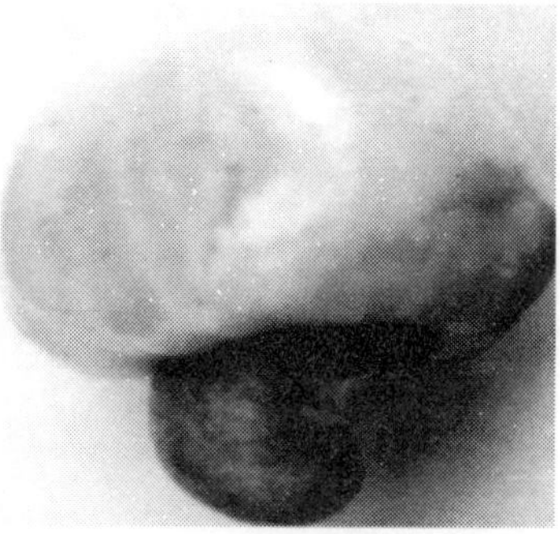

BPP-3

Year of release	:	1981
Centre	:	CRS, Bapatla
Hybrid/Selection	:	Selection
Canopy type	:	Open
Branching habit	:	Extensive
Flowering	:	March to May
Fruiting	:	April to June
Special character	:	Bunch bearer, small nut
Apple colour	:	Yellow
Juice percentage	:	67 %
Nut weight	:	4.8 g
Number of nuts/kg	:	208-210
Kernel weight	:	1.34 g
Shelling percentage	:	28.1 %
Export grade	:	W400
Mean nut yield/tree	:	11 kg

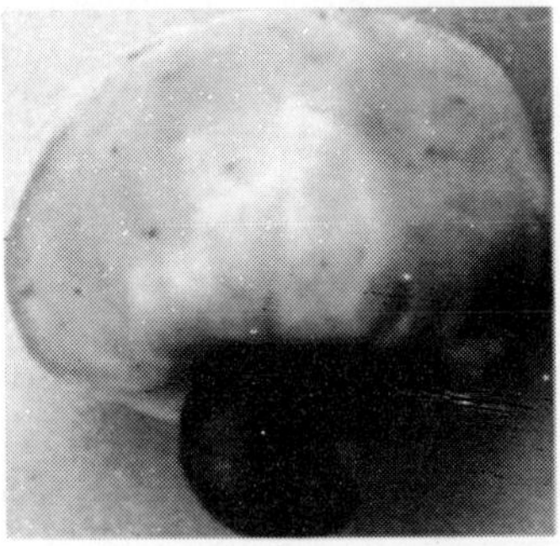

BPP-4

Year of release	:	1980
Centre	:	CRS, Bapatla
Hybrid/Selection	:	Selection
Canopy type	:	Open
Branching habit	:	Extensive
Flowering	:	February to April
Fruiting	:	April to May
Special character	:	small nut, low Shelling %
Apple colour	:	Yellow
Juice percentage	:	64 %
Nut weight	:	5.0 g
Number of nuts/kg	:	200
Kernel weight	:	1.15 g
Shelling percentage	:	23 %
Export grade	:	W400
Mean nut yield/tree	:	10.5 kg

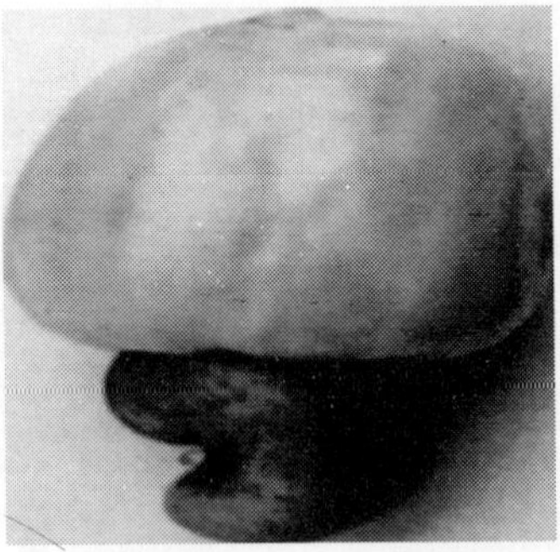

BPP-5

Year of release	:	1980
Centre	:	CRS, Bapatla
Hybrid/Selection	:	Selection (T.No.1)
Canopy type	:	Compact
Branching habit	:	Intensive
Flowering	:	February to April
Fruiting	:	April to May
Special character	:	Early bearer
Apple colour	:	Yellow
Juice percentage	:	64 %
Nut weight	:	5.2 g
Number of nuts/kg	:	192-200
Kernel weight	:	1.25 g
Shelling percentage	:	24 %
Export grade	:	W400
Mean nut yield/tree	:	11 kg

BPP-6

Year of release	:	1980
Centre	:	CRS, Bapatla
Hybrid/Selection	:	Selection (T.No.56)
Canopy type	:	Open/spreading
Branching habit	:	Extensive
Flowering	:	February to April
Fruiting	:	April to June
Special character	:	Long flowering duration
Apple colour	:	Yellow
Juice percentage	:	74 %
Nut weight	:	6 g
Number of nuts/kg	:	165-170
Kernel weight	:	1.44 g
Shelling percentage	:	24 %
Export grade	:	W400
Mean nut yield/tree	:	10.5 kg

BPP-8 (H2/16)

Year of release	:	1993
Centre	:	CRS, Bapatla
Hybrid/Selection	:	Hybrid
Parentage		T1 x T39
Canopy type	:	Compact
Branching habit	:	Intensive
Flowering	:	February to April
Fruiting	:	April to May
Special character	:	Early bearer
Apple colour	:	Yellow
Juice percentage	:	64 %
Nut weight	:	8.2 g
Number of nuts/kg	:	140-150
Kernel weight	:	1.89 g
Shelling percentage	:	27.0 %
Export grade	:	W210
Mean nut yield/tree	:	14.5 kg

Tamil Nadu Agricultural University, Coimbatore (T.N.) has released 3 high yielding cashew varieties. The characteristics of the released are given below.

Vridhachalam-1 (M-10-4)

Year of release	:	1981
Centre	:	RRS, Vridhachalam
Hybrid/Selection	:	Selection
Canopy type	:	Compact
Branching habit	:	Intensive
Flowering	:	January to April
Fruiting	:	February to May
Special character	:	Suitable for coastal region
Apple colour	:	Yellow with red tinge
Juice percentage	:	74 %
Nut weight	:	5.0 g
Number of nuts/kg	:	200
Kernel weight	:	1.4 g
Shelling percentage	:	28 %
Export grade	:	W320
Mean nut yield/tree	:	7.2 kg

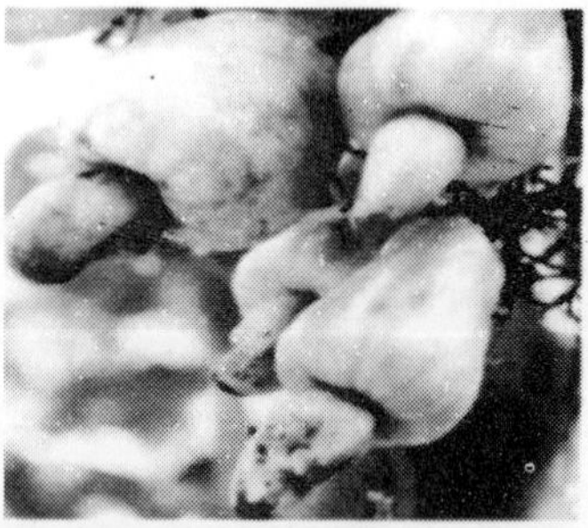

Vridhachalam-2 (M44/3)

Year of release	:	1985
Centre	:	RRS, Vridhachalam
Hybrid/Selection	:	Selection
Canopy type	:	Compact
Branching habit	:	Intensive
Flowering	:	Febuary to April
Fruiting	:	March to May
Special character	:	Compact canopy
Apple colour	:	Pinkish yellow
Juice percentage	:	82 %
Nut weight	:	5.12 g
Number of nuts/kg	:	190-195
Kernel weight	:	1.45 g
Shelling percentage	:	28.5 %
Export grade	:	W320
Mean nut yield/tree	:	7.4 kg

Vridhachalam-3 (M26/2)

Year of release	:	1991
Centre	:	RRS, Vridhachalam
Hybrid/Selection	:	Selection
Canopy type	:	Compact
Branching habit	:	Intensive
Flowering	:	January to April
Fruiting	:	February to May
Special character	:	Early flowering
Apple colour	:	Red
Juice percentage	:	72.8 %
Nut weight	:	7.18 g
Number of nuts/kg	:	140
Kernel weight	:	2.16 g
Shelling percentage	:	29.1 %
Export grade	:	W210
Mean nut yield/tree	:	11.68 kg

University of Agricultural Sciences, Dharwad or GKVK Campus, Banglore, Karnataka has released 6 cashew varieties through selection. The characteristics of these varieties are given below.

Ullal-1

Year of release	:	1984
Centre	:	ARS, Ullal
Hybrid/Selection	:	Selection
Canopy type	:	Medium spreading
Branching habit	:	Intensive
Flowering	:	November to April
Fruiting	:	February to May
Special character	:	Long flowering duration
Apple colour	:	Yellow
Juice percentage	:	64.2 %
Nut weight	:	6.7 g
Number of nuts/kg	:	145-150
Kernel weight	:	2.05 g
Shelling percentage	:	30.7 %
Export grade	:	W210
Mean nut yield/tree	:	16 kg

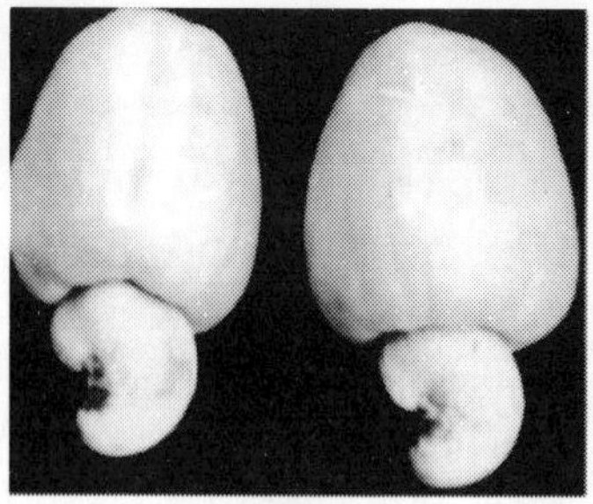

Ullal-2

Year of release	:	1986
Centre	:	ARS, Ullal
Hybrid/Selection	:	Selection
Canopy type	:	Med. spreading
Branching habit	:	Intensive
Flowering	:	Dec. to March
Fruiting	:	February to May
Special character	:	Short flowering duration
Apple colour	:	Yellow
Juice percentage	:	64.1 %
Nut weight	:	5.9 g
Number of nuts/kg	:	170-175
Kernel weight	:	1.85 g
Shelling percentage	:	30.5 %
Export grade	:	W320
Mean nut yield/tree	:	18.5 kg

Ullal-3

Year of release	:	1993
Centre	:	ARS, Ullal
Hybrid/Selection	:	Selection
Canopy type	:	Open
Branching habit	:	Intensive
Flowering	:	November to January
Fruiting	:	January to March
Special character	:	Short flowering duration
Apple colour	:	Dark red
Juice percentage	:	66.1 %
Nut weight	:	7.0 g
Number of nuts/kg	:	140-145
Kernel weight	:	2.10 g
Shelling percentage	:	30 %
Export grade	:	W210
Mean nut yield/tree	:	14.7 kg

Ullal-4

Year of release	:	1994
Centre	:	ARS, Ullal
Hybrid/Selection	:	Selection
Canopy type	:	Open
Branching habit	:	Intensive
Flowering	:	November to January
Fruiting	:	January to March
Special character	:	Short flowering duration
Apple colour	:	Yellow
Juice percentage	:	65.4 %
Nut weight	:	7.2 g
Number of nuts/kg	:	135-140
Kernel weight	:	2.15 g
Shelling percentage	:	30.8 %
Export grade	:	W210
Mean nut yield/tree	:	9.5 kg

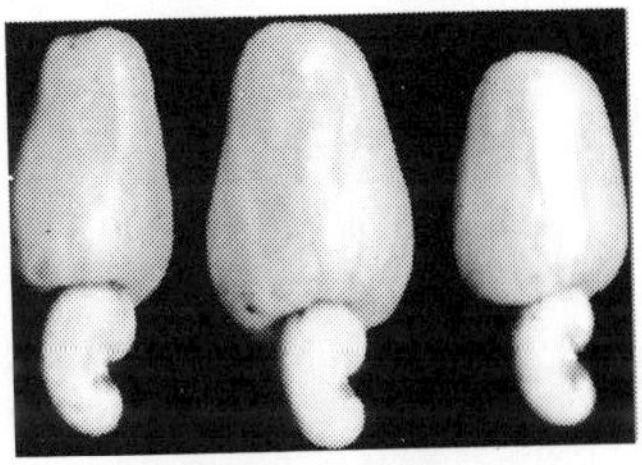

Chintamani-1

Year of release	:	1993
Centre	:	ARS, Chintamani
Hybrid/Selection	:	Clonal selection
Canopy type	:	Open
Branching habit	:	Extensive
Flowering	:	January to April
Fruiting	:	Febuary to May
Special character	:	Uniform nuts
Apple colour	:	Yellow red
Juice percentage	:	65.4 %
Nut weight	:	7.0 g
Number of nuts/kg	:	140-145
Kernel weight	:	2.10 g
Shelling percentage	:	31.0 %
Export grade	:	W210
Mean nut yield/tree	:	7.2 kg

UN-50

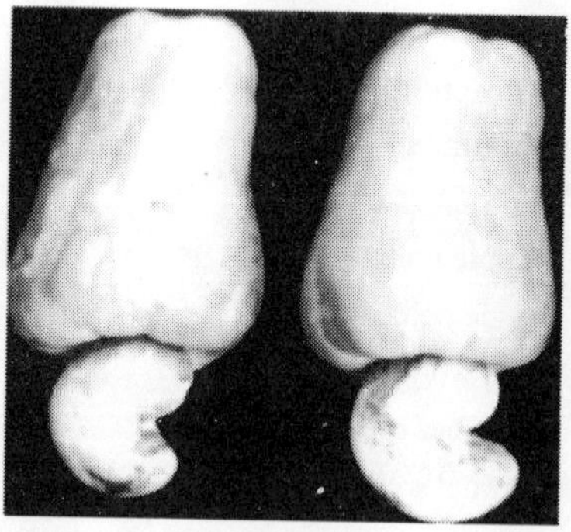

Year of release	:	1996
Centre	:	ARS, Ullal
Hybrid/Selection	:	Selection
Canopy type	:	Medium
Branching habit	:	Extensive
Flowering	:	November to January
Fruiting	:	February to May
Special character	:	Bold nuts
Apple colour	:	Yellow
Juice percentage	:	65.2 %
Nut weight	:	9.0 g
Number of nuts/kg	:	110-115
Kernel weight	:	2.24 g
Shelling percentage	:	32.8 %
Export grade	:	W180
Mean nut yield/tree	:	10.5 kg

Directorate of Cashew Research (former National Research Centre for Cashew), Darbe, Puttur, Karnataka released two cashew varieties. The characteristics of these varieties are given below.

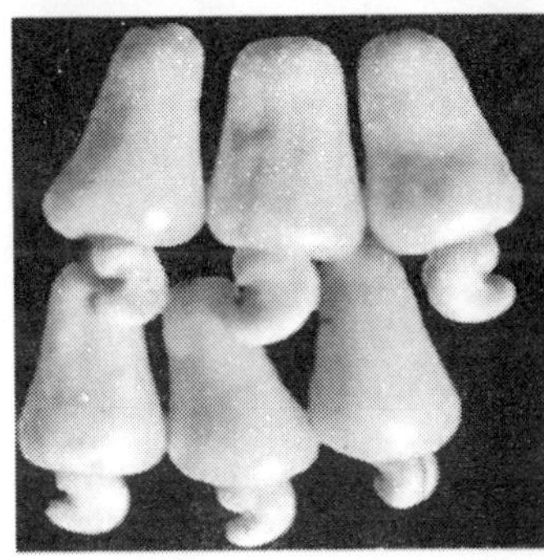

NRCC-1

Year of release	:	1989
Centre	:	DCR, Puttur
Hybrid/Selection	:	Selection
Canopy type	:	Medium compact
Branching habit	:	Extensive
Flowering	:	Dec. to February
Fruiting	:	March to April
Special character	:	Short duration
Apple colour	:	Red
Juice percentage	:	-
Nut weight	:	7.5 g
Number of nuts/kg	:	133
Kernel weight	:	2.10 g
Shelling percentage	:	25 %
Export grade	:	W210
Mean nut yield/tree	:	10 kg

NRCC-2

Year of release	:	1989
Centre	:	DCR, Puttur
Hybrid/Selection	:	Selection (2/9 Dicheria)
Canopy type	:	Medium compact
Branching habit	:	Intensive
Flowering	:	November to January
Fruiting	:	February to March
Special character	:	Short duration
Apple colour	:	Pink
Juice percentage	:	-
Nut weight	:	9.2 g
Number of nuts/kg	:	108
Kernel weight	:	2.15 g
Shelling percentage	:	29.0 %
Export grade	:	W210
Mean nut yield/tree	:	9 kg

Bidhan Chandra Krishi Vishawavidyalaya, Mohanpur, Nadia, West Bengal released a cashew variety through selection. The characteristics of this variety are given below.

Jhargram-1

Year of release	:	1989
Centre	:	RRS, BCKV, Jhargram
Hybrid/Selection	:	Selection (TN-16)
Canopy type	:	Medium compact
Branching habit	:	Intensive
Flowering	:	Febuary to April
Fruiting	:	April to May
Apple colour	:	Yellow
Juice percentage	:	63.5%
Nut weight	:	5 g
Number of nuts/kg	:	190-208
Kernel weight	:	1.5
Shelling percentage	:	30%
Export grade	:	W400
Mean nut yield/tree	:	8.5kg

Orissa University of Agriculture and technology, Sirpur, Bhubaneshwar, Orissa released a cashew variety through selection. The characteristics of this variety are given below.

Bhubaneshwar-1

Year of release	:	1989
Centre	:	OUAT, Bhubaneshwar
Hybrid/Selection	:	Selection WBDC-5 (V36/3)
Canopy type	:	Compact
Branching habit	:	Intensive
Flowering	:	January to March
Fruiting	:	March to May
Special character		Cluster bearing
Apple colour	:	Reddish yellow
Nut weight	:	4.6 g
Number of nuts/kg	:	217
Kernel weight	:	1.47g
Shelling percentage	:	32 %
Export grade	:	W400
Mean nut yield/tree	:	10.5 kg

ICAR Research Complex, for Goa, Ela Farm, Old Goa released a cashew variety through selection. The characteristics of this variety are given below.

Goa-1

Year of release	:	1999
Centre	:	ICAR Res. Complex, Goa
Hybrid/Selection	:	Selection (Balli-2)
Canopy type	:	Semi spreading
Branching habit	:	Semi intensive
Flowering	:	December to February
Fruiting	:	March to May
Apple colour	:	Yellow
Juice percentage	:	68 %
Nut weight	:	7.6 g
Number of nuts/kg	:	133
Kernel weight	:	2.2 g
Shelling percentage	:	30 %
Export grade	:	W210
Mean nut yield/tree	:	7 kg

Dwarf Cashew

The dwarf cashew tree is characterized by growth that begins earlier in the season, its smaller size and early productivity. Other characteristics that differentiate the dwarf cashew tree from the common type are: a compact and homogenous crown, with an average span reaching approximately twice the height; smaller, lighter green leaves; smaller stem diameter; initial branches closer to the ground; precocious flowering; a prolonged period of fructification; smaller nuts; and generally larger peduncles (Barros, 1988). Nut weight in natural populations varies from 3 to 10 g and the peduncle (cashew apple) from 20 to 160 g, thus showing a smaller variability in relation to the common type (Vieira *et al.*, 2005). The individual reproductive capacity of the dwarf type is also lower, with maximum known productions no higher than 65 g of nuts per harvest (Barros, 1995).

Purple Leaved Cashew

This cashew has got an ornamental value as its leaves are dark purple in colour.

Drought Tolerant Cashew Varieties

Drought tolerance study conducted at CRS, Madakkattara revealed that 3 varieties *viz*., Priyanka (H-1591), M26/2 and Vengurla-5 are tolerant to moisture stress condition (Salam and Peter, 2010).

Performance of Cashew Variety

Performances of the cashew variety depend upon the environment condition. The variety which has been released for that state may perform better in that state, but when grown in another state; its performance may or may not be similar or better. On the contrary, the same variety performs poorly with lesser yield and of inferior quality. Hence the MLT of all released varieties is being undertaken under AICRP-Cashew programme to check the performance of all released cashew varieties under certain environment. Planting early flowering varieties on slopes and late varieties in plains is beneficial.

□□□□

CHAPTER – 7

Propagation and Production of Planting Material

Cashew being a highly cross pollinated crop, planting of seedling is not recommended. In the past seed propagation was in vogue, which has resulted into large variation in plants due to highly heterozygous nature. However, it is seldom practiced now-a-days except to raise the rootstock materials. Selection of seed is very important for raising rootstock material. Seed nuts which are heavy and sink in water should be selected for raising seedling. Seed nuts should be collected from high yielding trees. Seed nuts collected from the current season's crop during March, April and May are the best for seed production as they have been found to give high percentage of seed germination. Studies conducted to investigate the time taken and germination percentage revealed that heavy nuts (sunken in water) gave 92 per cent germination on 14-17 days, whereas floating nuts took 14-22 days and gave 64 per cent germination (Krishnamurthy *et al.*, 1985).

Seeds may be sown in alkathene bags, containing three parts of red soil mixed with 2 parts of fine sand and a few holes for drainage and aeration of the soil. They take about 15-20 days for germination. Seedlings of 45-60 days are transplanted. When the older seedlings between ages of 2-6 months old are transplanted, the percentage of casualties is high. Seedlings of 7-8 months old could be transplanted more successfully than those belonging to the age groups of 2-6 months, if the plants are subjected to "stumping" treatment. In this treatment, 7-8 cm are cut at the top and root portions of the seedlings while lifting them from the nursery beds and prior to transplanting.

Sowing seeds 'at stake' at a depth of 6-8 cm is better than seeds sown erect with the stalk end upward and suture slightly slanting, at 45°. These seeds give better germination and produce plants in straight stem. Cashew seeds retain viability for eight months and give about 80 and 100% germination. Rao and Hasan (1957) reported that eight months old seedlings and older ones produced lateral roots with the taproot and fibrous roots. It was also stated that no lateral roots could be seen in the case of one or two months old seedlings and the production of lateral roots ranged from one to six. Therefore, investigations were undertaken to find out whether it would be possible to induce better root and shoot development by treating either cashew seed or seedlings with plant growth regulators (Shanmugavelu, 1985).

Among the various growth regulator treatments, GA was found the best recording 100 per cent germination, though all the growth regulators except CIPA gave fairly high percentage of seed germination. GA induced the maximum shoot length, 77.39 per cent increase over the control, in a period of 60 days after germination.

When the seedlings were sprayed with growth regulators at weekly intervals, IBA at 100 ppm induced root growth in 8 weeks. The induction of good root systems and shoot development caused by these growth regulators would help establishment of seedlings in dry arid regions.

Vegetative Propagation

Cashew being a cross-pollinated crop, the seedlings manifests variation. In order to bring uniformity in stand, duration of flowering, quality, yield etc., it is advantageous to resort to clonal propagation. In West Coast, air layering of shoots 4 cm in thickness was found successful. One year old shoots as well as current season's shoots were found suitable. The method consists of removing a ring of bark 0.5 cm wide and covering with moist sphagnum moss and wrapping with ISO gauge polythene film. Among the various rooting media like sphagnum moss, local moss, saw dust, wood shavings, coir fibre, coir husk dust and red soil mixed with leaf mould in equal parts, tried for air layering, wood shaving was the best.

A number of propagation methods have been tested for the multiplication of cashew. Air-layering was found to be one of the popular methods among growers. This technique, however, produced trees with poor anchorage as the root density was found to be low. It also resulted in poor field establishment and high susceptibility to cyclones and drought conditions. Air layering was therefore found to be unsuitable for commercial exploitation. In the case of mound layering too, the absence of a tap root was found to be disadvantageous. Epicotyl grafting was another method that had limitations due to high mortality at transplanting and incidence of collar-rot at the nursery stage. Soft-wood grafting developed at the research centres was found to be the most viable method of propagation that was commercially acceptable; it gave a success rate of about 80 per cent. The technique of soft-wood grafting described below is similar to epicotyl grafting except the difference in the age of the rootstock.

In a 19 years old trial at the Regional Cashew Research Station, Vengurla (Maharashtra) on the comparision of seedlings, air-layers and inarched grafts raised from a single mother tree. Vengurla-1, the yield of vegetatively propagated trees was better than the seedling progenies. However, the size of the trees as judged by the volume of canopy, the seedlings were more vigorous than the air-layers and inarched grafts.

Cashew is commercially propagated by soft wood grafting which has been standardized and commercially utilized for large scale production of plant material in the country. Selection of planting material is most important in cashew cultivation. Being a highly cross pollinated crop, vegetative propagation in cashew is mainly recommended on commercial scale to produce true to type planting materials. Softwood grafting is the only method which is commercially feasible and practically highly successful in cashew. Seedlings raised from the raw nuts should not be used as planting material as it gives lot of variation in growth and yield. Grafts are recommended for precocity, uniformity in quality and for higher yields.

Among the various methods tested, softwood grafting was found to be the best for vegetative propagation. Soft-wood grafting technique has revolutionized the availability of elite planting material of cashew in the country. It has also been shown that softwood grafting is feasible for commercial multiplication. Based on these results, India has been producing over 15 million grafts annually under both government and private sectors (Bhat *et al.*, 2010).

There are about 100 Regional Cashew Nurseries under public and private sector catering to the needs of graft requirement in addition to large number of small to medium cashew nurseries. About 50,000 ha of area are brought under cashew every year in the country by planting over 10 million cashew grafts at the rate of 200 plants/ha. This has enabled to increase area under cashew with recommended varieties both in traditional and non- traditional regions which already has large impact on the increase in production and productivity and will also enable the country to achieve self sufficiency in cashewnut production in next year's and beyond (Bhat *et al.*, 2010).

Softwood Grafting and Nursery Management in Cashew

I. Establishment of scion bank

i. Production of scion sticks

- The Scion Bank should be established with the recommended varieties of the region.

- The plants should be planted in a closer spacing of 4m x 4m, in order to get continuous supply of scions.
- It should be fertilized with recommended quantity of NPK during monsoon and tender shoots should be protected by timely sprays of insecticides.
- The flower panicles should be cut off in order to get more number of scions.
- The plants in the scion bank should be maintained by heading back the top to a height of 1.5 m from ground level and by pruning the dried branches.
- Pruning of trees may be carried out annually during September –October in the scion bank.

ii. Selection of scion and precuring

- Select non-flowered, 3-5 month old lateral shoots of current season's growth.
- The selected shoots should be about 10-12 cm long, straight, uniformly round with pencil thickness, brown coloured having dormant plumpy terminal buds.
- The top 4-5 leaves should be dark green in colour indicating proper maturity of the scion sticks.
- The selected scion sticks should be precured by clipping off leaf blades, leaving petioles.
- The scion sticks can be detached from the mother tree after 8-10 days, before they sprout and utilized for grafting.

iii. Collection of scion stick

- The precured scion sticks should be separated early in the morning to avoid desiccation and its length should not less than 10cm.
- Soon after separation from the mother tree, scion sticks should be dipped in water and placed in a polythene bag of 100 gauge thickness and brought to the nursery shed for grafting.
- The scion sticks wrapped in sphagnum moss cloth and placed in a polythene bag of 100 gauge can be kept for 3-4 days and used for grafting, if necessary.

iv. Raising of root stock seedlings

Selection of seednuts

- Collect fresh seednuts during the peak period of harvest in February-April and sun dried for 2-3 days.
- Medium sized nuts of 6-8 gm should be selected in order to get uniform and vigorously growing seedlings required for soft wood grafting.
- The seeds should be bold and the highest specific gravity of more than 1.0. The sun dried, graded seeds should be treated with Carbaryl (5g/kg of seeds) before storing them in gunny bags or polythene bags. The bags should be arranged on wooden planks in the store room.

Preparation of potting mixture

- Prepare the potting mixture in the proportion of one part red soil, one part sand and one part compost in heavy rainfall areas and one part red soil, one part compost in low rainfall areas and mixed with little quantity of rock phosphate (10g/2kg of potting mixture).

Filling polythene bags

- Use high density polythene bags of 25 cm x 15 cm size and 300 gauge thickness for filling potting mixture.
- Punch about 30-40 holes of 0.5cm diameter uniformly on the polythene bags for ensuring good drainage. The bottom corners of the polythene bags should be folded inside before filling potting mixture.
- Before filling the polythene bags place few dry leaves (leaf mould) and small pebbles at the bottom of the bag for better drainage.
- Fill the bags properly up to the brim with the potting mixture and arrange the bags in beds of 1000 (10 x 100) number in each.

Sowing of seeds

- The seednuts should be soaked in water for 12-24 hours before sowing in order to get good germination. After pre-soaking, nuts are then treated with Bavistin @ 1g/l water in order avoid further fungal infection.
- Pre-soaking of seeds results in earlier and higher percentage of germination.
- At the time of sowing, the soil should be moist and loosened in the polythene bag.

- Sow the seeds in the centre of the bag, stalk-end upwards, with a depth of not more than 2.5 cm and cover with little soil.
- Water the polythene bags immediately after sowing and daily during drier months. Excess irrigation should be avoided.
- The good and healthy seednuts usually germinate within 15-20 days after sowing.
- Seed bed may be mulched with paddy straw till germination takes place and partial shade may be provided during summer months.
- To control pests while seeds germinate, malathion 5% dust, or spraying of chloropyriphos (Durmet 20 EC @ 0.05%) should be applied.

Maintenance of seedlings in the nursery

- Daily watering is required during summer season.
- During rainy season, to avoid water stagnation and disease occurrence seedlings may be covered by polythene sheet.
- To get vigorous seedlings with single main stem, the side shoots arising from the leaf axils should be removed frequently.
- To prevent roots penetrating into the ground, grafted plants should be shifted frequently or placed on thick gauge black polythene sheets.
- To control tea mosquito and other sucking insect pests, Imidachloprid 17.8 SL @ 2.8 ml /10 litre of water may be sprayed.
- Select 45-60 days old seedlings as root stocks for grafting.

II. Soft wood grafting technique

Preparation of root stock

- Disinfect the grafting knives by dipping them in fungicide solution.
- Root stock seedling brought from the nursery beds should be kept in the nursery shed under shade for grafting.
- Remove the leaves by a sharp grafting knife leaving two pairs bottom leaves.
- At a height of 15-20 cm from ground level a transverse cut is made on the root stock and the terminal shoot is removed.
- A cleft of 6-7 cm deep is made in the middle of the decapitated stem by giving a longitudinal cut.

- A little portion of wood is removed from the inner sides of the cleft at the top, so that after grafting the joint will be perfect.

Preparation of scion

- Select a matching scion stick of the same thickness as that of the root stock of 10-12cm long by cutting off the excess portion at the bottom.
- Shape the cut end of the scion in to a wedge of 6-7 cm long by chopping off the bark and little portion of wood from two opposite sides.
- While preparing the wedge, the gum on the cut surface should not be disturbed and the cut surfaces should not be spoiled by touching with fingers.

Grafting

- The wedge of the scion is inserted into the cleft of the root stock to see that the cambial layers of both the root stock and the scion come in perfect contact with each other.
- The graft joint is secured firmly with 2.5cm wide and 30cm long polythene strip of 100 gauge thickness.
- A long and narrow HD polythene bag of 15cm x 12.5cm size and 100 gauge thickness is inserted on the grafted plant, tied at the bottom with a single knot in order to protect the scion stick from drying up.
- In places with high humidity the polythene cap need not be used.
- The freshly grafted plants are left in the nursery shed for about two weeks to encourage sprouting of the terminal buds.
- After two weeks, the polythene caps are removed gently and the grafts are shifted to open condition in the nursery. Within 3-4 weeks, 70-80% of the grafts will sprout.
- Graft becomes ready for planting within 5-6 months.

III. Seasonal influence on graft success

- The success of soft wood grafting is higher during the summer season February-April.
- During other season, graft success is slightly reduced due to non availability of proper scion sticks and unfavourable weather conditions.
- The average graft success throughout the year is 65-70%.
- Grafts need to be watered frequently depending on the season.

- Excess water needs to be drained by providing drainage holes in polybags.
- To prevent roots penetrating into the ground, grafted plants should be shifted frequently or placed on thick gauge black polythene sheets.
- During summer months (January-May) grafted plants should be protected by providing partial shade by erecting *pandal* of dry coconut fronds or green nylon nets (50%). Watering on alternate days should be done in summer.
- Bordeaux mixture spray (1%) may be given at 10 days interval during rainy season to control fungal infection of tender seedlings and grafted plants.
- Imidachloprid 17.8 SL @ 2.8 ml/10 litre of water may be sprayed to control sucking insects and leaf eating caterpillars as and when the damage is seen.
- The flower panicles produced by the graft should be removed as and when observed in the nursery.
- The polythene strip should be removed from the graft joint after four months of grafting.
- The bottom leaves on the root stock should be removed after three months of grafting when scion leaves turn bronze colour to green.
- Partial shade has to be provided to avoid sun-scorch by placing the grafted plants in a lath/screen house. Direct sunlight should be avoided as polythene bags tend to perish.
- When transporting grafted plants, terminal shoots and taproots should be protected.

Effect of Gamma Radiation Treatments to Raw Nut on Germination, Seedling Growth and Establishment

Khalatkar and Bhargava (1987) treated the physiologically similar nuts with radiation of 60 Co-Gamma Cell with 10, 20, 30, 40 and 50 kR. These were administered at 1.2 kR/min. Lower doses of gamma radiations had no effect on germination; however, 50 kR resulted in decreased germination (35%). Seedling growth was influenced considerably by 10, 40 and 50 kR. Branching was profuse in 40 kR plants. However, flowering and fruiting was profuse after four years. Plants from 10 kR treatment exhibited early flowering and fruiting, besides the nuts from 10 and 50 kR plants were comparatively bolder and weighed more than that of control. Thus, gamma irradiation of cashew with 10 to 20 kR improved the performance of seeds in the form of seedling growth and establishment.

Steps involved in softwood grafting technique is given below

Selection of seed seedling nuts (6-8 g each)

Raising of rootstock in nursery

Different growth stages of in nursery

Selection of Scion

Collection of scion sticks (3-5 months old)

Maintenance of Rootstock in nursery

40 days old seedling ready for grafting

Preparation of scion bud stick

Softwood grafting

New sprouts in tied graft

Softwood graft

Maintenance of grafts in nursery

8 months old cashew grafts ready for sale

Establishment of Cashew graft in field

CHAPTER – 8

Cashew Morphology and Tree Characteristics

Pattern of Growth Flushes in a Bearing Tree

Two or three peak periods of growth are usually observed in a bearing cashew tree, even though under favourable conditions the stray shoot growth may occur almost every month. The pattern of growth of a bearing tree consists of a generative flower flush (reproductive) and a vegetative flush. The vegetative flush, consisting of lateral shoots, always develops soon after the main crop has matured. Flowering is terminal and is universally preceded by the vegetative flush.

There are two types of branching in cashew, intensive and extensive.

Branching Habit

1) Intensive branching

Shoot grows to a length of about 25-30 cm and terminates in a panicle. Simultaneously 3- 8 laterals arise within 10-15 cm of the apex. Some of them may also terminate in panicles in the same flowering season and repeating same growth pattern. This gives a well covered busy appearance to the tree.

2) Extensive branching

Shoot grows to a length of 20-30 cm and rests. A bud sprouting 5-8 cm below the apex leads to further growth. This process of growth continues for 2-3 years without any flowering and produces a spreading tree.

Even though both kinds of branching are observed in all trees, one type dominates in a tree. High yielding trees have more than 60 per cent intensive branch, whereas low yielders have less than 20 per cent. Seventy five per cent of the shoots of the intensive types produce flowers as against 12 per cent in extensive types.

Fig. 10: Extensive

Fig. 11: Intensive

Leaf Structure

The leaves are alternate, simple, glabrous, obviate, round and pinnately veined. Leaves are blunt at the tip and slightly cone shaped on the twig. Normally, three leaf shapes are observed in cashew as per cashew catalogue *viz.,* oblong, obviate (club shaped) and oval (Fig.12), whereas the colour of young leaf *viz.,* red, yellow red, greenish yellow and purple (Fig.13). The young leaves are reddish brown which after turns to dark green when mature, within 20 days. The mature leaves are intense green on upper surface and opaque green on lower surface.

The leaf length varies from 10 to 20 cm and the width from 5 to 10 cm in a ratio of 1.8:1. The petioles are about 0.5 to 1.0 cm long. Short and flexible petioles are broader at the base with deep brown in colour. Each leaf is of about 20 pairs of parallel veins, slightly standing out on the under leaf. The bearing flushes have 6-8 leaves and only 4 leaves in non-bearing flushes. The bearing twigs have a more leaf area than the non-bearing ones.

Reddy *et al.* (1988) reported a maximum leaf length of 167.78 mm and a minimum of 88.48 mm. Rao and Hassan (1957) also reported that the number of leaves produced on the new leader shoots varied, either on the same tree or between different trees, and ranged from 3 to 14, with a mean of 9 per shoot.

Leaf shapes

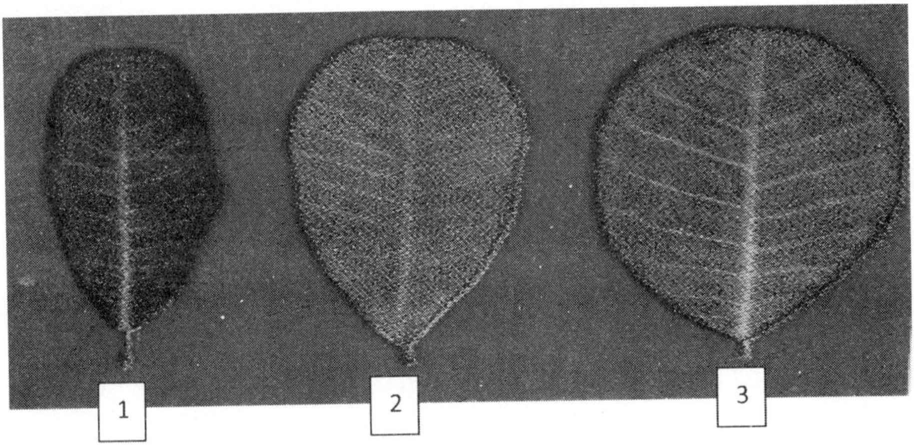

Oblong Obovate (club shaped) Oval

Fig.12: Leaf Shape

Colour of young leaves

Red Yellow Red Green Yellow Purple

Fig.13: Young leaves colour

Canopy

A tree can be visually rated as having upright and compact, upright & open and spreading type (as shown in Fig.14). In general, it is the branching habit of the plant that determines the canopy of the tree. In case of cashew, high lateral to leader ratio of branches measured per unit area is desirable.

In one square meter area of the canopy, a compact type tree will bear more than 29 lateral branches. These laterals are borne on about 5 to 6 leader branches. Leader branches are the laterals of the previous season. In a compact type, there should be more than 3 laterals for one leader on an average.

At Chintamani (Karnataka, India), Reddy *et al.* (1988) observed tree canopies ranging from compact to sparse; with diameters ranging from 2.5 to 6.1 m. Tsakiris and Northwod (1967) measured the canopy development of young trees in Tanzania. It appeared that under favourable conditions, young cashew trees grew at a rate of about 1m per year and the canopy diameter increased by 1.5 to 2 m per year for the first five to six years. Thereafter, growth did slow down. Shoot growth may occur throughout the year, especially when rainfall is well distributed (Ohler, 1979). Canopy shape may influence the yield per tree as cashew trees bear their fruit at the periphery and fruit production becomes almost nil on branches. Trees with a narrow and highly conical-shaped canopy would form a much larger surface area than a tree with a low spreading canopy. The latter is preferred for easy harvesting, especially when apples are to be harvested, before they fall on the ground (Ohler, 1979).

Source: "Catalogue of Minimum Descriptors of Cashew Germplasm Accessions-III" (Swamy et al., 2000).

Fig. 14: Tree habit

Roots

Cashew being a perennial tree, it has to remain continuously in a soil over a long period of time. It has a deep and spreading root system. Root distribution pattern of cashew depends on the age of the tree, the type of planting material, the soil type and its environment in which it is grown, level of nutrition, irrigation, etc. Majority of feeding roots normally reside in the surface layer of the soil. Adams (1975) showed that after emergence, the radical rapidly developed into a tap root, which started producing lateral roots four days later. The laterals also grew quickly at first and were produced progressively lower along the taproot as it elongated. The cashew tree has a taproot penetrating deeply into the soil with an extensive lateral root system (Ohler, 1979). Taproot of three half-years-old cashews extends up to 3.2 m (Tsakiris and Northwood, 1967). They also found that lateral spread of root varied with age of trees. Khader (1986) reported that 85 per cent thick roots and 45 per cent fine roots occur within 15 cm surface soil layer. Shallow soil depth restricts vertical spread of cashew roots (Vidhyadharan and Peethambaran, 1979). According to Wahid *et al.* (1989), the lateral root activity of adult cashew tree was traced up to 4 m and to a depth of 60 cm. Using ^{32}P soil injection techniques; they found that absorption was mostly from 0 to 15 cm soil layer than from deeper zone, suggesting that cashew is a surface feeder. They explained that an area of 2 m radius around the tree accounted for above 72 per cent of total active roots.

In shallow laterite soils, cashew tree develops an extensive root system and the majority of roots (89.3%) occur 300 cm laterally around the tree and 100 cm vertically. The pattern of decrease in root density in relation to lateral distance suggests that fertilizer application is to be done within 300 cm laterally around the cashew tree (Salam *et al.* 1995). The young plant put out strong and extensive root both sideways and downwards over an area double the size of the tree's canopy size within 1.5 year of age (Agnolony and Giuliani, 1977). The roots which emerge from the seed, the radical rapidly develops in to a taproot that produces laterals within 6-7 days and then shoot emerges, when the taproot has already more than 20 cm length. In Madagascar, Lefebvre (1969) and Andrianirina (1990) observed on various occasions that young trees of 2 to 3 months had taproots with a length of more than 80 cm and 5-month old trees had taproots of 120 cm. They also found that trees could utilise a large volume of soil because their roots grew not only vertically to a considerable depth but also in a large radius. In general, the root distribution studies indicated that 72 % of the active roots are located up to 15 cm depth and 2m radius (Wahid *et al.*, 1989). Further, the layers produced more lateral roots, whereas the seedlings and grafts produced the tap-root which grows directly downwards.

Vegetative flush

Under south Indian conditions, Rao and Hassan (1957) studied on growth flushes and described two distinct growth flushes. The flower emergence in cashew commences by the middle of September and continues until the end of February, the main season being October-November. Cashew vegetative shoots which appear in April-May develop panicles in about 92 per cent cases during the next flowering season, while those produced in December-January bear flowers in about 10 per cent shoots in the same year.

Inflorescence

The cashew tree normally starts flowering second year itself; however seedlings starts within three to five years. The inflorescence is an indeterminate panicle of polygamomonoecious type. Each branch of the cluster bears *in* terminal flower subtended immediately by two or more bracts. From the axils of these, bracts grow further to bracted flower stalks. The average length of panicle varies from 20 to 60 cm. The proportion of perfect flowers varied with types/ variety. Three different shapes of inflorescence (Fig. 15) and compactness of inflorescence (Fig. 16) as per cashew descriptor are as follows.

Narrowly pyramidal Pyramidal Broadly pyramidal

Fig. 15: Shape of inflorescence

1. Loose 2. Compact

Fig. 16: Compactness of inflorescence

Floral biology

The cashew tree produces both male and hermaphrodite flowers on the same panicle. There are more male flowers than hermaphrodite flowers produced and the ratio is generally 10 male to 1 hermaphrodite flower, however, it varies depending on the genetical, climatological and location specific. The sex ratio can also vary during the flowering period. In addition, more than 80 % of the hermaphrodite flowers drop before they produce a mature nut. In terms of timing, male flowers typically open earlier in the day, while hermaphrodite flowers open later in the early afternoon. The female part of the hermaphrodite flower is receptive to pollen from the male flower for about 6 hours, and for better nut set, pollination should occur within this time.

Structure of Flower

The flowers are small, white to light green at the time of anthesis, later turns to pink. Flowers are either bisexual or staminate and both occur intermixed in the same inflorescence. On the same tree, the perfect flowers are larger in size than the staminate. Cashew panicle is polygamous with male and hermaphrodite flowers. The flowers are pentamerous; calyx green, colour of the inside petal at the time of anthesis is either pure white or white with faint or distinct longitudinal streaks. The colour of the petal outside is white; anther rounded, pink or yellow, turning grey at the time of dehiscence; ovary pale yellow with a pink spot at the junction of the style, style pale white to creamy white, stigma needle shaped and creamy white. In a staminate flower, there are five sepals, five petals, stigma absent and 7-10 stamens of which one is exerted and the rest are staminodes and inserted (Fig. 17 & 18). In a hermaphrodite flower, there are five sepals, 7 to 10 stamens, of which, one is exerted and the rest staminodes and inserted (Fig. 17 & 18), the gynoecium consists of the style and stigma. The stigma and one stamen are protruding outside; the former is taller than the latter. The petals of a hermaphrodite flower are always broader than those of the male flower.

Cashew plant starts flowering during October to January. Flowering is the process in which the tree shifts from vegetative phase to reproductive phase. There are various internal and external factors which trigger the plant to flower. Flowering is a process extending gradually from shoot to shoot in a tree. There is wide variation in the duration of flowering from tree to tree and on an average; a tree completes flowering ranging from 45 to 90 days depending upon climate and irrigation facilities provided to cashew orchard.

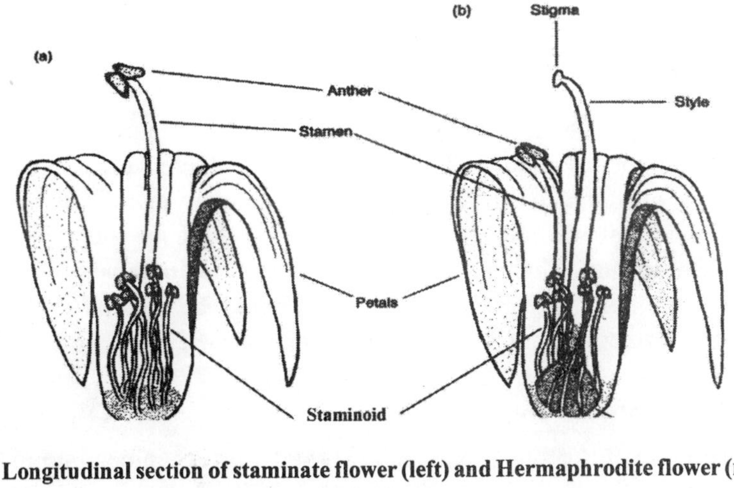

Fig. 17: Longitudinal section of staminate flower (left) and Hermaphrodite flower (right) of cashew

The floral structure and phenology of nine seedling cashew trees were studied for three years in north Western Australia. The hermaphrodite and male flowers were similar in structure except that the male flowers lacked a pistil and were smaller than the hermaphrodites. The stigma was covered by a thick cuticle, and the transmitting tissue narrowed toward the base of the style. Flowering occurred after the wet season during the driest and coolest period of the year. The trees produced many more male than hermaphrodite flowers, with the majority of hermaphrodites produced during the first three weeks of the flowering period, two weeks before the peak opening period of the male flowers.

Fig. 18: Hermaphrodite flower (left) and staminate flower (right) of cashew

There were two peaks of opening per day for both types of flower, one in the early morning and one around midday. Less than 40% of the hermaphrodite flowers set fruit, and this was followed by a high rate of premature fruit drop. There were significant differences between tree to tree in the length of the flowering period, the number of hermaphrodite and male flowers, the proportion of hermaphrodite flowers, the number of fruit set, the number of fruit drop, and in the number of mature fruit, on the contrary, there were no year to year or within-tree differences. In addition, there was a highly significant correlation between the number of hermaphrodite flowers and the numbers of both fruit set and mature fruit. Thus, there is scope for selection of superior genotypes, and that the number of hermaphrodite flowers can be used as a reliable and stable selection criterion (Wunnchit and Sedgley, 1992).

Number of flowers per panicle

Damodaran *et al.* (1966) found 200 to 1600 flowers under humid coastal conditions of Kerala. Raju (1979) recorded 4880 flowers under Bangalore conditions, while Patnaik *et al.* (1985), counted 43.77 to 115.80 flowers per panicle at Orissa. Reddy *et al.* (1985) stated that one of the various factors responsible for poor yields of cashew trees in India was the presence of a large number of male flowers. The presence of a high percentage of hermaphrodite flowers is a desirable characteristic of high yield varieties. They recorded that the total number of flowers per panicle varied from 201 to 643. Khan *et al.* (1988) recorded a high number of flowers per panicle up to 837 under Mangalore conditions. The difference in the number of flowers per panicle may be due to the difference in the age of trees, type and source of plant material as well as climatic conditions of that region.

Number of male (staminate) flowers per panicle

In India, the percentage of male flowers per panicle ranged between 25.82% and 96% (Morada, 1941; Rao and Hassan, 1957; Reddy *et al.*, 1985 and Ghosh, 1988a). There is a marked difference in the ratio of hermaphrodite flowers to male flowers per panicle. It ranges from 0.004 to 0.689 with the highest recorded in Bengal (Damodaran *et al.*, 1965; Patnaik *et al.*, 1985; Reddy and Rao, 1985; Ghosh, 1988a; Khan *et al.* 1988) (Table 20).

Number of perfect (hermaphrodite) flowers per panicle

In order to have a high fruit-set, a tree should possess a high percentage of perfect flowers. A positive correlation between yield and number of perfect flowers and each panicle possessed an average of 286.1 flowers of which, 199.8 were male and 86.3 perfect (Rao and Hassan (1957), Damodaran *et al.* (1965); Sriram (1970) and Parameswaran (1979). Srihari Babu (1981) stated that in case of high yielding trees, the hermaphrodite flowers should be average up to 45 per cent of total flower number. The number of flowers per panicle and the ratio of hermaphrodite to male flowers in various countries are shown in Table 20.

Sex ratio

The sex ratio is indicated in two possible ways: the number of hermaphrodite to male flowers, or the number of male to hermaphrodite flowers. The ratio varies between regions and majority produce more male than hermaphrodite flowers. In India, the ratio even varies with localities; but the figures showed little difference between male and hermaphrodite flowers produced (Table 20).

Table 20: Number of flowers per panicle and ratio male to perfect flower in different regions

Region	Number of panicle flowers	Perfect flowers/ panicle	% Perfect flowers/panicle	Male flowers/ panicle	% Male flowers/ to male	Ratio perfect	Research workers
Australia	413	32.1		410.8	-	1:12.80	Heard *et al.* (1990)
Jamaica	193 - 801	13-96	-	180 - 705	-	Up to 1:28	Ohlers (1979 b)
Madagascar	187	21	-	166	-	1:7.9	Fofifa (1981)
Senegal	1005	152	-	853	-	1: 6.61	Behrens (1996)
Tanzania	767	63-67	-	250-400		1:3.7-1:6.7	Northwood (1966)
India							
Karnataka	201 - 643	53.0-212.75	17.08-74.18	25.82-83	90-91	1:1.19	Reddy *et al.* (1989)
Kerala	200 -1600	-	0.45-24.9	-	-	1:1.45	Damodaran *et al.* (1965)
Orrisa	43.77-116	-	5.94-20.69	-	up to 96	-	Patnaik *et al.* (1985)
Manglore	329	13	2.27-65.2	-	36-95	-	Rao and Hassan (1957)
West Bengal	-	4.97-65.2		-	-	-	Ghosh *et al.* (1988 a)

Pollination, Fertilization and Fruit Development

In India, the flowers open between 9 am and 2 pm and the staminate flowers open earlier than the bisexual flowers (Damodaran, 1965). The stigma remains receptive for about 24 hrs before anthesis and continues to be so, for about 48 hrs after anthesis. Pollen fertility was 94 per cent in all types studied by Damodaran et al. (1965). Pollination is reported to be carried out by flies, bees and ants as well as by wind. Smith (1958) suggested that bees may be used to promote greater pollination. The occurrence of strong scented flowers and sticky pollen grains also emphasize the importance of insects over wind as pollinating agents.

Bees promote greater pollination because scented flowers and sticky pollen grains attract them. Similarly, the ants attracted because of the sugary substance oozes from the styler scar part of the tender nuts. The bagged inflorescence does not produce nuts unless it is hand pollinated or insects are allowed inside. Self-pollination is also possible, as nuts have developed in inflorescence from hand-pollinated and then bagged.

The cashew appears to have sticky pollen (as it is close with relative mango) because, when the anthers dehisce, the pollen remains attached to the anther lobes and is not easily blown away. The flowers are also strongly scented, which suggests that insects are more important than wind as pollinating agents.

A few sticky-tape traps were put into some trees to catch wind-blown pollen, but on examina-tion no pollen was seen. This again supports the view that wind pollination is of limited importance. At flowering time, there are large populations of flies and other winged insects capable of acting as pollinating agents and nearly always a large ant population which frequents the inflorescences. It is possible that a considerable amount of self-pollination is carried out by the ants as they crawl from flower to flower (Northwood, 1966).

Fruit setting

The cashew tree produces innumerable flowers of which, less than 10 per cent are perfect. Under normal conditions, nearly 85 per cent of the flowers are fertilized of which only 4-6 per cent reaches maturity, the remaining drop down at various stages of development. Pillai and Pillai (1975) reported maximum shedding (more than 40 per cent) at mustard stage, 20 per cent at pea nut stage, and 20 per cent at later stages. Northwood (1966) attributed fruit drop in cashew during the early stages of development due to physiological reasons, while the controversial reason suggested by Pillai and Pillai (1975) that insect attack also played an important role in immature fruit drop, apart from physiological reasons. Rao (1974) suggested that pollination, in nature was inadequate since 55 per cent fruit set was gained by hand pollination. Pappiah *et al.* (1978) increased the number of perfect flowers and enhanced the fruit set by application of growth regulators. IAA, IBA and NAA had negligible or no effect, while, 2,4-D spray alone had increased the shelling percentage and total yield per tree.

Rao (1956) observed only 3 % fruit set on the west coast of India, while Murthy *et al.* (1975) found 6 to 12 % on the east coast. Nawale *et al.* (1985) reported 8% to 26.6% fruit set under Kokan conditions. Patnaik *et al.* (1985) found that under normal conditions in Orissa (India) 11.9% to 54.5% of the total flowers set fruit while, 45.80% to 88.08% dropped off without fertilization. The possible reasons for a poor fruit set might be due to imbalances in the sex ratio, the condition of the pistil, and inadequate pollination and fertilization (Rao and Hassan, 1957; Nawale *et al.*, 1985). The final yield was, therefore, in proportion to the initial fruit setting. According to Reddy *et al.* (1985), high yielding cashew trees should have a high percentage of fruit set of more than 6%. Smith (1958) suggested that bee colonies be introduced into orchards to increase pollination and fruit set. Heard *et al.* (1990) confirmed the effectiveness of honey bees and native Australian bees as good pollinators that have a positive effect on fruit set and total yield.

Fig. 19: Nut sets per panicle of cashew at RFRS, Vengurla

Structure of the Fruit

Nut

The fruit consists of the epicarp, mesocarp, endocarp, testa and kernel. The epicarp forms the epidermis and is coriaceous. The cells are elongated with hollows frequent intervals. The mesocarp is the thickest of the three layers and is spongy and vacuolar. In the mesocarp, which has a honey comb structure, there are ducts fill with sticky, resinous corrosive oil, the cashew nut shell liquid (CNSL). The longitudinal section of cashew nut are shown in Fig. 20 below

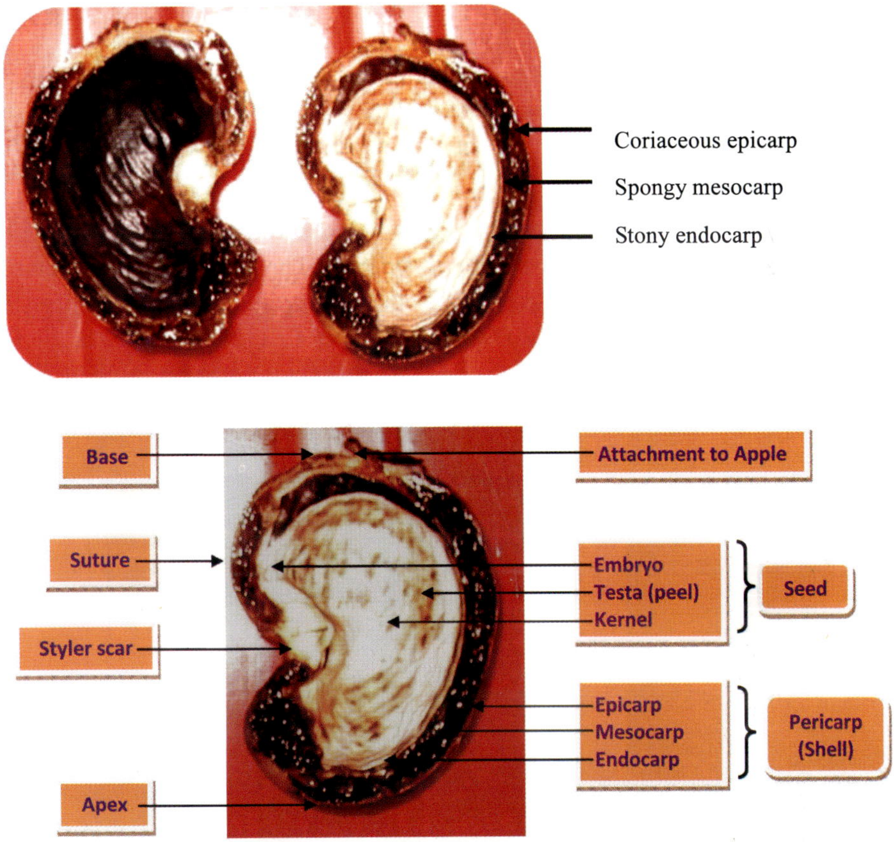

Fig. 20: Longitudinal section of cashew nut

The cashew nut is a kidney shaped drupaceous nut, greenish grey in colour. The nuts vary in size, shape and shelling percentage (Rao and Hassan 1957). A single nut is 3 to 5 cm in length and 2 to 3.5 cm in width. The weight of single nut usually varies from 3 to 20 g. The shelling percentage varies from 15-30. From the commercial point of view, the biggest possible nuts are preferred. The highest grade of export quality comes from bold nuts weighing above 10 g. From the Breeder's point of view, the medium size nuts are preferred. In this case, the scope of increasing the yield is more without considerable loss in quality.

Variability in the shape of cashew nut exists with variety. The normal shape of nut is that of kidney. A nut with oblong-ellipsoid shape is also observed.

Cashew apple

The cashew apple has a base at the point of attachment to the peduncle, ridges on its body and a cavity attached the nut. The nut remains snugly at one end of the swollen pedicel, known as the cashew apple. The apple contains 70-75% juice.

The fleshy peduncle, the cashew apple, is juicy and sweet when ripe. The apple varies in size, colour, juice content and taste. The yellow apple tended to be heavier, softer and less astringent than the red apples. Usually the apple is six to ten times heavier than the nut. The apple is a rich source of vitamin C and minerals.

The parts of cashew apple are shown in Fig. 21 below

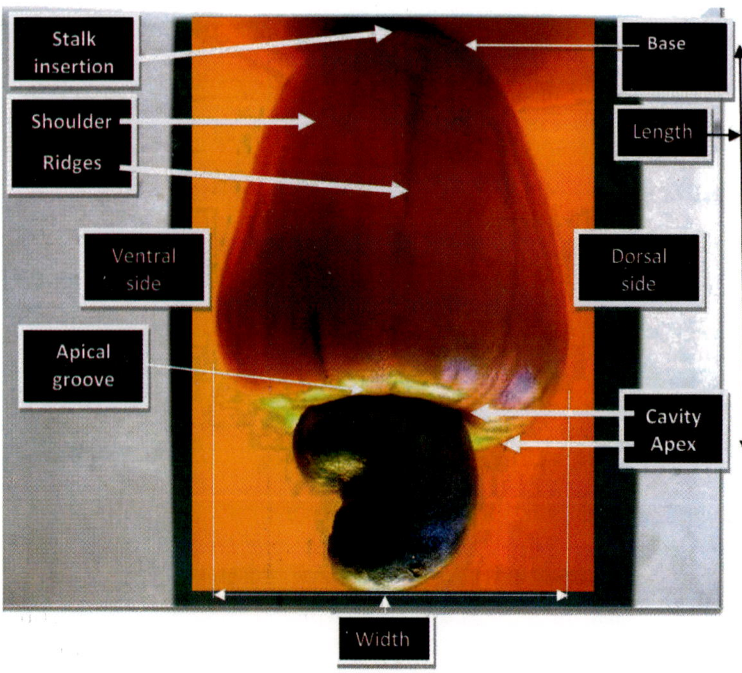

Fig.21: Cashew apple with nut

There are different shapes of cashew apple, namely; mostly cylindrical, conical-obovate, round and pyriform.

A High Proportion of Flowering to Non-flowering Laterals

Flowering in fruit trees is a complex phenomenon, including cashew. In high yielding types, out of the total number of lateral branches, a high percentage (more than 60 per cent) seem to be terminating in panicles. Perhaps, the flowering may be in 1-3 waves, or at one time synchronously. This character is being much influenced by the weather conditions also.

Early flowering

Precocity in flowering plays an important role in production of specific variety. The advantages of early flowering are (1) to produce early crop for the market, (2) escape from pests; (3) abundance of pollinators; which ultimately lead to better fruit set, higher production with minimal cost of production. This feature helps in adoption and cultivation of cashew in arid and semi-arid tracks of the country along with coastal belt.

A short flowering period

In view of the harvest to be made by manual labour and high cost of labour thus involved it is appropriate to make the choice of tree from among the genotypes that completes flowering in one flush. The range of this trait is from 45 to 110 days. A type that falls within 60 days span or below, coupled with high yield would be the right choice.

High bisexual flower percentage

Large number of male flowers is borne in cashew, in comparison to the bisexual flowers. The latter ones alone develop into fruits after pollination and fertilization. The narrow ratio of male: hermaphrodite flower results in to poor fruit set and final produce. Thus, wider sex ratio is desirable especially in cross pollinated crops like cashew.

Number of nuts per panicle

Bisexual flowers as high in number as 120, appear in panicles of certain trees. Even if a high degree of fertilization is assured, all fruits may not develop and reach to maturity. Many of them drop out at the pea-pod stage. A tree is not considered very high yielding, if it does not bear on an average 5-10 fruits per panicles.

Pollination and nut set

In cross pollinated crops, the pollination plays an important role in pollination and thereby fertilization. Cashew is pollinated by insects (native bees, honey bees, ants, flies, wasps). Although a large number of flowers are produced on a panicle, nut set is relatively low because of the low sex ratio and poor pollination. Sometimes, poor pollination can be caused by a lack of pollinating insects and in this case the inclusion of honey bees can help. Nutrition, water availability and pest damage can also influence nut set.

Cashew flowers are self-fertile, but probably not self pollinating, as bagged owners set no fruit (Northwood, 1966; Free and Williams, 1976; Masawe, 1994), but self pollination by hand improved fruit set (Northwood, 1966). The presence of scent, nectar, coloured petals and sticky pollen, are the important features which favour insect pollination. The relative importance of insect and wind pollination has, however, not been critically established. It would also add considerable confidence to devising future strategies, if further more extensive research could be carried out on yield performance on the different sides of the cashew tree across a range of locations.

Wunnachit *et al.* (1985) reported that, cashew trees produce four types of pollen from the large and small stamens of the hermaphrodite and male flower (HL, HS, ML, MS). Comparative studies were made of the grain number, structure, viability, vigour, and sugar and amino acid composition of the four pollen types. Anther and pollen grain numbers and dimensions of the four pollen types were similar, as were pollen structure and staining characteristics. The fluoresce in di-acetate test showed that the HL pollen had the highest percent fluorescence, and viability of all pollen types had declined by 48 h after anthesis. Following controlled hand pollination, the ML pollen had the highest capacity to germinate on the stigma and penetrate the ovule, followed by the MS, HL, and HS pollens. Glucose and fructose and 19 free amino acids were present in all pollen types, with higher levels in the hermaphrodite than in the male flower pollen. The results indicate that the pollen of the male flower is specialized for pollination and fruit set, whereas that of the hermaphrodite flower may be specialized for insect attraction.

The various workers suggested the use of combinations of floral quality characteristics like high pollen fertility, prolific flowering, and high flower sex ratio plus high cross-compatibility as reliable selection criteria for cashew yield improvement.

Factor Affecting the Flowering in Cashew

The cashew tree normally comes in to flowering in 3-4 years. Each flower panicle consists of 200-1600 flowers. Nearly 90 per cent of the flowers in a panicle are male and the rests are bisexual flowers and only 10 per cent of the bisexual flowers produce mature fruits. The panicle emergence on shoots takes place during October-January in India. The sex ratio is about 6:1. Panicles appeared in young tender flush. Variation in the flowering due to altitude and latitude, abiotic and biotic stress on cashew flowering are already discussed in chapter 2.

Optimum weather and flowering in cashew

Cashew may require relatively dry atmosphere with mild winter for better flowering. The mild winter may defined as "low minimum surface air temperature ranging between 15 and 20°C coupled with more dew nights having moderate dew" in the humid climates.

There is gradual increase in number of hermaphrodite flowers from lower to upper part of the panicle. The numbers of perfect flowers and fruits set increased from the proximal to the distal end with the maximum yield at the fifth or sixth nodes in a panicle of 8-10 rachises. The productivity of an entire inflorescence or a whole fruiting branch is related to its orientation with respect to the sun. The number of nodes and fruits set and the percentage of fruit retained is always found maximum in panicles located on the western and southern sides of the canopy. Tea mosquito damage is also found less on these sides of the canopy, but the physiological disorder of 'apple cracking' may be higher here than on the eastern and northern sides. An increase in temperature during April-May months may adversely affected on cashew apple in South-West direction. Direction-wise yield pattern in cashew as below in Fig. 22.

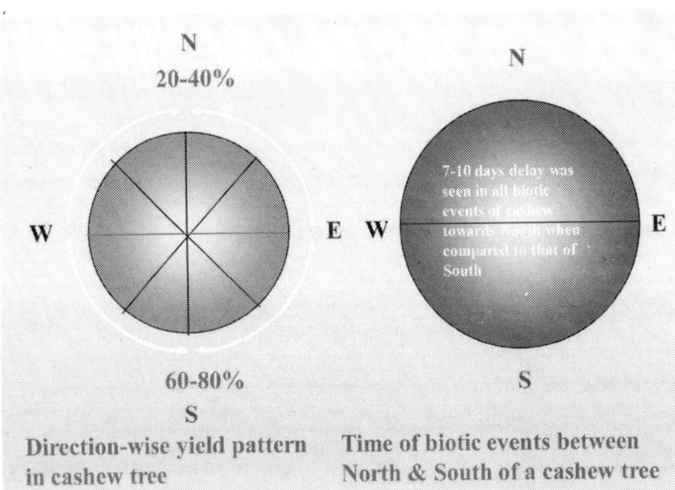

Fig. 22: Direction-wise yield pattern in cashew (*Source*: Rao and Gopikumar, 2010)

Flowering appears in two or three distinct phases and those appearing in the intermediate stage are generally the most productive ones. There are three phases of flowering as reported by Anon. (2002) as follows:

1. First male phase with 19-100 per cent male flowers
2. The mixed phase with 0-60 per cent male flowers and 0-20 per cent bisexual flowers, and
3. The second male phase with 0-7 per cent male flowers.

The mean duration of flowering is about 85 days in which the duration of first male phase is 2.4 days, mixed phase 69.3 days and second male phase 13.3 days (Anon., 2002).

Generally, each cashew flowering panicle performs the following steps depending upon the variety, geographical area, latitude and altitude, etc.

Mean duration of flowering - 85 days
↑
Duration of first male phase - 2 - 3 days
↑
Mixed phase - 69 to70 days
↑
Second male phase - 13 days.
↑
Fruit set to mature - 2 - 3 months
↑
First nut matures and then
↑
Development of apple start (Enlargement of pedicel)
↑
Harvesting period - 2- 3 months.

A period of 2-3 months elapses between fruit set and fruit maturity. The nut (fruit) reaches its maximum size during the first half of this period, whereas the fleshy cashew apple (enlarged pedicel) makes most of its growth during the second half of this period. The harvesting period lasts from 1½ to 3 months depending on the region and variety.

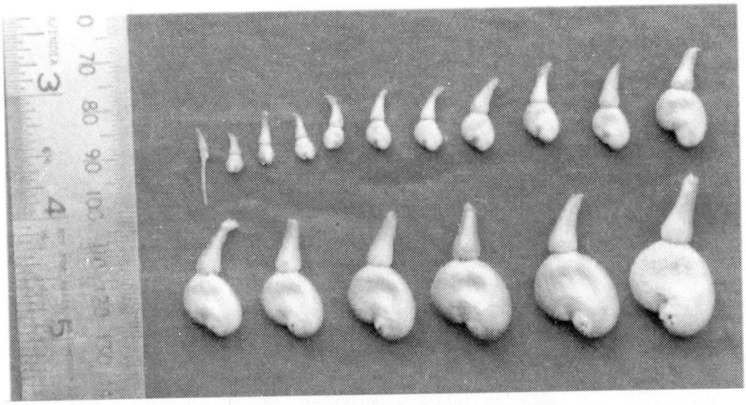

Fig. 23: 1 DAF (upper left) to 12 days after fertilization (DAF)

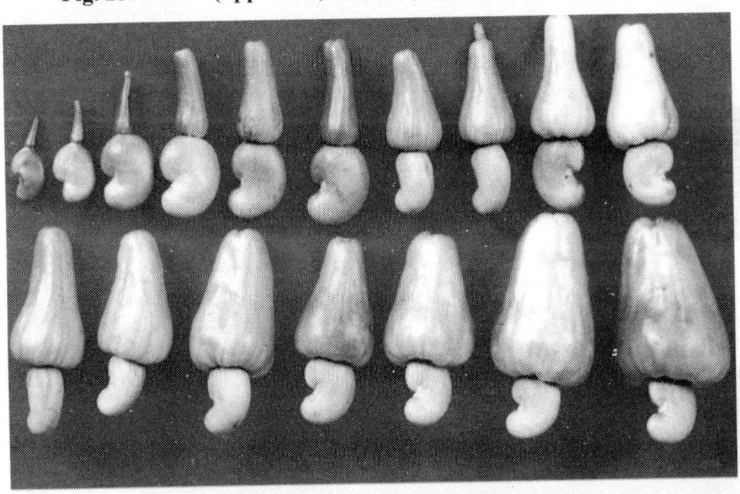

Fig. 24: 12 days after fertilization (Upper left) to maturity (Lower right)

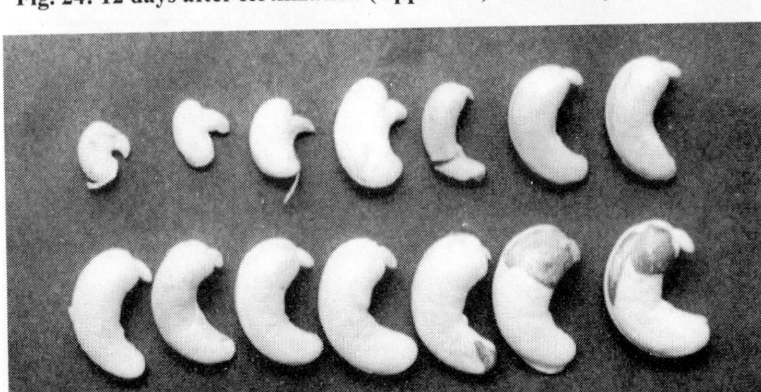

Fig. 25: Development of kernel in nut, 20 days after fertilization (Upper left) to maturity (Lower right)

CHAPTER – 9

Pollination Technique in Cashew

The emasculation and pollination in any crop depend on its floral structure and biology. Cashew is basically andromonoecious (polygamo monoecious) crop with staminate and hermaphrodite flowers appearing in the same inflorescence or panicle (Rao and Hassan 1957). The hermaphrodite/bisexual/perfect and male flowers occur intermediately in the same panicle in three or four phases viz., Male – Mixed – Male or Mixed – Male – Mixed– Male. Distinct phases depend on genotype and environmental factors. The tree shows considerable variation in the duration of different phases. The hermaphrodite flowers are bigger in size than staminate or male flowers (Damodaran *et.al.*, 1965). Male flower always exceed the bisexual flowers. The flowers produced early in a panicle are mostly male (Rao and Hassan 1957; Damodaran *et. al.*, 1965). Most of the perfect flowers are produced about one month after the first flower occurred in the panicle (Bhat *et al.*, 1998).

The flower is typically pentamerous and each hermaphrodite flower stands upon an obconic pedicel. The outer surface of the sepals and petals are pubescent with simple hairs. Flowers are creamy white to white in colour at the time of anthesis and red strips appear on it few hours after anthesis. The hermaphrodite and male flower are similar in structure except that male flowers are lacking in pistil and are smaller than hermaphrodite flower. Male flower have androecium which consists of one or two fully developed stamen (s) and 7 to 9 staminodes (Sterile stamens). The staminodes have only very short filaments and are closed in lower half of the open flower. In hermaphrodite flowers the anther level in mostly below the level of stigma. Depending upon flower type and level of anther, cashew flowers are classified into four group i.e. Male flower with long anther level (ML), Male flower with short anther level (MS), Hermaphrodite flower with long anther level (HL) i.e. just below the level of stigma and H.F. with short anther level (H.S.). In prefect flower, ovary is superior and unilocular. The base of perfect flower (H.F.) is mostly bigger in size than male flower.

The average length of panicle varies from 14 to 21 cm and the number of flowers per panicle varies from 200 to 1600 (Damodaran *et al.*, 1965.)

The staminate flowers generally open earlier than the hermaphrodite flower (Rao and Hassan1957). Majority of male flowers open between 9 a.m. to 11 a.m. while majority of hermaphrodite flowers open between 11 a.m. and 12 noon.

Out of 40 cashew varieties released in India, 25 are selection and 15 are hybrids. Regional Fruit Research Station, Vengurla has released 8 cashew varieties so far out of which two varieties i.e. Vengurla–1 and Vengurla–2 are selections and 6 varieties are hybrids (V-3 to V-8). Similarly two hybrids i.e. hybrid 320 and hybrid 303 are found to be very promising clones at this research station. The current average yield level of recommended cashew varieties in Maharashtra is around 1.5 to 2 tonnes/ha. Cashew plantation in Maharashtra State done before 1990 is of seedling origin due to which its average productivity is low. The present breeding strategy is to develop high yielding cashew varieties with high yield potential (more than 10 kg/tree) medium to bold nut (more than 8 gm) and high shelling % (more than 28%).

The hybridization programme was started at R.F.R.S., Vengurla in 1974. More than 2000 cashew hybrids have been evaluated; as a result 6 promising cashew hybrids were developed so far and at present more than 3000 new F_1 cashew hybrid seedlings are under evaluation. This large scale production of F_1 cashew hybrids was possible only because of this standardization of the pollination technique in cashew at this research station.

Vengurla is located at the sea shore and Regional Fruit Research Station is 3 Km away from sea shore and 8 m above MSL. It is at West Coast with high rainfall and humid climate. Under Vengurla condition stigma is receptive throughout the day after anthesis, however, maximum receptivity was found 2–3 hours after anthesis, dehiscence of anther starts from 10.30 a.m. which, however is influenced by genotype and weather conditions during the day hours. Generally high temperature induces early anthesis of cashew flowers whereas low temperature during working hours extend the period of anthesis. Similarly, frost during synchronized (mixed) flowering stage may extend the period of anthesis which results into low fruit set and yield.

The dehiscence takes place by means of longitudinal splits of the two anther lobes exposing the bluish grey pollen grains. Before anthesis of H.F. flowers (11 a.m. onward) of the selected female parent (on which hand pollination is to be carried out), staminate flowers from selected male parent should be detached from the panicles just after anthesis and before the anthesis of H.F. of same panicle because dehiscence of anthers start early after detachment. For hand pollination purpose, staminate flowers of the selected male parent are detached from panicle and kept under partial shady place for early dehiscence.

The existing procedure followed by the research workers in cashew pollination, a butter paper bag is used for covering the flowers or panicles after emasculation and pollination. The fruit set in this method is very low due to more weight of the paper bag and more of friction while putting the paper bags and also due to more impact of wind on panicle enclosed within the bag (Bhat *et. al.*, 1998).

Cashew flowers are very delicate and drop off quickly with slight friction if not handled carefully. Hence, it was felt that, the present pollination technique needs an improvement. Accordingly, this station has standardized the pollination technique.

The Procedure for Pollination Technique Involves the Following Steps

1. First select the male and female parent which are having wider diversity or according to the parents chosen by the breeder. The panicles that have flower buds which will open on the next day on the both male and female parents are selected. In case of the female parent tree, all opened flowers (both male & hermaphrodite flowers) and also the nuts formed, if any, by open pollination are removed from the selected panicles.

2. Next day between 8.00 a.m. to 10.30 a.m., all opened male flowers from the selected panicles in female parent tree should be removed. After 10.30 a.m., emasculation of the hermaphrodite flowers is carried out using ordinary pins before dehiscence from freshly opened hermaphrodite

flowers of the panicles. In the earlier method, stigma along with style is enclosed with butter roll / pantographic paper roll (Bhat *et al.*, 1998). However, in this method, each emasculated hermaphrodite flower is enclosed with butter roll paper (Fig.3) which is laborious and there is a chance of quick drop of hermaphrodite flower with slight friction. So, it was felt necessary to modify the earlier pollination technique. Accordingly, a method was developed in which complete panicle is enclosed within a cloth bags supported with rectangular/ rounded iron or alluminium wire cage (40 cm X 24 cm) of which one end is close and other end is open for insertion of the emasculated and pollinated panicle (Fig.7). This method gives the refined technique of pollination in cashew including procedure for emasculation, protecting the stigma from contamination and careful handling of the controlled hand pollination and finally getting more number of genetically pure hybrid nuts from selected panicles of the female parent. Similarly, another most important advantage of this method is, it protects the panicle and a tender nut from TMB as the complete panicle is enclosed in the cage. White cloth bag is used for insertion of rectangular/ rounded iron or alluminium wire cage. The cloth bag should have a size of 50 cm X 25 cm and sewed in such manner that one end is completely closed and other end is open for insertion of whole panicle inside the cage up to boot leaf. The cage should be closed with the help of rope. Care should be taken that tying portion of the panicle should be covered with cotton plug for avoiding damage to the panicle. After putting cotton plug on tying portion of the panicle, the complete panicle is inserted in the cloth bag (supported with cage) and then the bag is tied to the branch on the cotton plug. Due to more weight of the iron/alluminium cage, there is chance of breaking of panicle but this can be easily avoided by tying both ends of bag on other nearest branch of cashew tree with rope/string.

3. Then, freshly opened male flowers are collected in a petri dish from the selected male parent and the anthers are allowed to dehisce under partial shade (anther turns bluish grey when it dehisces because of the colour of pollen grains). Care should be taken to avoid direct sunlight on the male flowers kept in the petri dishes as cashew flowers will desiccates as they are very delicate. Separate petri dishes should be used for each pollen parent.

4. The bag is removed from the emasculated female parent and stigma is pollinated with pollen from freshly dehisced anthers of male flowers of the selected male parent kept in dishes (by holding male flower in hand and dusting the pollen grains on the stigma of hermaphrodite flowers.)

One male flower can be used for 2 hermaphrodite flowers (Bhat *et.al.*, 1989).

5. After the pollination process, complete panicle up to boot leaf is enclosed in the bag (supported with iron cage) and open end of the cage bag is tied to the panicle by tying the knot over a cotton plough in order to avoid further damage to the panicle with the help of thread and then both the ends of the bag are tied to the same or other nearest branch of the tree.
6. Each panicle must be labelled with zinc / wax coated labels showing the name of male and female parents and date of crossing. One female panicle is used for only one cross combination.
7. Record of the details of the cross combinations made should be kept such as, name of female and male parent, date of pollination, tree number, panicle number, number of hermaphrodite flower of female parent pollinated, etc.
8. All opened hermaphrodite flowers which are not used for pollination in female parent are removed daily. Further, opened male flowers of female parent may also be removed daily to keep the panicle clean. The above procedure (1 to 7) is continued for the next few days until 8-10 hermaphrodite flowers are pollinated in the selected panicle with the same selected male parent. Primegenic dominance operating in same cashew varieties may interfere in setting of crossed nuts (hybrid nuts) if numbers of pollinated flowers exceed the above mentioned limit i.e. 8-10 hermaphrodite flowers (Bhat *et al.*, 1998).
9. On the last day of pollination, all remaining flower buds are removed from the panicles of female parent used for pollination.
10. In this method, prophylactic spray of insecticide is not essential to the panicle. However, one to two prophylactic spray of insecticide helpful to reduce the incidence of TMB and other pests.
11. In this method, complete panicle is enclosed as such in the cloth bags after (last day of pollination) pollination, the matured hybrid nuts of the particular panicle are not lost or do not get mixed with the nuts of other panicles when cashew apple ripen and fall.
12. The hybrid nuts from the individual panicle (enclosed in bag) are collected when the apples are mature or fall in the cloth bag. The hybrid nuts are dried for 3 to 4 days in sunlight and stored in labelled cloth bags and in dry place.

13. The hybrid nuts are raised in poly bags (6' X 8') for germination. Hybrid seedlings are planted in the field for further evaluation. The new modified method of pollination technique in cashew is very easier to adopt and also give higher fruit set and fruit retention percentage in comparison to the earlier procedure of pollination.

Diagrammatical presentation of improved pollination technique in cashew is given below in Fig.26.

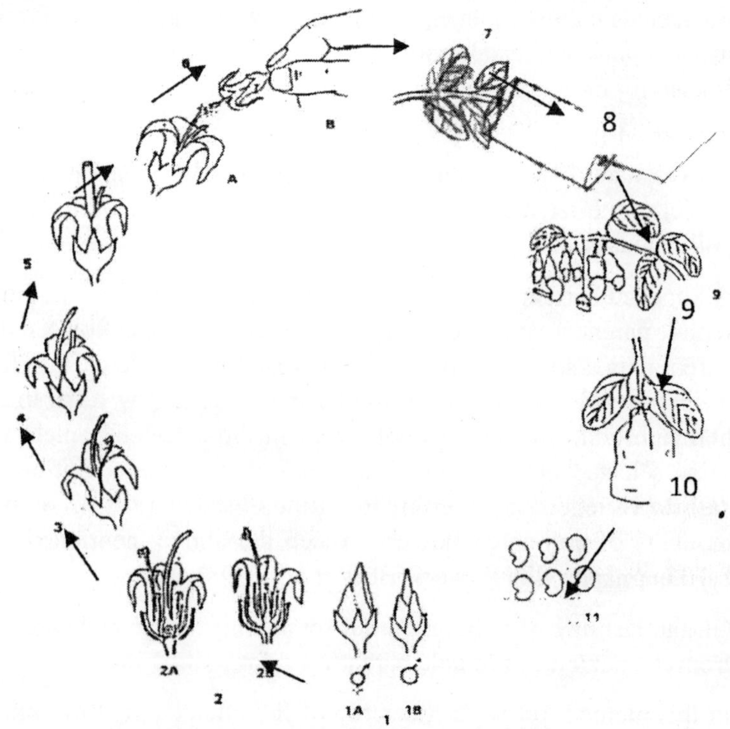

1. Flower buds1A Hermaphrodite flower bud1B Staminate flower bud
2. L. S. of flowers 2 A Hermaphrodite flower 2B Staminate flower
3. Hermaphrodite flower
4. Emasculation (Removal of anther)
5. In Earlier method, paper roll enclosed over stigma, before pollination. But in new improved technique instead of butter papers, complete panicle inserted in cloth bag supported with iron wire cage before pollination
6. Pollination (A X B)
7. Complete panicle inserted in cloth bag supported with iron wire cage after pollination
8. Bunch of hybrid nuts
9. Hybrid nut enclosed in a cloth bag
10. Hybrid nuts

Fig. 26: Pollination technique in cashew

□□□□

CHAPTER – 10

Establishment of Cashew Orchards

Selection of Site and Land Preparation

Cashew is a hardy plant; it thrives well on wide range of soil. While selecting the land for cashew, soils with water logging and excessive salinity of alkalinity should be avoided. Otherwise, soil depth, slope, stoniness, soil fertility and water availability seem to impose very little limitation. For establishing new plantations, land preparation should begin with the first pre-monsoon rains. The area should be cleared of bushes and other wild growth before digging pits for planting.

Planting Season

Planting of soft wood grafts is usually done at the beginning of monsoon season (June-July). Therefore, land preparation such as clearing of bushes and other wild growth, digging of pits for planting, should be done during pre monsoon season (April-May).

Spacing

A spacing of 7m × 7m or 8m × 8m is recommended for cashew (204 or 156 plant/ha). A closer spacing of 4m × 4m or 5m × 5m in the beginning and thinning out in stages and thereby maintaining a spacing of 8m × 8m after seven to eight years of planting can also be followed. This enables higher returns during the initial years and as the trees grow in volume, the final thinning is done. At Vengurla, highest yield/unit area was obtained from 5m × 5m square planting and highest nut yield/tree was in 8m × 8m square planting.

High Density Planting

It has been reported that high-density planting system in cashew is economical (Salam, 1997; Ghosh *et al.*, 2000). Maintaining tree-density of 625 ha^{-1} (4m x 4m) for the first 11 years, and diagonal thinning thereafter to reduce the population to 50%, resulted in maximum yield (Yadukumar *et al.*, 2001). High-density planting system in cashew doubled nut-yield during the first 10 years of planting (Yadukumar *et al.*, 2002). Salam (1999) reported that the varieties '44/3', 'Anakkayam-1' and 'H-1608' were suitable for high-density planting.

Fig. 27: Normal density planting (7m x 7m)

Fig. 28: High density planting (5m x 5m)

Fig. 29: Diagonal thinning in HDP after 10 years

Digging of Pits and Planting of Grafts

Normally cashew grafts are planted in the pits of 60 cm × 60 cm × 60 cm size. It is preferable to dig the pits at least 15-20 days before planting and expose them to sun so that the termites and ants, if any, which damage the roots of the grafts will migrate. The pits should be filled to 2/3 rd depth with a mixture of top soil, compost/farm yard manure (5kg) or poultry manure (2 kg) and rock phosphate (200 g).

Planting of grafts is done preferably during June-July by avoiding heavy rains. Usually six to twelve month old grafts are supplied by the Research Stations and registered private nurseries in polythene bags.

The soil in the centre of the filled pits is scooped out. The polythene bag of the grafted plant is removed carefully without disturbing the ball of earth. Then the ball of earth is placed in the centre of the pit where the soil was scooped out and it is covered with soil and pressed gently without disturbing root ball. Care must be taken to see that the graft joint remains at least 5 cm above the ground level at the time of planting. The grafted plant should be provided with a stake and tied loosely with a plastic thread immediately after planting to avoid breakage at the graft joint. Staking should be continued during second and third year of planting also. Immediately after planting, the basin around the graft should be mulched with green leaves. This suppresses weed growth and conserves soil moisture.

After Care

For the first two or three years, regular watering is necessary for the better establishment of the young grafts. The grafts are stalked and protected from scorching sun in summer by providing shade with coconut leaves and from cattle damage by fencing. The allies are ploughed to suppress weed growth. The basins of young plants may be mulched with dry leaves to conserve moisture in summer and also to suppress the weed growth in the basins.

Critical Management Practices

For the success in planting, the following management practices should not be missed.

- The planting material should be grafts, never use seedling for planting.
- Plant grafts (at least 6 months old) of varieties suited and recommended to the region.
- Adopt the pit method of planting (pit size not less than 60 cm X 60 cm X 60 cm).
- Transport grafts carefully to minimise transporting shock.
- Plant grafts at the onset of monsoon.
- Polythene tape around the graft union must be removed before planting.
- Staking and mulching of young plants are must after planting.
- Apply recommended fertilizers and manures as per a tree age basis.
- Irrigation during the initial years may be necessary for better establishment.
- Ensure gap filling in the second year.
- Remove branches emerging from the root stock regularly for first two years.

- Trained the plants during first 2-3 year to maintain straight single stem up to one metre height from bottom.
- Adopt recommended plant protection measures as and when required in time.
- Thinning of alternate plants should be done before severe competition sets in case of high density plantation.
- Remove the dead wood, infested shoots (insect/disease) once in a year after harvest to minimise unwanted load on tree.

□□□□

CHAPTER – 11

Integrated Nutrient Management in Cashew

Various studies have demonstrated that the cashew tree responds to the application of mineral nutrients, though the responses are significantly affected by plant age, the genotype utilized, the conditions of cultivation, soil and climate and of the crop management (Barros *et al.*, 1984; Ghosh, 1990; Ximenes, 1995; Bezerra *et al.*, 1999; Crisóstomo *et al.*, 2005). The crop develops well under semi-arid conditions, in which the problems associated with water availability and salt excess in the root environment are common. These two factors interfere with the acquisition of nutrients, in that the levels of nutrients in the plants are affected both by direct effects on absorption and transportation processes and by indirect effects associated with a reduction of growth of the various plant organs including the root system. The effect of salinity on plant nutrition depends on the types of salts that predominate in the cultivation environment which can result in a deficiency of essential nutrients and an excess of toxic ions. These changes under salt stress are accompanied by a loss of structural and functional integrity of the cellular membranes, a reduction in activity of various vital enzymes, and a diminished capacity of the roots to absorb nutrients (Zhu, 2001). On the other hand, water deficit reduces the extraction of nutrients from the soil, limiting the solubilization of nutrients in the root environment and altering the morphology of the root system (Kramer and Boyer, 1995). It should be added that the effects of excess salt and water deficit depend on various factors, including the species under study and the stage of development at which the plants are submitted to these.

Mineral Nutrition of Cashew Plants Under Stress

The concentrations of minerals in the leaves of adult cashew plants vary as a function of plant age, leaf age and mobility of the nutrient within the plant, season of the year, environmental conditions, and the genotype under study (Kumar *et al.*, 1982; Bezerra *et al.*, 1999). In an experiment performed by Kumar *et al.* (1982), great variations in the concentration of N, P and K were found in function of leaf age and season for the common cashew plant growing under a dry, non-fertilized regime. They also demonstrated that the process of fruit development causes a decrease in the levels of these nutrients in the leaf, which could be a consequence of ion translocation to the developing fruit. Dwarf cashew plants cultivated under dry and non fertilized conditions showed variations in leaf concentration of macronutrients along a cycle of production. The variations were associated with leaf age and the phenological changes of the plant over the course of a year (Bezerra *et al.*, 1999). Further they observed increasing levels of N, P, K, Mg and S up to the beginning of the reproductive phase (rainy season), and decreasing thereafter (with the exception of K which remained unvaried) during flowering and fruit development (dry season). The levels of Ca also tended to increase with the age of the leaf. Indeed, it was shown that the level of these nutrients was not altered in the leaves of a dwarf cashew plant that received higher doses of N and K fertilizers, and at times it was greater in nonfertilized plants. This could be the result of dilution, since fertilized plants presented greater canopy development (Richard, 1993; Crisóstomo *et al.*, 2005). It is also noteworthy that the levels obtained with fertilized plants (Crisóstomo *et al.*, 2005) were not different from those obtained with non-fertilized plants (Bezerra *et al.*, 1999), and this could also be associated with dilution and concentration effects, due to the influence of fertilization on plant growth. Nutritional stress in adult cashew plants also showed a similar complex response. Guilherme (2006) conducted a 12-month experiment in an area where the soil type Quartzarenic Neosol predominates using cuttings of the clone CCP 76 grafted on CCP 06 and that had been heavily pruned down to the stem before initiating the experiment. It was observed that after plants were irrigated with water containing different salt concentrations [0.5 (water control), 1.5, 3.0, 6.0 and 9.0 dS m^{-1}], only the levels of leaf Ca and Na were affected, while the changes in Mg, K, Cl and P were non-significant.

The effects of salt stress on some physiological and biochemical traits were evaluated in dwarf-cashew seedlings at the same developmental stage. Seeds were sown in trays containing vermiculite moistened with distilled water or with NaCl solutions having different electrical conductivities: 0.7, 1.8, 6.0, 9.8, 13.4, 17.4 and 20.6 dS m^{-1}. Salinity delayed and inhibited seedling growth and development, particularly in the shoot. Concentrations of Na^+ and Cl^-, but not of K^+, increased with increasing stress severity. With the exception of proline, concentration of organic solutes was only marginally affected by salt stress. Catalase activity in leaves increased slightly as a result of salt stress, whereas guaiacol peroxidase activity was induced only under low levels of salt. In contrast, activities of guaiacol peroxidase and ascorbate peroxidase increased dramatically in roots. Apparently, roots were better protected against oxidative damage than shoots, as judged from the decrease in lipid peroxidation in root tissues. In leaves, expression of 75 proteins, evaluated by 2D electrophoresis, was altered by salt stress: 35 of them increased their expression and three were apparently *de novo* synthesized. In roots, 69 proteins were modified by salt stress: 34 proteins increased their expression and two proteins appeared only in stressed seedlings. The changes in protein patterns were caused by the imposed salt stress rather than by a response to the developmental stage. Overall, these responses could play an important role in salt stress acclimation of cashew seedlings (Carlos *et al.*, 2008).

Fig. 30.1: Dwarf-cashew seedlings of clone CCP 06 on the 21st day after sowing. Seedlings were subjected to control (0 dS m^{-1}) and salt stress (0.7, 1.8, 6.0, 9.8, 13.4 and 17.4 dS m-1) conditions.

Source: Carlos *et al.* (2008)

Gas Exchanges in Cashew Plants Under Stress

The low availability of water in soil has both direct and indirect negative effects on the photosynthetic rate of plants. This could be due to stomatal closure that, in turn, reduces the availability of CO_2 inside the mesophyll cells, or due to non-stomatal limitations that diminishes the potential capacity of CO_2 assimilation (Lawlor, 2002; Flexas *et al.*, 2004). Nevertheless, stomatal closure appears to be the principal cause of reduced photosynthetic rates (Flexas *et al.*, 2004). Apparently, there is no inhibition of photochemical reactions or alterations in the pool of intermediary compounds of the photosynthetic carbon reduction cycle, at least within a change of up to 30% of the protoplast volume (Chaves, 1991; Pelleschi *et al.*, 1997). Under conditions of salinity, gas exchange is chiefly impaired as a consequence of altered physiology and biochemistry of stomata as well as due to effects of salts on photochemical events (Taiz and Zeiger, 2002). Salinity also impairs the transport of photo-assimilates and, consequently, the metabolism of carbohydrates, compromising the normal growth and development of plants (Kozlowski, 1997).

Nutrient Management

Cashew (*Annacardiun occidentale*) is a tree crop that follows an intermediate type of growth pattern. Since the tree thrives well even in adverse soil environments, it is believed that cashew does not require any fertilisers. This is not at all true. A large number of experiments conducted in India and abroad conclusively proved that cashew responds very well to applied fertilizers, except in very rich soils. Cashew being a regular bearer which puts forth flowering and sets fruit on the current season flush, the plant needs sufficient and continuous supply of nutrients for the growth of new flush, flowering, fruit set and development of nuts. The positive influence of major nutrients seems to be on vegetative growth, floral biology, fruit set, nut yield, shelling percentage and kernel grade.

Response to Nutrients

Information regarding the response of cashew to fertilizers is briefly indicated below.

a. Response to nitrogen

Nitrogen is the kingpin of nutrients. It is absorbed by plants in the form of nitrate and ammonical ions. Nitrogen is essential for the formation of proteins and it is an integral part of chlorophyll. An adequate supply of nitrogen enhances vegetative growth. When nitrogen becomes deficient, plant growth will be stunted and the leaves turn yellow. The yellowing or chlorosis usually appears first on the lower leaves and spread to the upper leaves later. Severe nitrogen shortage results death of leaf.

The response of cashew to applied nitrogen is tremendous and the same is observed universally. Nitrogen is the element that is absorbed by cashew in largest quantity (Salam, 1997). The growth and development of cashew is very much influenced by nitrogen application. Nambiar (1983) reported significant positive effects on various growth characters of cashew from Bapatla and Kumar (1985) from Karnataka. Increase in cashew yield due to N application was reported by several workers (Nair *et al.*, 1972; Lefebvre, 1973; Reddy *et al.*, 1982; Nambiar, 1983; Rao *et al.*, 1984; Veeraraghavan *et al.*, 1985; Ghosh, 1988b and Mathew, 1990). A review of literature indicates that the response of cashew to nitrogen application is universal, except in very rich soils.

b. Response to phosphorous

Conflicting reports are observable regarding the effect of P on cashew. Rao *et al.* (1984) observed no response to P application, in sandy loam soils. Similarly, Veeraraghavan *et al.* (1985) also could not observe any effect of phosphorous in laterite soils of Madakkattara. The main effect of phosphorous to increase the yield was limited to a dose of 25 kg P_2O per hectare, but when applied with nitrogen, phosphorous application increased yield up to a dose of 75 kg per hectare (Sawke *et al.*, 1985). Kumar (1985) reported positive influence of phosphorous on nut yield. Richards (1993) reported that soil P was a major limiting nutrient in P deficient soils of Australia. According to him, P application increased nut numbers and nut yield. A review of literature indicates that cashew is generally shy to show response to phosphorous fertilizers.

c. Response to potassium

Nambiar (1983) observed positive effects of potassium on cashew. Application of potassium increased the cashew nut production particularly in the presence of nitrogen (Lefebvre, 1973). Significant positive effects of potassium on growth and yield of cashew were reported by Ghosh (1988b) and Ghosh (1990). But Veeraraghavan *et al.* (1985) could not observe positive effect of potassium application in cashew. Kumar (1985) obtained linear response for potassium up to 150 g K_2O per tree, the highest level tried by him. A review of literature indicates that, next to nitrogen, the response of cashew is more to potassium application.

The response of cashew to the major nutrients is in the order of N, K and P. Cashew is a nitrogen lover and N and K nutrition is of greatest significance in enhancing the production and productivity of cashew (Salam, 1997).

Generally, little or no emphasis has been given to the economic evaluation of fertilizer use for cashew. Cashew is often grown on marginal soils and on wastelands mostly unsuitable for other economic crops. Nitrogen and phosphorous were found to be the most important nutrients during pre-bearing stage. At the bearing stage, K, together with N, is important. Application of fertilizers, their dosage, time and schedule under different agroclimatic zones has been standardized (Veeraraghavan *et al.* 1985; Sawke *et al.* 1985; Harishu Kumar and Sreedharan, 1986; Ghosh and Bose, 1986; Kumar *et al.*, 1997;

Latha *et al.*, 1994; Lenka *et al.*, 1998; Grundon, 1999; Grundon, 2001; Shingre *et al.*, 2003; O' Farrell *et al.*, 2002; Yadukumar *et al.*, 2003; Prasanna Kumar *et al.*, 2006; Salam *et al.*, 2008; Yadukumar *et al.*, 2008; Aikpokpodion *et al.*, 2009; O'Farrell *et al.*, 2010).

Rootstocks and grafts of cashew supplied with 150:20:100 ppm of N: P: K at the rate of 100 ml/plant/week had higher plant-height, stem-girth and number of leaves (Manjunatha, 2001). The ratio 2:1 of N:P is ideal for young cashew trees (Shi-Wenge *et al.*, 2005). Application of mineral nutrients in combination with organic fertilizers significantly increased plant height, dry weight of aerial parts and number of leaves in cashew seedlings (Lima *et al.*, 2001).

Application of Manures and Fertilizers

Cashew responds very well to fertilizer application. About 50 to 100 per cent increase in nut yield is commonly observed in this crop. Nitrogen is the key nutrient element for cashew nutrition followed by potassium. Cashew grows well in a variety of soil and climatic environments and nutritional dose depends on the soil and climatic environment in which it is grown. The productivity of the variety also decides nutritional dose. Separate fertilizer recommendations are available for every state. Right dose of fertilizers is to be applied at the right time following appropriate methods. Soil fertility based and productivity linked fertilizer recommendations are necessary for every environment especially when the climate changes under global warming situations.

Fertilizers and manures promote growth of the plants and advance the onset of flowering in young trees. Application of 10-15 kg of farm yard manure or compost per plant is beneficial. The fertilizer recommendation for Maharashtra for high yielding cashew variety is 1000 g N (2.2 kg urea), 250 g P_2O_5 (1.5 kg single super phosphate) and 250 g K_2O (400 g muriate of potash) for adult plant (4th year onward) per year. It is always advisable to use straight fertilizers instead of complex fertilizers. This is mainly due to the reason that the recommended dose of NPK for cashew is not available in any form among the complex fertilizers which are available at present in the market.

The ideal period for fertilizer application is immediately after the cessation of heavy rains, in a circular trench. While applying the fertilizers, it should be ensured that adequate moisture is available in the soil, so that the applied fertilizers are taken up by the plant. It is also suggested that fertilizers should be applied in two split doses during June-July and post monsoon (September-October) periods. However, if single application is done, it should be during the months of August. During first, second, third and fourth year of planting 1/4, 2/4, 3/4 and full dose of fertilizers respectively should be applied. From the forth years onwards, the full quantity is to be applied. The detail quantity of chemical fertilizer to be applied for different years is worked out based on 1000g N, 250 g P_2O_5, 250g K_2O (Table 21) and quantity of organic manure required to fulfil the total requirements of the nutrients on the basis of recommended chemical fertilizer dose is given in Table 22.

Table 21: Recommended doses of NPK fertilizers for cashew (g/plant)

Year/s after planting	FYM) (Basket	N (g)	Urea (g)	P_2O_5 (g)	Single (g) Super Phosphate	K_2O (g)	Muriate of Potash (g)
I year	1	250	500	63	400	63	100
II year	2	520	1000	125	800	125	200
III year	3	750	1725	188	1250	188	300
IV year	4	1000	2200	250	1500	250	400

Table 22: Quantity of organic manures required by adult cashew tree (kg/tree)

Manure	Quantity (kg/tree) required
Poultry manure	34.0
Farm Yard Manure (FYM)	80.0
Vermicompost	40.0
Green manure- Glyricidia	20.0
Biogas slurry	54.0
Water hyacinth compost	50.0
Castor cake	17.0
Groundnut cake	12.0
Pongamia cake	28.0
Cashew waste compost	96.0
Neem cake	19.0

About 72% of the total active roots of cashew reside within an area of 2 m radius around the tree. In laterite soils and soils of sloppy land in heavy rainfall areas, the fertilizers may be applied in the basins of individual trees in circular trenches. Circular trenches of 25 cm width and 15 cm depth prepared at a distance of 1.5 m away from the trunk for the fourth year onwards of planting, respectively. The trenches should be closed immediately after the fertilizer application and it is preferable to give a mulch of green leaves for enhancing the fertilizer use efficiency of the plant. Fertilizer application should be taken up only after weeding and cleaning the basins of individual trees within 2.5 m radius in order to avoid the competition for nutrients from weeds.

Placing of fertilizers at 1.5 m away from the trunk in the trenches of 25 cm width and 15 cm depth is suitable in high rainfall area while, in low rainfall area application in circular bands in 1.5 m width between 1.5 and 3 m from the trunk and incorporating in the soil was beneficial (Rao and Swamy, 2000).

Fig. 30.2: Fertilizer application in cashew

Response to Major Nutrients

Macronutrients (N, P, K) have appreciable effect on the absorption, translocation and distribution of nutrients in cashew. In general, low to moderate doses of any of the macro-nutrients may correct the deficiencies of one or more nutrients in cashew, by way of their enhanced uptake. The fertilizer experiments conducted at Regional Research Station, Vridhachalam (Tamil Nadu), in early years, indicated that application of nitrogen up to 500 g per plant (82.5 kg/ha) increased the yield 2 to 3 times more than unfertilized ones. The application of NPK fertilizers in a circular trench 25 cm broad and 15 cm deep at 2.5 m from the trunk appears to be useful in proper uptake of essential nutrients.

The role of macro and micro nutrients suggested by Mandal (2007) are as follows.

Nitrogen (N)

- Nitrogen is the major constituent of plants and due to high mobility, deficiency symptoms of nitrogen always appears first in older leaves.

Stem becomes thin and growth stunted. Light sandy soils of the coastal region where cashew is mostly grown, lacks in organic matter and is poor in nitrogen content. Nitrogen availability is also intimately connected with activity of soil microbes.

- It is reported that at low nitrogen, cashew seedlings (2 year old) were characterised with general yellowing of older leaves that soon spread to younger ones. The leaves were also smaller. Lack of nitrogen in cashew seedlings became visible 7 weeks after planting. Colour of the leaves changed gradually from dark to pale green and then to yellow, coinciding with stunted growth. The seedlings died after 4 months in case of severe nitrogen deficiency. Nitrogen content of shoots showed highly significant positive correlation with girth and height of seedlings
- In general cashew appears less demanding in its nutritional requirements than many other plants. However, nitrogen application significantly influences the yield of nuts. With increasing nitrogen levels, the concentration of N, P, K and Ca in all plant parts increases, but Mg (in roots), S (in stems), and Na (in leaves) gradually decline.
- It is reported from Madagascar that during the pre-bearing stage of cashew, nitrogen and phosphate were most important elements, but later, nitrogen in combination with potash become significant. An understanding of the mineral demands of cashew plant, the distribution and utilization of nutrients and the relationship among them, would be very useful in fertilizer programme, aimed at increasing yield and improving quality.

Phosphorus (P_2O_5)

- Phosphate is of special importance in germination of seeds, metabolism of seedlings, rapid development of roots and also ripening process of seeds. Heavy soils have high fixing power for phosphates.
- Phosphate deficiency results in stunted growth with dull-green foliage and by the fifth month, lower leaves of cashew seedlings wither and fall. Phosphate application while planting in pits or localized in the root region, is more effective than surface application.

Potassium (K_2O)

- Potassium deficiency symptoms may develop in cashew seedlings within 2 months. Lowest leaves turn yellow, starting at apex and along margins, showing necrosis. The symptoms spread rapidly from lower part of leaf towards the top. While potassium is applied to grown up cashew trees, the K content of leaves was peak at 14 days after application. K-deficiency occurs mostly on light sandy soil. Poor drainage decreases K absorption. However, the response to potassium was found only up to 250g per tree per year.

Calcium (Ca)

- Calcium is concerned in the activities of growing points and proper root development. Ca-deficiency is more prevalent in high rainfall regions where leaching is excessive. Cupping of young leaves with marginal scorching and death of growing points are due to Ca-deficiency, which could be pronounced within a month after planting in a calcium deficient soil. Application of lime (5 t/ha) and phosphorus (70 kg P_2O_5 /ha) increased the pH of sandy soil, and N and P content of the young leaves of cashew plants. Besides, lime application also improves the K content of young leaves.

Magnesium (Mg)

- Magnesium is a constituent of chlorophyll molecule, closely associated with energy supplying phosphorous compounds and is of importance in formation of seeds having high oil content. It is easily leached out and becomes deficient in coastal sandy soils, particularly during the wet season. It is highly mobile and hence deficiency symptoms appear first in older leaves with interveinal chlorosis / necrosis and then progress to younger ones. Magnesium deficiency occurs where calcium deficiency is also a problem and the deficiency also occurs by high potassium manuring. The deficiency can be corrected by foliar spray of $Mg\ SO_4$ @ 3 g /litre.

Response to Micronutrients

Zinc (Zn)

- Zinc is essential for formation of growth promoting and growth-regulating compounds in plants. Zinc deficiency is commonly called 'little leaf' or 'rosette'.
- This shortening of internodes, appears in acid soils and most of them are prevalent in sandy /calcarious regions.
- Zinc ions do not move readily in soil because of fixation with soil colloids.

- Hence foliar spray of Zinc (2 kg $ZnSO_4$ + 1 kg lime in 450 ml water/ha show quick response. Zinc with lime increases yield of nuts as well as growth of leaves and stem.

Boron (B)

- Boron is necessary for translocation of sugars. It regulates intake of water into the cell, acts as a regulator of K/Ca ratios helps reproduction of plants by promoting flower and seed production. It enhances tissue respiration and nitrogen metabolism.
- Boron availability decreases under dry soil conditions. It is readily leached through the soil and its deficiency seems to be more prevalent in sandy soils. Boron deficient plants fail to expand young leaves. The leaves are distorted and remain rolled, and the growing points die. It causes barren plants and poor seed development.
- It is suspected that gummosis in cashew might be due to boron deficiency. It is reported that 10 to 15 kg borax per hectare will provide adequate boron to balance the nutrient status of boron deficient cashew plants.

Manganese (Mn)

- Manganese is concerned with chlorophyll formation, acts as a catalyst in enzyme systems, often involving oxidation-reduction reactions. Mn deficiency leads to chlorosis and in excess, decreases the solubility of iron. In conjunction with Fe, Mn assists in synthesis of chlorophyll.
- The Mn toxicity symptoms are chlorotic mottling and necrotic lesions on leaves, mainly along veins. The necrotic spots enlarge as the leaf matures. Leaves are cupped with brown spotting of margins along with white mottle bands. The deficiency can be corrected by foliar application of $MgSO_4$ @ 3 g/litre.

Nutrient Removal

A 30 year old cashew tree removes 2.847 kg of N, 0.752 kg of P_2O, and 1.265 kg of K_2O through vegetative parts including root system, apples and nuts (Mohapatra *et al.*, 1973). Leaf N content of 1.51% in the month of April is considered optimum for high nut yield (Ghose and Bose, 1986). Macronutrient removal by cashew nuts and the cashew apple was N > K > Mg > P > S > Ca and K > N > Mg > P > S > Ca, respectively (Fragoso, 1999). Use-efficiency of N and K in cashew was 24.7% and 12.37%, respectively (Latha, 2000; Latha and Salam, 2001).

Table 23 a: Nutrient removal

Nutrient removal (30 year old tree) – Macronutrients

Plant parts	kg/tree		
	N	P_2O_5	K_2O
Leaf, stem & root	1.72	0.41	0.80
Fruit (155 kg)	0.37	0.12	0.28
Nuts (24 kg)	0.76	0.23	0.18
Total	2.85	0.76	1.26

Source: Mohapatra *et al.* (1973)

Table 23 b: Leaf analysis

Leaf analysis data (Kenya) – Macronutrients

Kind of tree	% of dry matter					
	N	P	K	Mg	Ca	S
Thrifty	1.98	0.21	1.69	0.20	0.09	0.15
Unthrifty	1.52	0.10	0.97	0.17	0.16	0.14

Source: Calton *et al.* (1961)

Leaf analysis data (Kenya) – Micronutrients

Kind of tree	ppm dry matter			
	Fe	Mn	Cu	B
Thrifty	45	95	16	9
Unthrifty	95	260	66	10

Source: Calton *et al.* (1961)

Table 23 c: Leaf analysis data (India)

Leaf analysis data (India) – Macronutrients

Season	% of dry matter		
	N	P	K
Before fruiting	1.41	0.09	0.63
After fruiting*	1.49	0.12	0.79
* fully matured leaves of previous season	1.19	0.06	0.46
*freshly matured leaves of current season	1.79	0.18	1.13

Table 23d: Fertilizer and stage of growth

Flowering response of young cashew to N and P fertilizer			
	Percentage flowering		
	At 2 ½yrs	At 3 ½yrs	At 4 ½yrs
With added N and P	35 %	87 %	100 %
Without added N and P	0 %	8 %	31 %

Source: Lefebvre (1970) http://www.pdf4free.com

Information on different nutrients elements removal from kernel, shell and cashew apple reported by different research workers are given in Table 24.

Table 24: Nutrient element concentrations in kernel, shell and cashew apple

Element	Concentration: On dry weight basis Apple, Kernel and Shell			Remark/source
	Apple	Kernel	Shell	
N (%)	0.616	2.793	0.343	*Source* : Mohapatra *et al.* (1973)
P (%)	0.085	0.410	0.005	
K (%)	0.390	0.455	0.175	
N (%)	1.21			*Source* : Nair *et al.* (1972); data of 1963-64
P (%)	0.25			
K (%)	1.38	-	-	
N (%)	2.87	4.52	2.07	*Source* : Beena Bhaskar (1992)
P (%)	0.09	0.23	0.03	
K (%)	2.07	0.79	0.54	
Ca (%)	0.15	0.34	0.12	
Mg (%)	0.09	0.06	0.05	
S (%)	0.09	0.16	0.03	
Fe (mg kg^{-1})	248	207	180	
Mn (mg kg^{-1})	27.5	25.1	34.2	
Zn (mg kg^{-1})	33.3	57.9	21.4	
Cu (mg kg^{-1})	13.5	23.3	4.7	
N (%)	0.91	3.20	0.50	*Source* : Richards (1993a)
P (%)	0.14	0.53	0.04	
K (%)	0.91	0.75	0.60	
Ca (%)	0.10	0.10	0.10	
Mg (%)	0.10	0.28	0.10	

Nutrient Concentration in Leaf

It is seen from the above research work that leaf nutrient status of cashew varied with physiological phases. The concentration of N, P and K in leaf was high at flushing and early flowering phase. The concentration of Mg and Cu in leaf was high during flowering and fruiting phase. The concentration of Fe in leaf was high during maturity phase. The concentration of S, Mn and Zn in leaf was high at post harvest phase.

The response of cashew to the major nutrients is in the order of N, K and P. Cashew is a nitrogen lover and N and K nutrition is of greatest significance in enhancing the production and productivity of cashew (Salam, 1997).

Factors Deciding Fertilizer Requirements

The nutritional dose of a crop depends on several crop and climate related factors in which the crop is grown. The level of crop productivity also influences the nutritional requirement. The important soil, climate and crop related factors that dictate the nutrient requirement of cashew is briefly indicated below:

a. Soil factors

In a given set if soil, the major soil related factors that influence the fertiliser requirement of cashew are the available nutrient (N, P, K, Ca, Mg, S, Fe, Zn, Cu, Mn, B, CI and Mo) status of the soil, the soil pH, soil organic carbon, the base saturation, and cation exchange capacity of the soil. In addition, the soil physics involving the structure, texture and water holding capacity also influence the nutritional dose. It is obvious that the fertiliser requirement will be low in a rich soil and high in a poor soil.

b. Varietal factors

The productivity of the variety is another important factor that influences the quantity of fertilisers. Varieties possessing higher yield potential should receive higher doses of fertilisers in tune with their higher physiological demand. Therefore, the production potential of the variety decides the fertiliser requirement.

c. Climatic factors

The most important climatic factor that influences the nutritional requirement of a crop is the quantum of rainfall and its distribution. The potential evapo-transpiration of the area is a major environmental parameter, which decides the soil moisture availability of the region. Cashew performs excellently well under low altitudes (less than 450 m) with a well distributed annual rainfall of

1500 to 2000 mm. The performance is good in humid tropical environments where maximum temperature does not exceed 35-40 ^{0} C and minimum temperature does not fall below 18°C. The humidity regime of 70 to 80 per cent without frost is good for better performance. The fertilizer requirement will be more in an environment with well distributed rainfall as it encourages production.

d. Type of farming

The nutritional requirement of a crop depends on the type of farming (rainfed or irrigated). Under rain fed conditions, the nutritional demand may be less compared to irrigate farming. Under irrigated farming, the crop productivity will be higher and cashew may demand higher nutritional dose.

Nutrient Recommendation

The recommendation of fertilizer dose for cashew in the major cashew growing states of India reported by Guruprasad *et al.* (2009) are indicated in the following Table 25.

Table 25: Fertilizer doses recommendations (g/tree/year) for cashew in different states of India

State	Fertilizer dose for adult tree (g)		
	N	P_2O_5	K_20
Kerala	500	125	125
	750	325	750
Karnataka	500	250	250
	500	125	125
Maharashtra	1000	250	250
Tamil Nadu	500	250	250
Andhra Pradesh	500	125	125
West Bengal	1100	1000	330
Orissa	500	250	250

Nutrition in relation to age

Cashew is a tree crop that requires at least seven years to attain the status of a fi'll-grown tree, with stabilized yield. During this period, the growth parameters in terms of height, girth and canopy spread increase gradually. Therefore, it is necessary to evolve fertilizer doses in relation to age. The fertilizer recommendation in relation to age of trees for different states, are given in Table 26.

Table 26: Fertilizer recommendation in relation to tree age

State	Percentage of fertilizer dose recommended for adult trees				
	1st year	2nd year	3rd year	4th year	5th year
Kerala	33%	66%	100%	100%	100%
Maharashtra	25%	50%	75%	100%	100%
Tamil Nadu	15%	30%	40%	60%	100%
Andhra Pradesh	0	25%	50%	75%	100%
Karnataka	12%	25%	50%	N-100% P & K 50%	100%
West Bengal	33%	66%	100%	100%	100%

Time of fertilizer application

Agronomically, for getting maximum efficiency of applied nutrients, it is necessary to apply the fertilizers at the physiological phase of the tree at which the internal nutrient demand is the highest. Beena *et al.* (1995) studied root activity of cashew in relation to phenological phases, employing ^{32}P soil injection technique, in the laterite soils of Kerala. Highest root activity and peak absorption of ^{32}P were noticed in flushing and early flowering stage, which extended from September to December. It is suggested that the onset of this phase is the most appropriate time for fertilizer application in a cashew orchard. Root activity was lowest in maturity and harvest phase (March - June). However, irrigation during this period enhanced ^{32}P uptakes by the roots. The seasons of fertiliser application for cashew, for various states of the country are shown in Table 27.

Table 27: The seasons of fertilizer application for cashew for various states

State	Month of application
Kerala	June-July and September-October
Maharashtra	August-September
Tamil Nadu	October-November
Andhra Pradesh	June-July and October-November
Karnataka	September
West Bengal	September-October

Method of Fertilizer Application

Fertilizers are to be applied in the tree basin where the feeding roots are concentrated, which depends on the age of the tree. It is generally believed that majority of the feeding roots reside under the canopy area, and as such fertiliser application can be restricted to this zone. For adult trees, fertilizers can be broadcast and incorporated over the entire tree basin (15 cm deep) within a radial distance of 2 to 3 meters within the drip line, leaving half a meter from the tree trunk. Instead, fertilizers can be placed in narrow trenches of 15 cm depth (taken 2-3 m radically around the tree), and covered with soil. For young trees, fertilizers can be broadcast and incorporated over the entire tree basin (10 cm deep) within the canopy area.

Method to increase fertilizer efficiency

For better efficiency, fertilizer application is to be done during the flushing and flowering season. The tree basin must be weed free before fertilizer application. The soil of the root zone should be moist to encourage absorption. Periods of heavy rain are to be avoided to minimize fertilizer loss. It is also necessary to apply fertilizers in the active root zone of cashew to improve efficiency. Dry spells between rains must be chosen for fertilizer application. Root injury should be minimum during fertilizer application.

Need for site specific recommendations

Cashew area in India is confined mainly to eight states (Kerala, Karnataka, Maharashtra, Goa, Andhra Pradesh, Tamil Nadu, Orissa and West Bengal). The present practice of fertilizer application is by adopting a common nutritional dose for the entire state. The soil fertility as well as the crop productivity varies to a greater extent between the cashew growing environments, the present practice of adopting a uniform fertilizer recommendation for the entire state, disregarding the variations in the level of soil fertility is irrational. The production potential of the variety also needs consideration. Hence, productivity linked and soil fertility based nutritional recommendations are to be evolved for every agro-climatic environment. Research efforts are being made in this direction.

Nutrient management in old cashew gardens

Most of the old gardens in the country especially public sector plantations are in highly neglected conditions and manuring is not done. The plant needs sufficient and continuous supply of nutrients for the growth of new flush, flowering, fruit set and development of nuts. The positive influence of major nutrients on vegetative growth, floral biology, fruit set, nut yield, shelling percentage and kernel grade have been studied in detail (Kumar *et al.*, 1993; 1995; 1997). Application of 500:750:250 g NPK/tree/year for old plantations was found to be optimum (Kumar *et al.*, 1993). Application during August-

September (west coast conditions) in circular trenches opened along the periphery of the 1.5-2m basin is beneficial. Foliar applications of urea (2%) with insecticides were found to increase nut yield per tree when sprayed at the emergence of new flush, flowering and fruit set (Kumar *et al.*, 1988). Studies on fertigation on cashew showed excellent response in terms of growth and nut yield besides better fertilizer use efficiency over conventional method of application of fertilizers (Kumar *et al.*, 1990).

Generally cashew crop is being cultivated on sloppy, eroded laterite and unfertile soils, whose physical, chemical and biological properties such as water holding capacity, gas permeability, nutrient supplying ability and microbial activities are at minimum. Therefore, there is a need to improve such soils to improve yield levels by proper soil management practices such as use of plant residues, augmenting bio-mass, composting, recycling plant residues to improve soil texture, structure, water holding capacity, soil biological activity and soil fertility levels. Thus, continuous addition of organic manures to the root zone helps in building up of soil fertility resulting in improvement of yield levels. Recycling of plant nutrients through cashew leaf litter available in old cashew plantations along with other organic wastes through collection and ex-situ decomposition has resulted in obtaining very good compost within 120 days. However, use of phosphorus, urea, cellulose decomposing and phosphorus solubalising micro organisims in the process of compost preparation has produced nutrient enriched compost within 90 days (Kumar and Hegde 1999).

Green leaf manuring

Glyricedia is potential green leaf manure that can contribute substantial quantities of organic N to cashew nut based production systems. Glyricedia may be raised as an inter crop or can be grow successfully as a live fence on borders of cashew orchard, specifically to add N for cashew.

Biofertilizers

Application of *Azospirillum*, *Azotobacter* and VAM increased germination percentage of nuts and plant growth, and reduced the incidence of fungal diseases in the nursery (Kumar *et al.,* 1998; Ramesh *et al.,* 1999). Inoculation of *Azotobacter* resulted in higher root growth (Oblisami *et al.,* 1985) and yield (Singh, 1997) in cashew. Among VAM, *Acaulospora laevis* and *Gigaspora mosseae* were better symbionts for inoculating cashew with (Lakshmipathy, 2000). Anantha Krishnan *et al.* (2004) reported that among VAM, *Glomus fasciculatum* was superior in terms of increased shoot-length, internode number, number of leaves, stem diameter, root length and root number, under nursery conditions. Sivaprasad *et al.* (1992) also reported that *Gomus fasciculatum* to be more effective at enhancing growth and P uptake of cashew plants. VAM (25 g/bag) was helpful for better graft-uptake at the time of grafting (Sridhar *et al.,* 1990).

Organic recycling

Cashew plantations have a vast potential for organic biomass for recycling. Availability of cashew leaf-litter from plantations of different age groups (10 to 40 years) ranged from 1.38 to 5.20 t ha^{-1} (Guruprasad *et al.,* 2009). About 5.5 t ha^{-1} available cashew- biomass waste can be converted into 3.5 tonnes of compost or vermicompost and helps to meet nearly 50% nutrient requirement of cashew plants (Yadukumar and Nandan, 2005; Yadukumar, 2007).

Composting

Composting is largely a biological process in which microorganisms of aerobic and anaerobic nature decompose organic matter and lower the carbon nitrogen ratio of the substrate. Compost is prepared from vegetable and animal refuses collected in the farm or in towns or villages. The available refuses in the farm are collected and stored till they form sufficient mass for compost making. A trench of suitable size, (4-6 m long, 2-3 m broad and 1-1.5 m deep) is dug, the accumulated refuse is well mixed and a layer 30 cm in thickness is spread all along the length and breadth of the trench. This layer is well moistened by sprinkling cow dung slurry and water over it. A second layer (30 cm thick) of the mixed refuse is then spread. The process is repeated till the heap rises to a height of 45 cm to 60 cm above ground level. The top is then covered with a thin layer of earth. After three months of decomposition, the mass is taken out of the trench and formed into a conical heap above the ground, moistened with water, (if necessary) and covered with earth. After another month or two, the manure will be ready for application to field (Salam and Peter, 2010).

Foliar Application of Growth Regulators, Pollination Attractants and Nutrients for Improving Productivity in Cashew

Foliar spray of plant growth regulators

Foliar spray of growth regulators Planofix, Nutron, IAA, IBA, NAA, 2,4-D and Ethrel were found effective in increasing total number of flowers, hermaphrodite flowers, sex ratio, fruit and yield per panicle and also improved physico-chemical composition of apples and nuts (Ghosh, 1988; Singh *et al.,* 1992 and Kumar *et al.,* 1994)

Gajbhiye *et al.* (2007) reported that for increasing the yield of cashewnut, two sprays of 10 ppm Ethrel, first at flushing and second at flowering stage. This is in agreement with the investigations in cashew by Mariappan *et al.* (1995) and Gavankar *et al.* (2007). The highest nut yield with Ethrel® spray has been reported by Mohan and Rao (1995) and Gajbhiye *et al.* (2007). However, the mean nut weight was not significantly influenced by spray treatments. Konhar and Mech (1988) reported highest fruit retention in cashew with Nutron 500 ppm, followed by Etherl 50 ppm and Planofix 45 ppm. Similarly, Kumar *et al.* (1994) found Ethrel®50 ppm, NAA25 ppm and 2, 4-D 4 ppm to be most effective in improving sex ratio and yield.

Foliar spray for pollination attractant in cashew

Fred and Melinda *et al.* (1998) reported that cashew nut trees are consistently ant-visited throughout the year, with the ants attracted to a large number of extra floral nectaries on the leaves, inflorescences, flowers, and developing nuts. The commercial productions of cashew nut, for example, in India, Brazil and East Africa, consistently apply pesticides, especially insecticides, in large monoculture plantings. Each year prophylactic spraying begins with the first flush of new leaves, continues through flowering, ending at about mid-nut development. They surveyed for ant diversity in sprayed and unsprayed cashew monocultures of various sizes and ages in Sri Lanka, India, and Malaysia to document the ant-cashew relationship and to explore the potential of ants replacing chemical pesticides in insect control. Using for-profit, commercial-size plantations as examples, they presented information that cashew has a strong potential for arthropod-dependent protection from pests and suggested important habitat considerations for encouraging ants within cashew plantings.

Naturally, black and red ants are seen to visit the flower frequently (Mandal, 2007). Cashew pollen is heavy and sticky. Several workers have reported the role of ants in cashew pollination (Northwood, 1966; Elsy *et al.*, 1986; Salunke and Kadam, 1995).

Pollination in cashewnut is major constraint for effective fruit set and fruit retention, attracting pollinators the ants by various methods during blossoming is found effective in cashew. A trial was conducted to study the effect of certain ant attractants on yield of cashewnut during 2001-2004 at Regional Fruit Research Station, Vengurla-416516 (Maharashtra). An adult 14 years old plantation of cashewnut was selected where the yield level was low. The treatments included post flowering sprays of jaggery (2.5 and 5%), glucose (0.5 and 1%), dried fish extract @ 500g in 10 litre water (fully dried fishes are grind in mixer and make a powder. The fish powder is filtered with the help of muslin cloth. Then mixed, fined dried fish powder @ 500 g in 10 litre of water and make a dried fish extract of 5%, cashew apple juice (1%) and establishment of red ant colony under cashew tree (The natural houses/colonies of red ants were brought as such from nearby plots of cashew plantation and were kept near the trunk base of cashew tree as one of the treatment). The foliar application of dried fish extract (5%) was found to be the best followed by jaggery (5%) and glucose (1%) for yield improvement as well as for the net return was concerned. The technique can be a component in a package for organic cashew production.

Foliar Nutrition in Cashew

A brief history of foliar fertilization

Foliar fertilization, that is nutrition through leaves, is a very effective technique of supplementary fertilization. The first reports on foliar application of mineral nutrients in plant production date back to the second half of the 18th century (Weinbaum, 1988). Particularly many studies on the uptake of mineral nutrients and their translocation within a plant were carried out after the Second World War.

Griss (1844) reported the use of iron sprays to control chlorosis in wheat was probably the first published report on foliar absorption of mineral nutrients. Downing (1869) noted the effect of dormant K sprays on promoting growth of fruit trees. Hamilton *et al.* (1943) were the first to recognize that apple leaves may absorb nitrogen in appreciable quantities. This finding was followed by extensive research on the use of foliar urea sprays on apple and other fruit trees.

Beginning in the early 1950s, scientists in numerous laboratories pursued research on the mechanism of absorption of foliar-applied minerals and growth regulators. The employment of radioisotopes and development of new laboratory techniques allowed the separation of the penetration process into its components. Penetration of ions through the cuticle and absorption by the mesophyll cells could be studied as individual processes. Such studies also provided knowledge on epicuticular waxes and leaf cuticles as barriers to penetration and on the relationship between leaf metabolism and ion absorption. It is not possible to mention here all those who contributed to this body of knowledge. Nevertheless, S. H. Wittwer, M. J. Bukovac, and W. Franke are the most outstanding pioneers in this area of study and must be mentioned (Swietlik and Faust, 1984).

Foliar application of nutrients was also being studied as a means of supplying nutrients to the fruit. Calcium and P sprays, especially, are used extensively to reduce the occurrence of physiological disorders of the fruit. Recently, it has been recognized that root uptake must be maximized in order to obtain the most benefit from foliar sprays.

Foliar nutrition is accepted by fruit growers worldwide. The two decisive factors in its favour are the reliability and rapidity of tree responses to foliar sprays.

The ability of plant leaves to absorb water and nutrients was recognized approximately three centuries ago (Fernandez and Eichert, 2009). The application of nutrient solutions to the foliage of plants as an alternative means to fertilize crops such as grapevine agriculture which was noted in the early 19th century. Following this, research efforts were applied to try and characterize the chemical and physical nature of the plant foliar cuticle, the cellular physiology and structure of plant leaves as well as focusing on potential mechanisms of penetration by foliar sprays. With the advent of firstly fluorescent and then radio-labelling techniques in the first half of the 20th century, it became possible to develop more accurate methods to investigate the mechanisms of leaf cuticular penetration and translocation within the plant following foliar application of nutrient solutions (Fernandez and Eichert, 2009; Fernandez *et al.*, 2009; Kannan, 2010).

The nutritional status of plants may also play a key role in the uptake of foliar application as many nutrients may have a synergistic effect on one another's absorption (Fisher and Walker, 1955). Foliar applications have been used in agriculture for many years, and it is a very useful tool to apply nutrients and chemicals directly to the leaves and fruit (Swietlik and Faust, 1984). A balanced nutrient status may be an alternative approach to increase SAR in fruit production systems.

Plant leaves are organs largely specialized in capturing light and CO_2, which play a critical role in plant survival and productivity (Sinha, 1999). While the basic role of the leaf is strongly conserved (Press, 1999), a great diversity in its morphology, size, surface topography and other features has been observed in the plant kingdom (McLellan, 2005). Such diversity is linked to the water transport capacity of the leaf and some features can be modulated by patterns of water availability during leaf expansion (Sack and Holbrook, 2006).

For all fruit and vegetable crops, the major pathway for mineral nutrient uptake is via the roots. Nutrients applied to the leaves can be absorbed and utilized by the plant; however, for nitrogen, phosphorus, and potassium the quantity absorbed at any one time is small relative to the larger levels required for growth by the plant. Foliar application of these three nutrients cannot be expected to supply the total amount required for crop production.

Cashew is often grown on marginal soils and on wastelands mostly unsuitable for other economic crops. Erroneously, the cashew is considered to require low levels of plant available nutrients because many plantations are found in soils of low natural fertility to which no fertilizers are applied. Balanced plant nutrition is a component of the intensive mechanised production system needed to grow cashew profitably (Chacko *et al.*, 1990). Nitrogen (N) is possibly the most important nutrient because of its impact on growth and yield, and because it can easily cause adverse environmental effects through misuse (soil acidification and offsite contamination). It increases vegetative growth, the duration of flowering, average nut weight, yield and kernel recovery when applied during the main period of vegetative growth (Sawke *et al.*, 1985; Ghosh and Bose, 1986; Harishu Kumar and Sreedharan, 1988; Ghosh, 1989).

In the recent past, considerable progress has been made in developing proper method of foliar feeding of the trees. Formerly it was thought that the lower surface was more efficient in absorbing solutions because of greater number of stomata on the underside. This apparently is not the case, because the upper surface of some leaves can absorb as much phosphate as the lower surface, even though it has only one seventh as many stomata. It is not easy to appreciate the potential absorbing surfaces involved. The leaf area of a 12-year-old mango tree, for example, may have about 10 times the land area it occupies (Bose *et al.*, 1999).

The absorption of nutrients by leaves is generally greater during the day time. Other factors which favour absorption and transport of nutrients include high humidity, proper temperature, and incident radiation, a carbohydrate supply and vigorous growth. Chemical and physical properties of the nutrient spray, such as pH, carrier ions and surface active agents, also exert a great influence.

Leaf characters like thickness of the cuticle, number of discontinuities (interspersed pectinaceous areas), hairiness, waxiness or smoothness of leaf surface, and other surface properties that affect adherence influence nutrient intake. Young, rapidly expanding leaves absorb more nutrients per unit area than do fully mature ones. Generally speaking, the more vigorous the plant, the more nutrient can be absorb through the leaves. However, not all plants have the same capacity of absorbing nutrients from foliar sprays (Bose *et al.*, 1999).

Foliar sprays should be applied either with a pressure sprayer or with one of the specially designed fertilizer spray guns. The trees should be sprayed until the nutrient solution begins to drip from the leaves. Most of the concentrations that are safe to use as foliar sprays can be combined with organic pesticides such as Malathion and Rogor. This enables the orchardist to feed the tree and control pests and diseases in one operation.

Where problems of nutrient fixation exist, foliar application proves the most effective means of fertilizer placement. So far the most important use of foliar sprays has been in the application of micronutrients. This is because micronutrients are required in small amount and also due to the possibility of their becoming fixed in the soil. Foliar application of major nutrients should be considered as a supplement, not as a substitute for the regular soil application of fertilizers, although foliar application of urea has been reported to be successful in a number of fruits, such as citrus, guava, apple and many others. Spraying results in more rapid absorption of nitrogen than soil application. Pineapple can receive as much as 75% of their N-fertilizer needs as urea sprays and about 50% of their supplemental phosphorous and potassium by foliar fertilization. Potassium sprays have also been found useful for several fruit crops like pineapple, citrus and guava (Bose *et al.*, 1999).

Nitrogen applied in the form of urea is most readily absorbed. After the urea in solution is absorbed into the leaf, it is split up into ammonia and carbon dioxide by means of the enzyme urease. Sodium and potassium are two other elements readily absorbed by leaves, and they are among the most highly mobile once inside the leaves. Phosphorous, chlorine, sulphur, zinc, copper, manganese, iron and molybdenum are less mobile in the order given. Calcium is absorbed but does not move out of the leaves. Despite this fact, calcium and magnesium can be applied successfully to the leaves of some plants to overcome deficiency of these elements.

The rate of movement of nutrients, such as nitrogen, phosphorous and potassium absorbed by leaves has been estimated as 2.5 cm in 5 minutes. The leaves are not the only tree parts capable of absorbing nutrients. The bark, buds, petioles and flowers also absorb nutrients up to some extent. Phosphorous and potash applied to the bark of apple and peach trees move into the branches 50 to 60 cm from the points of application within a 24-h period. Most absorption, via the bark, occurs when the nutrient is applied just before the buds of temperate tree crops begin to swell in the early spring. The nutrient solution applied to the bark can be 10 times stronger than that applied to the leaves, without causing any injury to the tree. Since tropical trees are mostly evergreens and they are never devoid of leaves, the application to the bark of such trees has limited scope.

Foliar applications prove a valuable means to complementing root treatments and an alternative to supplying nutrients to plants in special situations such as at peak nutrient demand times or in conditions of low soil availability (Weinbaum, 1988; Brown, 1999; Lester *et al.*, 2006). The need to supply high concentrations to ensure that foliar uptake takes place is more dramatic with macronutrients (N and K), which are required by plants in greater amounts which are not likely to be met only by foliar treatments (Johnson *et al.*, 2001).

Foliar spray uptake is determined by many plant-environmental characteristics which will ultimately have a direct influence on factors such as leaf morphology, structure, position, sun exposure or the rate of physiological processes in the plant. The prevailing environmental conditions during plant growth have a direct influence on the leaf surface in terms of cuticle thickness or amount and composition of epicuticular waxes (Bird and Gray, 2003; Koch *et al.*, 2006). Environmental factors such as relative humidity (RH), temperature (T) and light play an important role in relation to both the behaviour of the spray solution on the leaf surface and the subsequent foliar uptake and distribution process (Ramsey *et al.*, 2005).

Plant nutrient management can influence flowering, fruit set, fruit size and the amount of vegetative growth and other plant characteristics. By carefully choosing the components of fertilizer programme, the grower can nudge a crop toward earlier, heavier fruit set (Abd-Allah, 2006). Foliar feeding has been used as a means of supplying supplemental doses of minor and major nutrients, plant hormones, stimulants and other beneficial substances. Observed effects of foliar fertilization have included yield increases, resistance to disease and insect pests, improved drought tolerance and enhanced fruit quality (Havlin *et al.*, 2005; Tariq *et al.*, 2007).

Urea has become the leading form of N fertilizer worldwide. Urea, a naturally occurring compound, can also be made by reacting carbon dioxide with ammonia at high temperature and pressure. Its high N content (46% N) makes urea economical to produce, transport, and deliver to the farm. Foliar application of urea can be extremely beneficial in some circumstances for plants.

Several cereal, vegetable, and perennial crops respond favourably to foliar applications of urea with increased growth, yield, and quality. These benefits can include boosting N concentrations, reducing N losses through leaching and denitrification, and supplying N when root uptake is limited Mikkelsen (2007). However, foliar-applied nutrients may be directly absorbed by plants (without the buffering effects of the soil). The modern N fertilizer industry produces urea that is remarkably safe, consistent, and effective for enhancing plant growth. Urea has many properties that make it the most commonly used N fertilizer in the world. Biuret toxicity problems are generally rare, but special attention should be made for fertilization of especially sensitive crops.

Application of some nutrients (not all) through foliage can be from 10 to 20 times as efficient as soil application. However, this efficiency is not always achieved in actual practice due to weather extremes, application of the wrong spray mix, or of the right mix at wrong time. Foliar fertilizations are often timed to coincide with specific vegetative or fruiting stages of growth and the fertilizer formula is adjusted accordingly. Therefore, judging what foliar materials to apply and at what plant stage to spray with soil applied organic and inorganic fertilizers are important principles to make best uses of this technique (Yaseen *et al.*, 2004). A properly formulated foliar spray particularly amended with appetizers/bioactive materials/bio-stimulants and surfactants increases uptake of nutrients from the soil (Yaseen *et al.*, 2004) because foliar fertilization causes the plant to pump out more sugars and other exudates from its roots into the rhizosphere (Marschner, 2003). Based on the studies made by Ankaiah and Rao (1983) concluded that spraying four per cent urea on three weeks old leaves in cashewnut will be of great advantage. They also suggested that cashew leaves at the age of third week are having more nitrogen absorbing capacity as urea in cashew. When the leaves are four week old, there was less nitrogen content most probably may be due to heavy cutimization at this age. Further leaf scorching was also noticed in all the concentration of urea at the age of one week. It was universally agreed that the rates of absorption in the young leaves are greater but in cashew very tender leaves (one week old) not suitable for greater foliar nutrient absorption.

Balanced plant nutrition is a component of the intensive mechanised production system needed to grow cashew profitably in Australia (Chacko *et al.,* 1990). While cashew responds to fertilization (Grundon, 1999), nitrogen (N) is possibly the most important nutrient because of its impact on growth and yield, and because it can easily cause adverse environmental effects through misuse (soil acidification and offsite contamination). It increases vegetative growth, the duration of flowering, average nut weight, yield and kernel recovery when applied during the main period of vegetative growth (Sawke *et al.,* 1985; Ghosh and Bose, 1986; Harishu Kumar and Sreedharan, 1988; Ghosh, 1989).

Cashew being a regular bearer which puts forth flowering and sets fruit on the current season flush, the plant needs sufficient and continuous supply of nutrients for the growth of new flush, flowering, fruit set and development of nuts. The positive influence of major nutrients on vegetative growth, floral biology, fruit set, nut yield, shelling percentage and kernel grade.

Cashew responds well to fertilisation. Of the primary nutrients, nitrogen has been found to be most important for cashew. Young cashew plants respond well to nitrogen and phosphate application up to pre-bearing stage but at the bearing stage, potash in combination with nitrogen is most important. The response of cashew to applied nitrogen is tremendous and the same is observed universally. Nitrogen is the key nutrient element for cashew nutrition followed by potassium. Nitrogen is the element that is absorbed by cashew in largest quantity (Salam, 1997). The growth and development of cashew is very much influenced by nitrogen application. Nitrogen promotes early growth and development of young cashew plantation by producing more leaves and shoot facilitates early flowering and produces more female flowers which reflects in an increased yield. The beneficial effects in growth and productivity of cashew kernels are manifested only if phosphorus and potash are provided along with nitrogen. Phosphorus increases girth, rate of leaf production and mostly promote root growth. Magnesium is also essential for the absorption and translocation of phosphate in cashew. Adequate supply of potash improves the precocity of flowering, increase number of flowers and setting percentage, thereby increased cashew nut production. It improved the nut characteristics, particularly nut size including nut quality. A combination of K and N has showed positive effect on the number of flowers per bunch and increase in yield (Mandal and Rethinam, 1994). Foliar feeding of urea has been a success in many fruit crops and it is practiced commercially on pineapple in Hawaii. Foliar application of elements is easy and economical too but no serious efforts have been made to standardize the requirements, time and technique of feeding (Chundawat, 2002).

Cashew is a very important plantation/horticultural crop of India but neither receive any management, nor inputs and thereby resulting in lower productivity. The crop is seldom fertilized and as such the average yield per tree will be declining gradually from year to year. However, research results showed that positive tremendous response can be obtained through regular application of fertilizers and improved management practices which could give two to three fold yield increases. Cashew requires regular fertilizers application to ensure early and high yield in new plantation and regular high yield from mature trees (Rupa and Bhat, 2010). The challenge for cashew crop will be to improve yields in marginal lands and sloppy area where the soil is exposed to degradation strongly limits the growth and production of cashew.

Cashew tree being hardy and drought resistant are often grown on poor marginal soils and also in waste lands which are considered mostly unsuitable for other economic horticulture crops. However, it performs better in well drained, brown forest soils, red sandy loam and light coastal soil with high water holding capacity and rich in organic matter. Cashew prefers slightly acid soil of pH 4.5 to 6.5, with low calcium (Ca) content.

In India, the cashew is cultivated on soil that is infertile, leached, acidic and sometimes containing excess changeable aluminium (Al) (Falade, 1984 and Badrinath *et al.*, 1997). Falade (1984) concluded that the physical and chemical characteristics of the soil influence the height of the plant as well as the diameter of the crown, and the morphology of the root system. He also concluded that light textured soils, free from stones and without an impervious layer or horizon within the top 100 cm are the best for growing cashews.

In West Coast cashew mostly planted in lateritic soil. Red and lateritic soils are generally acidic and low CEC, low to moderate base saturation. Due to intensive leaching and presence of high amount of iron and aluminium oxides, the soils show deficiency of nutrients, such as N, P, and K causing nutrients imbalance (Rupa and Bhat, 2010). Similarly, Cashew is mainly grown in high rainfall area of West coast especially on hill slopes. Under these heavy rainfall areas, most of applied nutrients losses due to leaching and availability of applied nutrients might be low during critical stage that may affect the growth and yield of cashew. Plant uptake and loss of nitrogen through leaching and erosion in cashew farm calls for regular fertilization of the soil in order to maintain nitrogen equilibrium within the soil system. Therefore, foliar feeding of nutrient in addition to recommended dose of manure and fertilizers may play the major role in increasing the cashew yield by optimum uptake of nutrient with minimum losses and reduce nutrients imbalance. Foliar applications of nutrients can supplement the soil supply and thereby reduce the quantity of fertiliser needing to be applied to the soil (El-Otmani *et al.*, 2002).

Besides this, in each growth cycle, the nutrients are removed from the soil to supply the vegetative parts of the plant (leaves, branches, trunk and roots) and for export in the harvested fruits and pseudo fruits. Thus plant growth and satisfactory harvests are only possible by replacing the nutrients exported by the harvested parts. Mohapatra *et al.* (1973) A 30 year old cashew tree removes 2.847 kg of N, 0.752 kg of P_2O_5 and 1.265 kg of K_2O through vegetative parts including root system, apples and nuts.

Vishnuvardhana *et al.* (2002) observed that the largest cashew yields were obtained with 1000, 250 and 250 g/plant/yr of N, P_2O_5 and K_2O, respectively, but economically, 500, 250, 250 g/plant/yr N, P_2O_5 and K_2O produced the best results. Ghosh (1990) concluded that the weight of the nut, number of nuts, height and vigour of the plants were increased and reached a maximum with 600 g N/plant/yr.

The existing recommended dose of manure and fertilizers in cashew is the application of 1000g N, 250g P_2O_5, 250g K_2O and 40 Kg (4 Basket) FYM per plant per year for cashew tree from 4th year onwards. However, some researchers found that productivity can be increased by foliar spray of macro and micro nutrients. Foliar sprays of nutrients (urea 2 to 4%; DAP 1%; orthophosphoric acid; $ZnSO_4$ 4%; Cu 0.3 to 0.6%), at emergence of the flush, panicle initiation and fruit-set stages, ensure better fruit-set and enhanced nut yield (Sapkal, 2000; Anonymous, 2010; Ghose, 1988; Ankaiah and Rao, 1983). Cashew yellow leaf spot occurrence in low soil pH (4.5 - 5.0) could be corrected by foliar sprays of Mo salts (Subbaiah *et al.,* 1986).

An experiment was conducted to access the efficacy of water soluble nutrients (fertilizers) with reference to growth and flowering in cashew cv. Vengurla-4 at Agriculture Experimental Station, Paria, Tal. Pardi (396 145), Dist. Valsad (Gujarat) during the year 2011-12 and 2012-13 under rainfed conditions with eleven treatments *viz*., 19:19:19 (1% and 2%), 0:52:34 (1% and 2%), 13:0:45 (1% and 2%), 0:0:50 (1% and 2%), urea (2%), $MgSO_4$ (3%) and control with three replication in randomized block design. Uniform eight year old trees of cashew variety Vengurla-4 planted at 8m x 8m were selected for experimentation. The recommended dose of fertilizers was applied to all the treatments including control. All the water soluble nutrients (fertilizers) sprays were applied at flushing, flowering and fruit set. Production of laterals (m^2), panicle (m^2), number of male, perfect and total number of flowers per panicle were noticed significantly maximum when plant sprayed with urea 2% and 19:19:19 2% during both the years and in pooled analysis also. Significantly higher sex ratio was found under the treatments urea (2%), 19:19:19 (1 & 2%) and 13:0:45 (1 & 2 %) in cashew during both the years of study. Minimum duration of flowering in cashew cv. Vengurla-4 has been recorded in urea 2%, which was about 15 days earlier than control (Gajbhiye *et al*., 2014).

Similarly, foliar application of urea (2%) and T_2 19:19:19 (2%) was found to be best for improving fruit set per panicle, fruit retention (%), nut weight, apple weight, yield (kg/tree & t/ha) and shelling (%) of cashew cv. Vengurla-4 under South Gujarat conditions. Foliar feeding of urea and 19:19:19 (2%) directly to the metabolite sites considerably enhanced yield by increasing laterals/panicles per m^2, per cent of perfect flowers, fruit set, retention of nuts up to harvest and also increasing nut size and weight (Gajbhiye *et al*., 2013).

1 Foliar Nutrition

1.1 Uptake of foliar applied nutrients

According to Franke (1967) mineral nutrient uptake by leaves involves three steps. After mineral nutrients are deposited on the leaf surface they (1) penetrate the cuticle and epidermal walls by means of diffusion; (2) are adsorbed on a surface of the plasmatic membranes, and (3) pass through the plasma membranes and enter the cytoplasm.

The process of mineral nutrient uptake by plant leaves is firstly via penetration of the cuticle and epidermal walls through diffusion. Penetration of the cuticle is hampered by the structure of the cuticle. The cuticle is made up of two layers. The outer layer consists of cutin with epicuticular waxes, while the inner layer is made up of cellulose and pectic substances, encrusted with cutin. The epicuticular wax is hydrophobic, while the cutin is made up of hydrophilic polyesterified hydroxyl fatty acids (Swietlik and Faust, 1984).

Diffusion of the substances through the cuticle is affected by temperature and the concentration gradient across the cuticle. Diffusion of organic compounds is higher than that of inorganic compounds. Kannan and Wittwer (1965) reported that urea diffused at a rate of 10 to 20 times higher than inorganic compounds. When the nutrients penetrate the cuticle, movement occurs via apoplastic or symplastic transport towards the vascular tissue. The nutrients are then loaded into the phloem, a process which is energy dependent, and subsequently transported out of the leaf or fruit as sources. Alternatively, the nutrients may also be actively transported across the plasma lemma to the leaf cells where incorporation into organic compounds may occur which would allow for easy transportation throughout the plant (Haynes and Goh, 1977).

Nutrients or foliar sprayed chemicals may also enter the leaves via the stomatal pores. The stomatal pores are in fact cuticular invaginations and come into direct contact with foliar applications (Norris and Bukovac, 1968). Trichomes are another path by which foliar applied nutrients may enter the leaves; their effect on absorption is a function of the leaf age and plant species (Hull *et al.*, 1975). In addition to these structure that grant access to foliar applied minerals, the presence of pores and canals in the cuticle of the apple fruit that may facilitate uptake of foliar applied nutrients was demonstrated in a study by Miller (1982).

1.2 Factors that affect the uptake of foliar applied nutrients

1.2.1 Environment

The environment can have different effects on the absorption of foliar applied nutrients because it impacts on the development of the cuticle and directly affects physiological processes in the plant (Flore and Bukovac, 1982). Leece (1978) found that as the season progressed, the secondary wax structure increased on the abaxial side of plum leaves. This was ascribed to the associated increasing light intensity that prevailed towards the end of the growing season. Results from this study thus suggest that the absorption of foliar applications early in the season may be more effective, where leaves are formed under environmental conditions and where the thickening of the cuticle may not yet have occurred. In addition, higher temperatures increased both the formation of waxes and leaf expansion in this study on plums, which in turn lead to less wax per unit surface as leaf expansion is faster than the formation of waxes (Leece, 1978). Although such a cuticlar structure would allow for better absorption, the stomata are usually closed under higher temperatures, which in turn will lead to a reduction in stomatal absorption. Foliar applications during high temperatures and light intensity conditions may also damage the leaves by raising the concentration of the foliar chemicals on the leaves, which in turn cause physical damage (Swietlik and Faust, 1984).

1.2.2 Leaf anatomy

Fisher and Walker (1955) showed that phosphate (P) absorption was better in younger than in older apple leaves. This may be due to a less developed cuticle structure in the younger leaves. However, Leece (1978) showed that wax structure development in plum leaves was not determined by the physiological age of the leaves, but rather by the phenological stage of the leaf within the growing season. It was determined that leaves of the same physiological age, where some developed earlier in the season, had no abaxial waxes opposed to those that developed later in the season. The lower surface of a leaf has been observed to absorb greater quantities of product than the upper surface of a leaf, mainly due to a thicker cuticle associated with the adaxial side of the leaf (Cook and Boynton, 1952).

Among fruit tree species *Prunus* leaves are regarded as less efficient in foliar absorption than are apple or citrus leaves (Impey and Jones, 1960; Leece, 1978).

Cashew absorbed 14^C urea through leaf as well as roots in the molecular form. It is likely that after absorption, urea may be metabolized. The ability of cashew to absorb urea through leaf suggests the possibility of feeding N through foliage to this crop (Salam and Kamalan, 1993).

1.2.3 Fruit crop

The degree of absorption of foliar-applied nutrients differs between different fruit species. Plum leaves are less effective at absorbing foliar applied nutrients than apples and citrus leaves (Swietlik and Faust, 1984). This phenomenon can be attributed to the difference in leaf structure, as plum leaves have a wax layer on both sides of the leaf. Additionally the guard cell are also covered with wax, and as mentioned in section 1.1 these natural openings play an important role in the absorption of foliar applied nutrients (Leece, 1978).

1.2.4 Interactions between nutrients

The nutritional status of plants may also play a key role in the uptake of foliar application as many nutrients may have a synergistic effect on one another's absorption (Fisher & Walker, 1955). One example being $MgSO_4$ sprays, found to alleviate magnesium (Mg) deficiencies in apples with an adequate nitrogen

(N) status. The chemical formulation of mineral nutrients within the foliar application is also an important factor that affects nutrient absorption. A chelated form of mineral nutrients has also been showed to increase absorption, and has been attributed to the increased mobility of these nutrients in this formulation (Kannan and Wittwer, 1965).

1.2.5 Solution pH

The absorption of some nutrients by leaves is pH dependent. The optimum absorption of Ca_2^+ for example, applied as a $CaCl_2$ solution, occurred at a pH 7 in sweet cherry fruit (Lidster *et al.*, 1979). The optimum absorption of urea by apple leaves have been reported between pH 5.4 and 6.6 (Cook and Boynton, 1952).

1.2.6 Surfactants

Surfactants have been developed to increase absorption of foliar-applied nutrients by lowering the surface tension of the applied solution and by reducing the contact angle between the leaf surface and the liquid (Leece, 1978).

Mechanisms of Penetration of Nutrients into the Plant

The role of stomata in the process of foliar uptake has been a matter of interest since the beginning of the 20th century. However, in 1972, it was postulated that pure water may not spontaneously infiltrate stomata unless a surface-active agent to lower surface tension below 30 mN m^{-1} is applied with the solution (Schonherr and Bukovac, 1972).

As a consequence of this, most investigations were subsequently carried out on cuticular membranes isolated from adaxial (upper) leaf surfaces of species in which enzymatic isolation procedures could be conducted, e.g. from poplar or pear leaves. Utilizing this system, it was found that cuticles are permeable to water and ions as well as to polar compounds (Kerstiens, 2010). Furthermore, the occurrence of two distinct penetration pathways in the cuticle, one for hydrophilic and another for lipophilic substances, has been suggested (Schonherr, 2006).

The proposition that stomata could also contribute to the foliar penetration process was re-assessed by Eichert and co-workers at the end of the 1990's and subsequently validated (Eichert and Burkhardt, 2001; Eichert and Goldbach, 2008; Fernandez and Eichert, 2009). At present, the quantitative significance of this pathway and the contribution of other surface structures such as lenticels to the uptake of foliar applied solutions remain unclear. Since its first recorded use in the early 19th century (Griss, 1844), foliar fertilization has been the subject of considerable controlled environment and field research and has become widely adopted as a standard practice for many crops. The rationales for the use of foliar fertilizers include: 1) when soil conditions limit availability of soil applied nutrients; 2) in conditions when high loss rates of soil applied nutrients may occur; 3) when the stage of plant growth, the internal plant demand and the environmental conditions interact to limit delivery of nutrients to critical plant organs. In each of these conditions, the decision to apply foliar fertilizers is determined by the magnitude of the financial risk associated with the failure to correct a deficiency of a nutrient and the perceived likelihood of the efficacy of the foliar fertilization (Fernandez *et al.*, 2013).

Furthermore, foliar fertilization is theoretically more environmentally friendly, immediate and target-oriented than soil fertilization since nutrients can be directly delivered to plant tissues during critical stages of plant growth. However, while the need to correct a deficiency may be well defined, determining the efficacy of the foliar fertilization can be much more uncertain (Fernandez *et al.*, 2013).

According to Franke (1967) mineral nutrient uptake by leaves involves three steps. After mineral nutrients are deposited on the leaf surface they (1) penetrate the cuticle and epidermal wall by means of diffusion; (2) are adsorbed on a surface of the plasmatic membrane and (3) pass through the plasma membrane and enter the cytoplasm.

The cuticle consists of three layers (Fig. 31), namely (from the external to the internal surfaces of the plant organ), the epicuticular wax layer (EW), the cuticle proper (CP) and the cuticular layer (CL) (Jeffree, 2006). The EW layer is the outermost and most hydrophobic component of the cuticle. The CP that lies beneath the epicuticular waxes contains mainly cutin and/or cutan and is by definition free of polysaccharides (Jeffree, 2006). The CL is located under the CP and consists of cutin/cutan, pectin and hemicelluloses that increase the polarity of this layer due to the presence of hydroxyl and carboxylic functional groups. The middle lamellae and pectin layer (ML) is situated beneath the CL. Variable amounts of polysaccharide fibrils and pectin lamellae may extend from the cell wall (CW), binding the cuticle to the underlying tissue (Jeffree, 2006).

The cuticle is a waxy layer covering leaves, flower petals and fruits. Its primary function is to prevent moisture loss from plants and to protect the plant from injury. It is typically thicker on upper leaf surfaces than on lower leaf surfaces and is impermeable to aqueous (water) solutions; oily solutions will penetrate the cuticle more readily. Cracks can occur in the cuticle and can serve as a port of entry for foliar sprays. On young, immature leaves the cuticular layer is not as well developed as an older leaves. The stomata of plants are located almost exclusively on the leaves. These pores are only a few millionths of an inch in diameter and their primary function is to breathe for the plant. They allow carbon dioxide, the building block of plants, to pass into the leaf and here the chlorophyll molecules, using sunlight for energy, convert the carbon dioxide molecules into simple sugars. The stomata also allow the by-product of this reaction, oxygen, to escape from the plant. A third function of the stomata is to allow the escape of water vapour that evaporates from the inside of the leaves to cool the plant (Wittwer and Teubner, 1959).

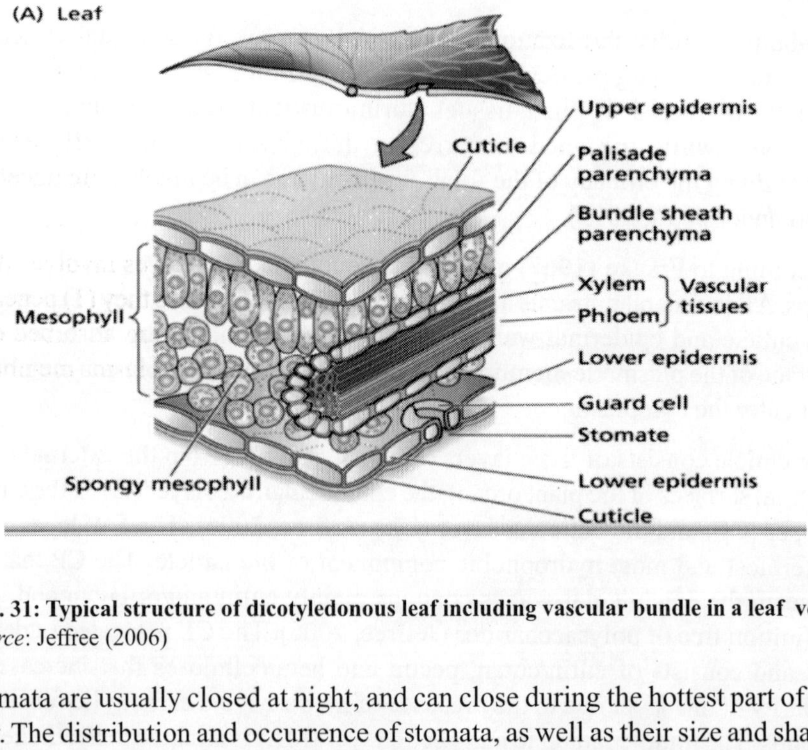

Fig. 31: Typical structure of dicotyledonous leaf including vascular bundle in a leaf vein.
Source: Jeffree (2006)

Stomata are usually closed at night, and can close during the hottest part of the day. The distribution and occurrence of stomata, as well as their size and shape, varies widely from one plant species to another. Broadleaf crops and tree crops commonly have most of their stomata on the lower leaf surface while grass species may have about the same number on both surfaces. As examples, small white beans commonly have 40,000 stomata per square centimetre on their lower surface and only 3,000 on their upper surface (Wittwer and Teubner, 1959).

Lipophilic, apolar compounds have been proposed to penetrate cuticles by a solution-diffusion process. The mechanisms of penetration by hydrophilic, polar compounds are not fully elucidated yet (Fernandez, 2013).

Stress situations during growing stages can be readily and effectively overcome with foliar fertilizers (and it is well-known how deeply stress influences growth, and consequently also yields, of our crops!) (Levitt, 1972). Supplementary and compensatory FF is more economical than root fertilization due to the higher degree of applied nutrients utilization (via the leaf usually approx. 85 %, compared to a mere utilization of 10 %, max. 30 % and only in exceptional cases 50 % via the root) which of course makes the nutrients more efficient (Wittwer, 1963).

Foliar feeding constitutes one of the important milestones in the progress of agricultural crop production. As a natural phenomenon of nutrient uptake, it has existed with all forms of plant life from their beginning. The marine algae and most other aquatic plants live in a one phase environment from which all the necessities of life, including minerals, are absorbed by the entire plant surface.

With land plants, adapted to a two-phase environment, there is the separation into top and root. The root is a specialized organ for absorption and anchorage. The aerial shoot, on the other hand, while adapted as an organ for photosynthesis will also function in the absorption of water and dilute solutions of both organic (usually pesticides) and inorganic (usually minerals) substances. It is suggested that if we apply foliar fertilizers to the cashew leaves in soluble forms, as much as 95 per cent of what is applied may be used by the plant (Anon., 2006). If we apply a similar amount on the soil, we find about 10 per cent of it to be used.

The quantity of soil-applied nitrogen (N) failing to get into a crop can range from 85 per cent in the worst case to 38 per cent in the best. By contrast, foliar-applied N can be up to seven times more efficient than soil-applied N. Other benefits of foliar-applied N include lower application rates (higher efficiency), plus the relative ease of obtaining timely, uniform applications. A combination of soil-applied and foliar applied N is the best management practice to reduce the potential for nitrates in groundwater Dixon (2012). Tukey and Marczynski (1984) also suggested that a combined soil and foliar fertilization should be recommended in plant production to increase both plant productivity and yield quality.

According to Marschner (1995) absorption of mineral nutrients by plants is species dependent. Picchioni *et al.* (1995) showed that the rate of B absorption by apple leaves was two to three times higher than that of pear, plum and sweet cherry. They further stated that differences in the uptake rates of mineral nutrients by leaves of different plant species result not only from the specific cuticular membrane structure, but also from various amounts of ectodesmata on a leaf surface. Wojcik (2004) reported that absorption rate of mineral nutrients by above ground plant parts considerably differ not only among plant species but also between varieties within the same species. Cuticular waxes embedded in the cutin/cutan matrix were found to be responsible for the barrier properties and diffusion of non electrolytes. Neutral, non-charged molecules penetrate the cuticle by diffusion and dissolution in cuticular waxes (Schonherr, 2000).

It is well documented that the lower leaf surface takes up mineral nutrients more rapidly than the upper side and it is accepted that the absorption rates of most mineral nutrients by young leaves are greater than those of old ones. Leaf ability to take up mineral nutrients also is dependent on the nutritional status of a plant (Wojcik, 2004).

According to Alexander (1986) there is a significant relationship between the ability of leaves to absorb mineral nutrients and the plant development stage. He listed the development stages of selected plant species having high needs to particular mineral nutrients. Besides it, foliar sprays of a given nutrient are most successful when applied at plant development stages requiring high amounts of this nutrient. Weinbaum *et al.* (2002) reported that foliar fertilization treatment is successful for perennial fruit crops with deep rooting systems since soil surface application of most fertilizers has little effect on rapid improvement of plant nutrition.

At first, this is rather surprising since plant surfaces are covered by the cuticle which has been considered impenetrable in the past; however, during the last few decades more evidences provided that absorption can take place through the surfaces of aerial plant parts. Foliar feeding in the broad sense involves absorption of nutrients by all above-ground plant parts.

An agriculture that aims for a high level of output per unit area must be based on a fertilizer application concept that provides the optimum supply of nutrients. Foliar fertilization is used throughout the world as both a preventive and a curative measure as it is able specifically to compensate for nutrient deficiency.

Foliar application of nutrients is gaining importance in many crops and countries. It is of routine use in horticulture and its potential in the most important agri-cultural crops is rising. Intensively grown crops like cereals are routine sprayed several times during the season for different purposes, like pesticide treatment and growth regulation. In such cases foliar fertilizer may be combined with pesticide application, provided that the foliar fertilizer formula is compatible with the pesti-cide applied. Such mixtures between foliar fertilizer and pesticide no doubt improve fertilizer cost effectiveness and can even increase the activity of the pesticide (Alexander, 1986).

Crops are characterized by growth stages needing a particular supply of all or of certain nutrients to reach optimum yields. Plant demand is generally at its maximum during the period of exponential growth. During such critical stages leaves show a particularly high efficacy for absorbing nutrients. Further, in any vegetative period there are times when the nutrient uptake from the soil is impeded to such an extent that there is no guarantee that all the parts of the plant will receive an adequate supply of nutrients. A good example is dry weather. It is obvious that a minimum level of soil moisture is necessary if foliar nutrition is to be successful. When there is a prolonged period of dry weather, even foliar fertilization is no longer able to exercise a favorable influence on the yield. On the other hand, during short times of dry weather foliar fertilization may be the only measure-particularly in critical growth stages-by means of which the full yield can be ensured at all (Alexander, 1986).

Schönherr (2000, 2001) proposed the following general principles: (i) since ionic compounds in solution are surrounded by water molecules, a diffusion pathway parallel to the lipophilic pathway must occur; (ii) air relative humidity (RH) could be a key factor in determining the rate of penetration of solutions; (iii) given the importance of RH, active ingredients, either hygroscopic or with low points of deliquescence, will promote foliar penetration; (iv) concentration would serve as driving force for penetration, and the absolute amount which penetrates would probably be proportional to concentration; (v) smaller molecules should penetrate faster than larger ones; (vi) cations and anions will penetrate in equivalent amounts; and (vii) effective wetting agents can greatly increase the rate of penetration of electrolyte solutions. In contrast, stomatal uptake may resemble diffusion in bulk water, and therefore, it should not be restricted by the factors mentioned above (Eichert *et al.*, 2006).

For maximum stomatal entry, nutrient sprays must be applied when the stomata are open, early morning applications are the best. Also, there is less evaporation during the early morning thus giving a better chance for maximum uptake by leaves. Timing is keeping this point in mind spray so did both in regard to time of the season and time of the day a critical factor in foliar spray. High relative humidity during the time of application will also enhance uptake by minimizing evaporation (Gary and Grigg, 1999). Foliar sprays may be effective only during "critical stages" of plants growth cycle and must be applied during or shortly before the critical period to be effective. Since immature foliage does not have well-developed cuticular layer, application of nutrient sprays when there is a significant amount of young foliage present will enhance cuticular entry. Factors that affect foliar absorption include relative humidity, temperature, pH of the nutrient solution, age of leaf, concentration of the nutrient solution, difference in the nutrient compounds (formulations), use of surfactants and addition of non-nutrient facilitating or carriers-mediated agents (Gary and Grigg, 1999).

Plant leaves are organs largely specialized in capturing light and CO_2, which play a critical role in plant survival and productivity (Sinha, 1999). As the third organ of the higher plants after the shoot and the root, the leaf is essentially used for photosynthesis and transpiration. The leaf blade is flat in shape for this purpose. A bi-facial leaf has an adaxial assimilation parenchyma above an abaxial spongy mesophyll and the vascular bundles in the form of the ribs and has an epidermis on both sides. These have stomata either only on the underside or on both sides of the leaf, which enable gas to be exchanged for photosynthesis and respiration as well as releasing water vapour in stomatal transpiration. The fact that the leaf with its epidermis can also function as an organ that absorbs and excretes water and substances dissolved in it. Compared with the uptake of nutrient salts through the roots, however, foliar fertilization is doubtless a much more speedy way of supplying microelements in particular. It can, however, also be used to satisfy acute needs, with lower concentrations of macro elements (Wolfgang, 1986).

The supply of nutrients via roots is restricted in some soil conditions such as high pH, high lime content and often inadequate water supply. A particular advantage of foliar nutrient application is that it supplies nutrients directly to the various metabolizing parts of the plant without the possibility of antagonism due to cations or deposits caused by reactions with anions (Doering and Gericke, 1986).

A large part of the nutrients applied remain unused by the fruit tree and leach into deeper soil layers, causing soil and groundwater contamination. Groundwater contamination ranks among the most serious problems facing the world today since it entails risks for the supply of drinking water in the future, also leading to other environmental degradation problems such as disturbance of biological processes and prevention of the normal development of the flora and fauna of aquatic systems (Kang *et al.*, 2008). Due to the aforementioned reasons, in the cultivation of apples and other fruit crops, scientists give priority to fertigation and foliar nutrition, as well as types that significantly improved the efficiency of fertilizers, with minimum adverse environmental impact (Amiri *et al.*, 2008).

Although fertigation is generally seen as a preferred type of fertilization, foliar nutrition has been used very often lately in fruit production. The main advantage of foliar nutrition is that it can be used under very limited root nutrient uptake conditions (root damage, long periods of dry weather, and stagnation of water on surface of soil). Furthermore, the degree and rate of nutrient utilization through foliar fertilizers are higher than those through soil fertilizers, particularly at low temperatures (Wojcik, 2004; Pfeiffer *et al.*, 2008). Additionally, this type of fertilization largely contributes to environmental protection by reducing undesirable nutrient leaching from the soil, an accompanying manifestation of most other fertilization types (Dong *et al.*, 2005; Totten *et al.*, 2008).

For all fruit and vegetable crops, the major pathway for mineral nutrient uptake is via the roots. Nutrients applied to the leaves can be absorbed and utilized by the plant; however, for nitrogen, phosphorus and potassium the quantity absorbed at any one time is small relative to the larger levels required for growth and development by the plant. Foliar application of these three nutrients cannot be expected to supply the total amount required for crop production.

The largest problem with foliar feeding is that, plant is not able to completely replace the nutrition through the roots, primarily due to plant requirements for some elements (N, K, P) is considerably higher than their concentrations in the foliar fertilizers. As part of adequate nutrition, fruit trees should be frequently sprayed, which would however lead to an increase in production costs (Murtic *et al.*, 2012). Although it cannot be used as a uniform model of fertilization, foliar fertilization can serve as an excellent supplement to other fertilization types (Doring and Gericke, 1986)

Gonzalej *et al.* (2010) reported results of their research, provide strong evidence that specific nutrients applied to the foliage, including flowers, can efficiently meet crop nutrient demand and stimulate specific metabolic and physiological processes resulting in increased yield of commercially valuable large size fruit and, hence, grower net income. They also reported foliar fertilization strategies were all cost-effective and have the potential to reduce nutrient accumulation in the soil, run-off water, surface water and groundwater. With the increasingly high cost of fertilizer, replacing soil-applied fertilizers in part with properly timed foliar fertilizers provides growers with a tool to reduce fertilizer use and expense while increasing yield and revenue and protecting the environment.

Foliar fertilization is most effective when soil nutrient availability is low, top soil dry, and root activity during the reproductive stage is decreased. Foliar fertilization is also successful in increasing content of fruit Ca 2^+ and cereal grain protein. It is proposed that this treatment should be recommended in the integrated plant production because it is environment friendly and increases productivity and yield quality (Wojcik, 2004).

Foliar application is based on the principle that the nutrients are quickly absorbed by the leaves and transported to different parts of the plant to fulfill the functional requirement of nutrition. It is an ideal way evading the problems of nutrient availability. This method is highly helpful for correction of macro and micro elements deficiencies to restore disrupted nutrient supply and overcome stress factors limiting their availability. This method has been commercialized in a number of fruit crops like citrus, pineapple, guava, *etc.* Foliar sprays are found excellent supplements to soil application as they provide rapid absorption of nutrients for immediate utilization in metabolic activities besides ensuring economization of fertilizer use.

The efficiency with which foliar applied macronutrients are utilized depends on the mobility of the specific nutrient throughout the entire plant, mobility comprising long distance transport especially phloem transport as well as the symplastic transport. Potassium and nitrogen are examples of nutrients showing high mobility and when taken up by leaves they can be rapidly distributed throughout the entire plant. Supply of Nitrogen in the foliar application through KNO_3 and urea has shown betterment in growth irrespective at non-saline as well as saline rooting medium (Jabeen and Ahmad, 2012). It is commonly believed that the optimal pH values of spray solutions for the maximum uptake of most mineral nutrients are within the range of 3.0-5.5 (Kannan, 1980). Acidic foliar sprays can penetrate leaf surfaces more effectively, but it is possible for a negative effect to occur when too much acidity is present. Each type of organic acid has its own pH disassociation range with the mineral as the pH drops (increased acidity) (Jabeen and Ahmad, 2012).

Bondada *et al.* (2001) reported that young leaves (0.25 month and 1 month old) absorbed a greater percentage of N than the older leaves (3 month and 6 month old). Epicuticular wax concentration increased and 15N absorption declined with leaf age. Nitrogen deficient leaves (1.80% N) had greater wax concentration and lower N absorption than N sufficient leaves (2.60% N). Maximum N absorption from foliar urea sprays occurred at night through the abaxial surfaces of young leaves with sufficient N.

At present, India is the second largest producer of fertilizer- nitrogen and the third largest producer of phosphate fertilizers in the world. Potash is totally imported. As regards consumption, India is second only to China in nitrogen and phosphorus (Prasad, 2012).

The use of water soluble fertilizers in different crops is meagre in India while these are mainly used in developed countries. It has been reported that water soluble fertilizer consumption in USA was about 17% of the total fertilizers used in all the crops in 2009. In India, these fertilisers mainly used as foliar application or applied through drip irrigation system (fertigation). Water soluble fertiliser can contain various elements which are necessary for such as N, P, K, Ca, Mg S, and the nutrients content can be designed according to crop requirement (Bhaskaran and Subrahmanyam, 2012). They also enlist the summary of foliar nutrition trials with productivity improvement in various crops as reported by different SAUs, in that 30 % increased in cashew yield over control (reported by Tamil Nadu Agriculture University, Coimbatore) and 76% increased in mango yield over control (reported by University of Agriculture Science, Bangalore).

Use of foliar feeding in fruit production largely contributes to environmental protection as it reduces the risk of soil and groundwater contamination with undesirable mineral substances. In view of the current goal of agricultural policy to pave the way for sustainable agriculture and, therefore, fruit production, foliar feeding combined with adequate basal dressing can be recommended as a fertilization model which, if properly employed, can satisfy the criteria with regard to environmental protection and fruit tree nutrient requirements.

The most efficient or economical use of fertilizers is essential because of their high cost as well as concerns about pollution. Foliar feeding is one way to achieve this goal because nutrients can be applied directly to the site of their metabolism, and are therefore not subjected to various losses as occurs with soil application. If application can be combined with plant protection measures and growth regulators, its cost can be reduced. Moreover, for a rapid response and correction of deficiencies, foliar sprays of nutrients have been used singly or in combination. Foliar feeding accelerates flowering, fruiting and improves fruit quality, and is also recommended for rainfed conditions.

Foliar sprays are useful to maintain optimum nutrient concentration in the plant during the growing cycle by optimizing the movement of nutrients. Foliar fertilization is better than conventional soil fertilization, (i) in soils having an acute shortage of nutrient supply, (ii) nutrients either due to their total absence or due to trace concentration, are immobilized on account of unfavourable soil conditions, (iii) under soil nutrient imbalances i.e. having an unfavourable influence on root absorption for an optimal growth and (iv) restricted nutrient uptake through the plant roots (Dixit *et al.*, 1979; Srivastava and Singh, 2003).

Mineral feeding is one of the main tools to use improve fruit yield and quality (Tagliavini and Marangoni, 2002). All nutrients play an important role in improving growth and fruiting via set encouraging cell division (Nijjar, 1985). Nitrogen is essential to the cashew tree for vegetative growth and fruit production. It increases leaf chlorophyll levels and photosynthesis, hence promoting shoot growth, flowering and fruit production. Providing cashew trees with additional supplies of nitrogen at critical phenological stage of flushing, flowering and fruit set has proved beneficial, increasing growth and yield. It also increases the ability of the cashew tree to utilize other nutrients.

Foliar spraying has been used as a means for preparation of supplements doses of macro and micro nutrients, plant hormones and other useful substances. Observed effects of foliar nutrition have included yield increases, resistance to diseases and insect pests, improved drought tolerance and enhanced crop quality. Plant response is dependent on species, fertilizer form, concentration and time of application, as well as the stage of plant growth (Kuepper, 2003). He further suggested that foliar spray should be done when wind is minimal. This is especially important with finely atomized sprays because they drift readily. Absorption is increased when sprays also reach and coat the undersides of leaves. This is where most of the plant's stomata are located. Always delay foliar fertilization until air temperatures drop to 80°F or below. Absorption at higher temperatures is very poor because plant stomata are closed. Some of the most effective foliar feeding is done late in the evening or in the early dawn, when temperatures are right and wind is minimal. Absorption is further enhanced when weather conditions are humid and moist. The presence of heavy dew on the leaves facilitates foliar feeding.

Foliar fertilizers are being used in fruit crops that contain various macro nutrients. Foliar fertilizers are known to immediately deliver nutrients to the tissues and organs of the crop. Recently new generation specialty fertilizers have been introduced exclusively for foliar feeding and fertigation. These fertilizers have different ratios of N, P and K, are highly water soluble and so amenable for foliar nutrition. Currently, foliar fertilisers are commonly supplied to commercial fruit and vegetable crops (Zhang and Brown, 1999).

Foliar spray are found excellent supplement to soil application as they provide rapid absorption of nutrients for immediate utilization in metabolic activities beside ensuring economization of fertilizer use (Ravishankar *et al.*, 2010).

CHAPTER – 12

Canopy Management in Cashew

Canopy management is the manipulation of tree canopies to optimize the production of quality fruits. The canopy management, particularly its components like tree training and pruning, affects the quantity of sunlight intercepted by trees, as tree shape determines the presentation of leaf area to incoming radiation. An ideal training strategy centres around the arrangement of plant parts, especially, to develop a better plant architecture that optimizes the utilization of sunlight and promotes productivity. Light is critical to growth and development of trees and their fruits. The green leaves harvest the sunlight to produce carbohydrates and sugars which are transported to the sites where they are needed – buds, flowers and fruits. Better light penetration into the tree canopy improves tree growth, productivity, yield and fruit quality. The density and orientation of planting also impact light penetration in an orchard. Generally, in close planting, quicker shading becomes a problem. An east-west row orientation results in more shading as compared to the western and southern orientation of trees. Strong bearing branches tend to produce larger fruits. The problem of a fruit grower is initially to build up a strong and balanced framework of the trees, then equip them with appropriate fruiting. Obviously, pruning in the early years has to be of a training type to provide strong and stocky framework with well spaced limbs or any other desired shape.

The canopy architecture is one of the factors that decide the nut yield of cashew. For obtaining higher yield, it is necessary to optimize the plant density along with canopy management. In case of high/ultra high density planting models, the major controlling factors that decide the nut yield are the extent of incoming radiation and percentage radiation intercepted by tree canopies which could be enhanced by proper canopy management practices. For commercial cashew cultivation, the natural form and shape of the trees are to be modified through the practice of pruning to achieve the targeted yield.

Cashew a fast growing woody perennial is generally characterized by its spreading branches of erratic growth which forms irregular shaped canopy (Ohler, 1979a, b). Canopy manipulation and containment by way of pruning as an orchard management horticultural technique has not been attempted so far. But in a plantation for sustainable crop, proper management of canopy from the beginning of orchard life is essential to prevent the decline of the yield beyond 10 to 15 years (Quinlan and Tobutt, 1990; Stassen, *et al.*, 1999). Thereby, not only shape of the cashew tree is irregular but also poor yield is realized in long run. In fact, canopy shape manipulation in cashew by possible means for better crop yield was not felt essential in the past. However, high density planting or ultra high density planting with precocious varieties and their canopy development and management is the most important horticulture technique and is one of the possible means of achieving the targeted yield in several horticultural crops (Saroj and Nayak, 2013). It is well proven that the small trees with proper canopy exposure to solar radiation can give better fruit yield (Singh, 2010) and thus it is essential mat the efforts should go to develop canopy management technique for ease of orchard management and higher productivity in cashew too (Saroj and Nayak, 2013).

For cashew, the normal optimum spacing recommended is 8m x 8m which would be totally occupied by the plant within 8 to 10 years unless the canopy is maintained by trimming and thinning. Therefore, to ensure a better canopy structure and frame work for a productive orchard. Under high density planting system better ground coverage by plant canopy and interception of light (up to 80 percent) during early period of orchard life definitely has added advantages in enhancing the crop yield per unit area (Yadukumar and Nayak, 2000). In normal spaced plantations the ground coverage was minimum and the light interception was less than 50 per cent up to 9 to 10 year after planting and thereby resulting in the lower yield per unit area (Anon., 1998). In high density system, dropping of dry leaves/unit area was more and thus, increasing to soil nutrients, conservation of soil moisture and in checking the weed growth compared to normally spaced plantations.

Principles of Canopy Management

Canopy management refers to the manipulation of tree architecture to optimize the production of quality nuts. An ideal training strategy centres around the arrangement of plant parts, to develop a better plant architecture that optimizes the utilization of sunlight and promotes productivity. The green leaves harvest the sunlight to produce carbohydrates and sugars which are transported to the sites where they are needed. Better light penetration into the tree canopy improves tree growth, productivity, yield and fruit quality. The density and orientation of planting also impact light penetration into the canopy. Generally, in close planting, quicker shading becomes a problem. An East-West row orientation results in more shading as compared to the Western and Southern orientation of trees. The initial job of a plantation owner is to build up a strong and balanced framework of the trees (Dora and Kabita, 2013). Some of the basic principles in canopy management are:

- Maximum utilization of light.
- Avoidance of built-up microclimate congenial for diseases and pest infestation.
- Convenience in carrying out the cultural practices.
- Maximizing productivity with quality produce.
- Economy in obtaining the required canopy architecture.

The selection of training methods also depends upon variety, resource situations and growing needs. Pruning, an essential horticultural technique for the improvement of crop yield in several horticultural crops is not well understood in cashew (Saroj and Nayak, 2013). Pruning of the deadwood alone helps in improvement of 30-40% nut yield in cashew (Khan *et al.*, 1987); while, Mohan and Singh (1989) reported that the nut yield can be doubled by leader shoot pruning. In other tree crops like apple, pruning helps in inducing dwarfing and thereby better size and shape to the canopy and same has been recommended in high density planting to restrict the tree size and to improve the quality of the produce (Mika, 1986). Pruning also helps in changing the pattern of branching and flushing in several other crops like guava, ber, grapes, apple, pear and several ornamental crops. As the older cashew plantations available in the country have become senile, they need rejuvenation to improve the yield performance so as to achieve self sufficiency in cashew nut requirements (Saroj and Nayak, 2013).

From the beginning, proper framework should be developed by adopting appropriate training methods. The sprouts coming from the root stock portion of the graft, that is, from the portion below the graft joint, should be removed frequently during the first year of planting. If it is allowed to grow, it usually results in the death of the grafted scion and only the root stock seedling will grow. Studies conducted at National Research Station, Puttur suggest that initial training and pruning of young cashew plants during the first 3-4 years is essential for providing proper shape. The trees are shaped by removing the lower branches and the water shoots which come up from the base during the first 3-4 years. Therefore, little of no pruning is necessary. The plants should be allowed to grow by maintaining a single stem up to 0.75-1.00 m from ground level (i.e. lifting the canopy skirt to a height of 0.75-1.00 m). This can be achieved by removing the side shoots or side branches gradually as the plant start growing from the second year of planting. Weak and criss-cross branches can also be removed. Branches growing unwildy may also be cut off. Proper staking of the plants is required to avoid lodging due to wind blow during the initial years of planting. Initial training and pruning of the cashew plants facilitates easy cultural operations such as terrace making, weeding, fertilizer application, nut collection and plant protection, most importantly prevention of Cashew Stem and Root Borer infestation by easily swabbing the trunk portion with Carbaryl (0.2%) and chloropyriphos (0.2%).

In older cashew plantations, removal of dried/dead wood, criss-cross branches, water shoots *etc.* should be attended to at least once in 2-3 years. This allows proper growth of the canopy and receives adequate sunlight on all the branches, pruning of cashew plants should be done during August- September and the cut surfaces are smeared with Bordeaux mixture paste (10%) to avoid fungal infection.

The flower panicles emerging from the grafts during the first and second year of planting should be removed (deblossoming), in order to allow the plant to put up good vegetative growth. The plants are allowed to flower and fruit only from the third year onwards.

Pruning provides a definite form and shape to the tree. Pruning of dead wood and criss-cross branches can increase yield by 30-40% (Khan *et al.*, 1987). Training indirectly assists in ease of other operations such as weeding, manuring and hoeing (Satpathy, 1988). Results of pruning on 28-year old trees revealed that trees with three branches pruned recorded the highest number of panicles/sq. m (39), highest number of flowers/panicle (588.70) and fruit-set to an extent of 14.42%, while unpruned trees recorded only 7.75% increase in yield (Panda, 1990). Under Jhargram conditions, pruning of leader-shoots during July enhanced the number of productive laterals, increased the number of bisexual flowers per panicle, fruits per panicle and yield per tree (Chattopadhyay and Ghosh, 1994). Leader-shoot pruning doubled the yield in cashew (Mohan and Singh, 1989). Pruning treatment increased the number of laterals/leader but did not affect duration of flowering and harvest (Mohan and Rao, 1995). Leader-shoot pruning at least once every 2-3 years helps boost nut yield (Nayak, 1996).

Fig. 32: Tree luxuriantly grow after pruning

Canopy Management in Cashew

To achieve this initially grafts were allowed to grow with a clear single stem up to a height of 75-100 cm by removing all the lateral sprouts. Thereafter, branching is allowed in all the directions in different whorls up to a height of 4-5 meters and subsequently, the central leader is detopped to ensure a globular canopy shape. However, trimming of branches and removal of criss-cross and low spreading branches was recommended to maintain the canopy size and shape (Nayak, 1996). When the plantations were raised with closer spacing of 4m x 4m or 5m x 5m or with closer spaced hedge row system the plants tend to remain lanky and straight growing or crowded if they were not shaped properly. This requires efforts form the first year of orchard life as the subsequent early thinning of some plants can be prolonged and maximum benefit of high density system can be reaped. Studies however, are to be conducted with various training systems e.g. central leader system, modified leader system, multi leader system, open vase system, closed vase system, spindle shape system, cone shape system and palmette system under different spacing and system of planting as attempted in various tree crops (Saroj and Nayak, 2013).

Plants should be meticulously trained from the very beginning so as to harvest the maximum benefit of high density system of planting and avoid thinning later on. Canopy management in cashew plantations of different growth stages are discussed below:

1. Canopy management in newly established cashew plantations

Different methods of training employed to shape the cashew trees to desired architecture at initial stages of plantations. There are three different pruning systems suggested by Nayak and Bhat (2010) are as follows:

i) Modified leader system

In this system, side sprouts on the leaf axils of young graft are removed periodically during the first year as and when arise and a clear single stem (trunk) of 0. 5 to 0.7 m from the ground level is maintained and later the trunk is allowed to branch in all the direction. The central leader is detopped at a height of 3 to 3.5 m and a clear semiglobular canopy canopy be forced to form. Height of detopping may be decided depending on the spacing allotted to the plants. Less spaced plants are detopped at lower height. The canopy needs annual maintenance by minimum trimming of the over growth after the harvest of the crop. This kind of canopy helps in tapping maximum sunlight and helps in reducing the dead wood and water shoot development. The system is well suited for the plants spaced at a closer spacing of above 5m x 5m or less (Singh, 2010).

Fig. 33: Initial training by removing the side sprouts

Fig. 34: Modifed leader system of training

Fig. 35: Open centre system of training

Source: Nayak and Bhat (2010)

II) Open centre system

In open centre system, plant is allowed to grow up to a height of 30 cm to 45 cm height from ground with a clear single stem and then the main stem is pinched off in this system. The lateral shoots in all directions are encouraged to grow and form a vase shape. Canopy shape is maintained by minimum trimming annually. This shape helps in flowering and setting of nuts both in inner and outer surfaces of the canopy and covers the allotted space faster. For the plantations having wider space (8m x 8m) the system is adoptable but the system cannot be adopted in very closely planted plantations (Singh, 2010).

III) Pruning to bush shape

This training system can be adopted in closely spaced plantations and need to be attempted from the initial years of planting. In closely planted plantations under high density system canopy development within the manageable size is most essential. Plants can be prune to bush shape at a height of 0.75 to 1.00 m. Under high density system the plants of pruning responsive varieties such as Ullal-1, K-22-1. NRCC Sel-2 and VRI-3 showed encouraging yield performance. Yield performance of bush pruned plants was superior even at a closer spacing of 2.5m x 2.5m for initial several years and thus a yield target of more than 4 tones/ha can be attempted to achieve. In this the fruit bearing branches will be orthotropic in growth behaviour and healthier than plageotropic fruit bearing branches of unpruned plants (Nayak and Bhat, 2010).

Table 28: Growth and yield performance of 5 year old variously trained plants

Training system	Height of plant (m)	Canopy spread (m²)	Yield/plant(kg)
Central leader system	3.9	5 x 5	2.9
Modified leader system	3.5	5 x 5	3.4
Open centre system	3.4	5 x 5	3.8
Bush shape system	1.5	2.5 x 2.5	2.7
Control	3.9	5 x 5	3.1

Fig. 36: Bush shape pruning in cashew
Source: Nayak and Bhat (2010)

Canopy Management in Productive Orchards

Cashew responds very well and gives higher yield when exposed to bright sun light. The well maintained cashew plants need annual pruning and trimming to shape the plants so as to tap maximum sunlight leads to better photosynthesis. In a plantation the intermingling branches with neighbouring trees need to be trimmed every year and a clear gap of minimum one foot may be maintained for tapping the intermittent light. Depending on the spacing of the plantation the height of the tree canopy is regulated so as to overcome the shading effect of the plants over the neighbouring plants. For example, height of the plants spaced at 5m x 5m be contained at 2.5 m height. Similarly, for the plantations of 8m x 8m spacing the plants is detopped at 4m height. While attending the annual pruning the criss-cross branches, the dead wood and the branches which touch the ground can also is removed by pruning (Nayak and Bhat, 2010).

High Density Planting in Cashew

Adoption of high density planting is the best method to increase the productivity of cashew. This however, needs dwarf and compact cashew varieties which are not available at present in the country. A few recommendations were made on high density planting in cashew with different spacing's such as 5mx 4m, 6m x 4m and 5m x 5m or 4m x 4m depending on the fertility status of soil. The studies were taken up based on the popular cultivars prevailing during the days and those results may not be relevant today in all the localities. Pruning and diagonal thinning is recommended in high density planting system after 10 years of planting. Generally, in the initial one or two year's frequent removal of side sprouts and the sprouts corning from the root stock portion is required. From third year onwards frequent pruning is to be avoided and restricted to only shape maintenance as it may extend the juvenile phase of plants. Pruning and trimming, if required may only be attended once in two or three years as yearly pruning has no additional advantage on yield. Pruning and trimming must be done between May-June as soon as the harvest is over to avoid possible adverse effect on the ensuing yield (Yadukumar *et al.*, 2002). Such a recommendation was available for high density plantations as the slow growing and pruning responsive cultivars were not known at the time. The studies which were conducted at the time were with vigorous growing types of cashew without emphasis on pruning and canopy management techniques (Saroj and Nayak, 2013).

In high density system, dropping of dry leaves per unit area was more and thus increasing in soil nutrient status, conservation of soil moisture and in prevention of weed growth was observed. Under high density system better ground coverage by plant canopy and interception of light (up to 80%) during early period of orchard life definitely has added advantage in enhancing the crop yield per unit area. In normal spaced plantations the ground coverage was minimum and the light interception was less than 50% up to 9 to 10 year after planting and thereby resulting in the lower yield per unit area.

Ultra High Density Planting in Cashew

In the recent past, studies on ultra density planting of cashew revealed that the yield per unit area can be enhanced in multifolds over the traditional practices of growing cashew at wider spacing. Growing at a spacing of 3m x 3m or 2.5m x 2.5m can accommodate 1111 plants or 1600 plants/ha and even an average yield of 2 kg per plant can bring more than 2-3 tonnes per ha. The identified pruning responsive and precocious varieties such as NRCC Sel-2, Ullal-1, VRI-3, K-22-1, V-4 and Ullal-3 have shown superior performance under this technique and are under experimentation (Saroj and Nayak, 2013). The early results on the trials with ultra density planting of precocious cashew varieties revealed that per unit yield of cashew nuts can be enhanced in multifold (Nayak, 2011). Large scale trials of participatory approach in farmers field also was successful and an average yield of one kg/plant could be realized in the very first fruiting season after planting (1.5 year old plants) under irrigated and well managed orchards (Saroj and Nayak, 2013).

Pruning System for Ultra High Density

Under high density orchard system, the canopies needs to be contained within allotted space which can be done by forcing them to grow like bushes rather than trees. In the absence of dwarf and high yielding clone or dwarfing rootstocks or chemicals known to induce dwarfing of cashew trees, the only option remained for containing the canopy is by pruning (Nayak and Swamy, 2010). In high density system, plants tend to grow lanky and erect for want of more sunlight resulting in drying of lower branches. Therefore pruning and training needs to be attempted from the initial years of planting and needs to be maintained by annual pruning.

Timely pruning of exhausted growth induces productive growth which can help the plants to remain more productive. Immediately after the harvest of the crop every year the total growth need to be pruned of at height of 1.0 to 1.5 M from ground level. However, the initial frame work of primary branches should not be disturbed. The best season of pruning was found to be May - June in West Coast region and a Bordeaux (1%) spray soon after pruning takes care of the possible fungal invasion. As the new flushes tend to remain juvenile pest control be given utmost importance. The resultant new flushes always grow in orthotropic in nature and as a result they will have less competition for light and space among themselves. Only a few selected slow growing varieties have responded well to the technique (Saroj and Nayak, 2013). The growth of pruned plants of a pruning responsive varieties indicated much superior performance (Table 29) over the plants spaced at higher spacing without much pruning (Nayak *et al.*, 2008). Under ultra high density system the yield can be maximized by accommodating of more number of plants by canopy containment.

Table 29: Intensity of resultant flushes (lateral shoots) in pruned plants

Variety	Pruned trees		Unpruned trees	
	Total no. of flushes/m² area	No. of flowering laterals/m² area	Total no. of flushes/m² area	No. of flowering laterals/m² area
VRI-1	26	15	22	20
Ullal-1	29	26	21	17
Ullal-2	26	18	19	16
VTH30/4	21	6	16	14
NRCC Sel-1	24	2		12

Potential Benefits of Ultra Density Planting

Different cashew varieties such as NRCC Sel-2, Vengurla-4, Ullal-3 and VRI-3 when tried under ultra density planting at 2m x 3m spacing, the yield of 1st harvest at the plant age of 1.5 years was up to 1.20 kg in NRCC Sel-2 under irrigated situation in rest of the varieties the average yield ranged from 1.00 kg -1.10 kg/plant. The ultra density planting with proper pruning technique can produce the anticipated results in terms of yield and the target of 3-4 tonnes/ha can be achieved and the nut requirement of the country be met in a short span of years (Saroj and Nayak, 2013). The quality of the produce in the newly emerging flushes was always better than unpruned plants. The leaf numbers, number of flushes, leaf area and leaf positioning was always better in pruned plants and thereby the enhanced quality of nut and apple in terms and size and colour and nut filling (Nayak *et al.*, 2013).

Photosynthetic Efficiency

The net photosynthesis (Pn) had reached a minimum level in high tree density plot during 6th, 7th and 8th years. This was attributed to mutual shading. The differences in Pfd (Photon flux density) and Pn between the tree densities were significant ($P<0.05$) in this study although differences between the three years (6, 7 and 8 years) were not significant (Balasimha and Yadukumar, 1993). Subsequently on the 9th, 10th and 11th year Pn at the top positions of the canopy of trees in high tree density and low tree density were similar. In low tree density plots (156 trees/ha), the Pn values for the middle height of the canopy were significantly more than high tree density plot. Pn increased in middle position of the canopy from 5.98 in 9th year to 7.48 μmol/m²/s in the 11th year in high tree density plot. This was mainly due to pruning of the overlapping branches to maintain 80 percent light interception in the 9th and 10th year and due to thinning in the 11th year. On the other hand, overlapping branches were removed (8 kg dry weight per tree) each year (8, 9 and 10th year) to limit light interception to 80 per cent (Saroj and Nayak, 2013).

The yield per tree did not increase considerably particularly in the tenth year. The effect of pruning of overlapping branches on 10th year was not evident as the ground coverage and light interception increased to more than 90 and 80 per cent respectively in the same season. Incidence of less tight (Pfd) in the middle position of the canopy (927 and 998 μmol, resulting in reduced photosynthesis (5.98 and 6.98 μmol $co_2/m^2/s$) may be the reason or yield levels not improving even after pruning of overlapping branches on tenth year. When total tree population was reduced to 312 trees/ha in high density plot, the Pn in the remaining trees increased considerably (11th year) both at top and middle height positions of the canopy. This is reflected in the increased yield obtained one year after thinning in 12th year (Yadukumar *et al.*, 2001).

Dried Branches and Weed Biomass

Dried branches due to shading under different tree density in 7th year varied from 0.90 t/ha in the low tree density plot (156 trees/ha) to 3.25 t/ha in high tree density (625 trees/ha) plot. During 11th year it varied from 1.24 t/ha in low density plot to 1.76 t/ha in high tree density plot. Drying of branches was more prevalent in the high density plots due to mutual shading of branches. The biomass of dried branches in high tree density plots was 10 times more than widely spaced trees (low tree density plot) indicating shade effect and consequent drying of live branches in the lower portion of canopy.

Weed biomass collected from the low tree density plots during 7th year was significantly higher (11.80 and 10.71 t/ha) than the high tree density plots (2.55 t/ha). This is due to maximum light interception in high density plots and minimum light interception in low density plots. The trend was similar in the 11th year but, the total weed biomass was 1/3rd in low (2.92 t/ha) and high tree density plots (0.83 t/ha) compared to those in the 7th year (Yadukumar *et al.*, 2001).

Average Root Density

A minimum of 0.7 kg of total roots in the high tree density (625 trees/ha) and a maximum of 1.20 kg of total roots in the low tree density (156 trees/ha) were present per cubic meter of soil volume (Yadukumar *et al.*, 2001). The corresponding gross weight of roots was 7.45 kg/ tree in high tree density (625 trees/ha) and 18.04 kg/tree in low tree density (Saroj and Nayak, 2013). However, root density variation with spacing was also reported in coffee and areca nuts, as with wider spacing always had maximum total root weight compared to trees with closer spacing's.

Soil Moisture

In the high tree density plot (625 trees/ha) moisture content was higher up to 60 cm depth but decreased at lower depths of 60-90 cm. Higher moisture content at 0-30 cm depth of the soil may be because of heavy shading over the ground and the thick mulch layer of heavy leaf deposit on the ground surface. The weed occurrence was also minimum in these plots resulting less competition for moisture. Soil moisture content at lower depths was less by 3.9 per cent in the case of high tree density plots compared to low tree density plot indicating more utilization of moisture at lower depths by roots during peak summer season. This may be due to more density of roots at higher depth in plot where trees were closely spaced there by extracting more available moisture from lower depth (Yadukumar *et al.*, 2001).

Management of High Density Plantations

There is an intensive management required in high density plantations from the beginning of the site improvement, soil working, selection of proper root stock and variety, planting, initial training, deblossoming, thinning, irrigation, nutrition, mulching, pruning, plant protection, harvesting etc as it differ from many aspects of normal planting of orchards. Due to closure planting, the consideration of resource sharing, high leaf litter, more transmission of pest and diseases, less availability of light to ground floor etc. should also be taken into consideration (Saroj and Nayak, 2013).

Nut Yield

The work done at two research stations in East Coast (Jhargram, West Bengal) and West Coast (Vengurla, Maharashtra) of India on high density plantations (5m x 5m) indicated that the cumulative yield of 9.0 and 5.14 tonnes/ha respectively can be obtained from 10 and 19 years old cashew orchard in comparison to normal spacing of 8m x 8m with the yield level of 2.59 and 2.12 tonnes/ha respectively (Saroj and Nayak, 2013).

These studies clearly indicate that canopy management in cashew plantations of different categories (new, productive and senile plantations) pays rich dividends. Hence, appropriate steps should be taken to generate awareness among the cashew growers on the subject of canopy management. This would definitely bring in desirable results and enhance the productivity level of cashew in the country.

□□□□

CHAPTER – 13

Rejuvenation of Old Cashew Orchard

Pruning in young tree during initial stage help to form a strong frame work for better growth and development of the tree. However, on long term way, the tree becomes huge with gigantic canopy results in to problem of majority of agronomic operation, plant protection and harvesting of crop. Top working provides an opportunity to control the growth of tree and makes the tree productive with manageable canopy, in old and senile orchards.

Top working is a technique evolved to rejuvenate the unproductive and senile cashew trees. Poor yielding trees of 5-15 years of age can be successfully rejuvenated by top working. The unproductive trees are to be headed back to a height of 1.00 m from the ground level in October-November. The stem should be sawed off to avoid stump splitting. Soon after beheading, the stumps and cut portions should be given a swabbing with carbaryl 50% WP (50 g each per litre of water) or Bordeaux paste. Sprouts emerge 30-45 days after beheading. Sprouting will be profuse in young trees. So, it is better to keep only 4-6 healthy shoots on beheaded tree spreading in all directions and similarly the sprouts which come out from ground level (most probably water shoots) also removed frequently. On 30-45 day old shoots, graft scions of high yielding varieties following soft wood grafting technique. On each tree, 4-6 grafts are to be done to ensure at least 3-4 successful grafts per tree. The best season for grafting is March-April. Thinning of the extra shoots arising from the stumps should be done to obtain better growth of the grafts. Removal of sprouts below the graft joint and removal of polyethylene strip from the graft joint should be done. Top working is simple and can be done by farmers themselves after getting proper training. The top worked trees start yielding right from the second year of top working. Early bearing can be treated as one of the best advantages of this technique. The major disadvantage associated with top working is the huge casualty of trees due to stem borer attack. Intensive care and management to ward off stem borer is essential. As such adoption of top working on a larger scale would be difficult.

Uneconomical cashew trees top-worked with high-yielding Ullal-1 variety resulted in a four-fold increase in nut yield per tree within five years (Kumar, 1990). Wherever populations of cashew stem and root borer (CSRB) are maximum, top-working technology may not be suitable, unless regular follow-up action is taken to manage the incidence of CSRB (Swamy, 1995).

Beheading of tree

Result of non thinning of flush

Thinning of flush

Coppice grafting on 4-6 selected flush

Rejuvenated cashew tree

Fig. 37: Rejuvenation technique in cashew

Fig. 38: 20 years old rejuvenated cashew orchard at AICRP-Cashew Vengurla centre

Rejuvenation of Old Trees by Pruning

The normal optimum spacing recommended for cashew is 8m x 8m. There will be crowding of canopy by overlapping branches in 8 to 10 years of age depending on the nature of variety, unless the canopy is maintained by trimming and thinning. Plantations which have overlapping branches will lead to dead wood development and decline in yield of cashew in the later years. Such plantations can be rejuvenated through canopy redevelopment by limb pruning.

The redevelopment of canopy is necessary in the older plantations, where the canopies are overcrowded resulting in reduced in yield. This is possible by heading back (limb pruning) of the exhausted canopy at 1.0 - 1.50 m height. Immediately after beheading the trees, the pruned wood needs to be removed from the plot so as to prevent the damage by shot hole borers. Remaining portion of the trunk and exposed part of roots is to be swabbed with 0.2 per cent chloropyriphos at least 2-3 times after beheading till new canopy is developed. Regular checking for the damage by Cashew Stem and Root Borer (CSRB) is to be done and managed. The cut ends of the branches should be smeared with 10 per cent Bordeaux paste to prevent the gummosis and entry of any pathogens. The best season for limb pruning is May-June i.e., immediately after the harvest of the annual crop (Nayak *et al.*, 2008).

The new flushes arise from the dormant buds on the trunk. These new sprouts form the canopy within a period of 6-7 months depending on the variety. Yield levels during the first season after pruning are slightly lower while the yield from second season onwards is better than the unpruned plants. Yield performance of the 4 varieties during first 3 years after limb-pruning is given in Table 30. It was also observed that cashew varieties such as Ullal-1, Ullal-2, Chintamani-1, K-22-1, NDR-2-1, Kanaka, V-4 and NRCC Sel-2 responded well to limb pruning. The cashew trees after rejuvenation either by beheading or top working are more prone for CSRB. The old cashew plantations can be rejuvenated as explained above (Nayak and Bhat, 2010), provided they are managed well by taking care of individual plants to prevent damage by CSRB (*Plocaederus ferrugineus*).

Table 30: Yield performance of rejuvenated trees

Varieties	Yield/tree (kg)		
	1st year	2nd year	3rd year
VRI-1	1.62	3.19	8.01
ULLAL-1	2.95	4.42	8.84
VTH-30/4	1.81	4.61	9.30
NRCC Selection 1-1	0.72	1.23	8.10
Unpruned (Control)	3.25	3.01	5.59

Phytosanitation should be adopted to manage the rejuvenated plants. The newly developed canopy needs to be protected from Tea Mosquito Bug (*Helopeltis antonii*) and leaf beetles as they are also more vulnerable for the damage in the initial 5-6 months of canopy development. Care should also be taken to treat the cut wounds with 10 per cent Bordeaux paste during the beginning of monsoon season to prevent from any fungal infection and gummosis. If the plants are properly protected the canopy redevelopment can help in enhancing the nut yield to the fullest potential of the plant after 2-3 years. The limb pruned trees be pruned or trimmed as and when required in subsequent years. It may be noted that the yield levels of limb pruned trees are much better than the freshly planted cashew grafts of comparable years due to better root system. Pruning operation should be taken up only after complete picking of nuts which coincides with temporary quiescent stage in cashew (May-June in West Coast) (Nayak and Bhat, 2010).

Rejuvenation by Top Working

Rejuvenation of old and unthrifty cashew trees by top working, by grafting with scions of better variety has been developed in early eighties at Agricultural Research Station, Ullal, under University of Agricultural Sciences, Bangalore (Rao, 1985; Guruprasad *et al.*, 1988; Khan *et al.*, 1988) and subsequently modified as per local requirement and popularized by National Research Centre for Cashew and several other cashew research centers in the country.

Though the top working is a promising technology to rejuvenate the old senile plantations of seedling progeny in a short period, the technology could not be adopted on a large scale due to the problem of CSRB which damages and kills the top worked trees. The technology is being recommended only for the homestead gardens of young age which can be properly monitored and maintained.

Conditions and Points to be Considered Before Top Working

Tree should be within 15-20 years age. As far as possible it should have smooth bark and should not have many crevices on the bark at collar region. The colour of the trunk should be brown or ash coloured, but not black.

An experiment conducted at NRCC on top working, very good success was recorded. In a plantation of young tree of 8 years age of variety Vengurla-2 when top worked with the scion of Vengurla-4 to replace the variety in scion bank, 88 trees out of 94 top worked trees could be successfully saved and managed (Nayak and Bhat, 2010).

The expenditure involved and the yield performance for initial two years is furnished in Table 31. Studies on large plot trials of top working conducted both at NRCC and Karnataka Cashew Development Corporations, revealed that utmost care is required to protect the top worked trees from the devastating pest of cashew stem and root borer (Nayak and Bhat, 2010).

Table 31: Performance of 94* top worked trees; expenditure involved in top working and maintenance

Year	Expenditure involved in top working and maintenance (Rs)	Nut yield per Plot (kg)	Income per plot (Rs)
1996-97	3098	-	-
1997-98	2688	37.8	1512
1998-99	2756	184.2	8389
Total	8542	222.0	9801

*Out of 94 trees top worked 88 trees survived. Yield and income are from plot of 88 survived trees.

Pest Management in the Rejuvenated Trees

The trees which have been top worked need to be checked for any symptoms of CSRB pest incidence right from the first week after top working at 15 days intervals. Normally the fork regions and the cut ends (if uncovered) will have the pest entry and exudation of the fine frass material in small quantity. These spots of the bark should be chiseled cautiously so as not to make more damage onto the bark of the top worked trees and the tunnel made by the grubs, which bore into the bark, can be traced out by the freshness of frass. The young grubs remaining inside need to be removed out and killed. In case the frass comes out from the root zone then the soil at that spot need to be dug out and the infested root (Nayak and Bhat, 2010).

□□□□

CHAPTER – 14

Irrigation and Water Management in Cashew

In the field of agriculture/horticulture, various irrigation systems are being employed to fulfil the water requirement for better growth and development of the tree. Since, cashew is hardy plant which can sustain in rainfed conditions of warm and humid region; it has very good potential to produced higher nut yield. Of the various irrigation methods, drip irrigation system is best method in the areas where natural resources of water is limiting factor. It is well known fact that this system has many advantages.

During the first year after planting, protective irrigation is recommended at 15 days interval during winter and at 8 days interval during summer season. At each irrigation, 25 litres of water to be given per graft. The studies on mulching and irrigation were under taken with a view to determine fruit retention. It was observed that black polythene mulch with irrigation @ 60 liter of water per tree at 15 days interval showed beneficial effect on increasing fruit retention (66.15%) over control. Cashew is grown under rainfed conditions. However, supplementary irrigation during summer months, especially during January-March at fortnightly intervals @ 200 litres/plant proved to double the yield. This is evident from the research results obtained at NRC-Cashew, Puttur (Karnataka).

In India, cashew experiences severe moisture stress from January to May, adversely affects its flowering and fruit set. Cashew responds well to drip irrigation. Irrigating the crop at 60-80 litres of water/tree once in four days through drip after initiation of flowering till fruit set and development can lead to significant higher yields. Since the cashew is predominantly grown under rain fed condition, it is essential to conserve the water as well as soil. In flat to medium sloppy lands, terraces need to prepare with crescent bund for efficient utilization of the available water during pre- and post-monsoon periods.

Considering the heat web attack during March-April, it is suggested to provide the micro irrigation to the cashew orchard. It will protect the plants from heat web and will reduce the fruit drop. In addition to this, it will maintain good crop health, proper soil condition which will lead better development of nuts and will improve the quality of the produce.

Under drip irrigation, better plant growth, more number of flowers per panicle and enhancement in the yield with plastic mulch has been reported by Mishra *et al.* (2008). Further, they reported that the highest nut yield (1.42 t/ha) was recorded under 0.8 V drip irrigation with mulch as compared to other treatments. An increase of 108 % in nut yield and 122 % in net seasonal income were recorded in case of 0.8 V drip irrigation with mulch as compared to conventional ring basin irrigation. The benefit cost ratio was also found to be highest (3.1) for the same treatment. Mulch alone could increase the yield by 16% even in the absence of drip.

Fertigation saved 50% fertilizer requirement and doubled cashew yield (Richards, 1993; Yadukumar and Mandal, 1994). Irrigation at 200 litres of water per tree once in every 15 days after flowering during summer months increased cashew yield. Irrigating cashew with 60-80 litres of water per tree every four days through drip after flowering until fruit-set and development, in combination with application of 750:187.5: 187.5 g NPK per tree led to significantly higher yields (Yadukumar and Rejani, 2008; Yadukumar *et al.*, 2009). Several researchers have reported superiority of fertigation in terms of higher scion production and nut yield (Nawale *et al.* 1985; Yadukumar and Mandal, 1994; Kumar *et al.*, 1998; Blaikie *et al.*, 2001, Ingle *et al.,* 2005). Response to irrigation varied among cashew genotypes (Oliveira *et al.*, 2006). Dwarf clones did not respond well to irrigation (Oliveira *et al.*, 2003).

Effect of Irrigation on Photosynthetic Rate of Cashew

Photosynthesis of two-year-old cashew trees cultivated in containers, under irrigation, was greater than those cultivated under the same conditions but without irrigation (Bezerra *et al.,* 2007). The similar behaviour was observed by Blaikie and Chacko (1998) for stomatal conductance to water vapour such that a close relationship between these two variables was found. In an experiment with irrigated dwarf cashew plant clones cultivated in the experimental field of Paraipaba, Ceará state (North Eastern Brazil), a region whose annual precipitation varies between 600 and 1200 mm, concentrated in the first four months of the year, Oliveira *et al.* (2006) found an increase in the production of nuts of irrigated plants of around 70% (the mean over seven years). Increase in yield was strongly correlated with the increase in quantity of nuts produced per plant. Schaper *et al.* (1996) in northern Australia also found significant differences in photosynthetic rates of irrigated and non-irrigated cashew trees, with higher differences during October, when the level of water in the soil was at its lowest and the plants were in full fruit.

Irrigation and Water Management for Cashew

1) *In- situ* soil and water conservation trenches

Most of the cashew plantation is to be done on a mountain slopes results into soil erosion and nutrients loss that directly affect the growth and yield of the tree. To avoid this, it is necessary to prepare a stretch of 2 m plain land (bench terrace) near the base of the cashew tree that prevents the soil erosion. Similarly for getting the good growth and yield of cashew, a trench of 2 m length, 45 cm depth and 50 cm breadth should be prepared in between the two cashew tree on upper side. This trench will help to reduce the soil erosion and conserved the soil moisture.

Reduction in peak runoff and increase in recession time and groundwater recharge due to soil and water conservation practices have been reported by several workers (Singh *et al.,* 1989; Deshmukh *et al.,* 1992). Black polythene mulch was helpful in conserving soil moisture (Nawale *et al.,* 1985). Using coconut coir pith as soil mulch in cashew plantations resulted in 14.15% higher water retention and suppression of weeds up to 73.52% (Kumar *et al.,* 1989). Badhe and Magar (2004) reported that trapezoidal shaped staggered trenches (230 ha^{-1}) of 4.5 m length, 0.60 m top-width, 0.30 m bottom-width and 0.30 m depth were effective in reducing runoff and for conservation of soil and nutrients. Rejani and Yadukumar (2010) showed that proper soil and conservation technique like modified crescent bund, or coconut husk burial has increased the mean moisture content significantly (15.6 % and 16.1% respectively, on dry basis compared to 11.5% dry basis in control in March), reduced annual runoff 25% and 19% respectively, compare to 37% in control, reduced the soil loss (68% and 58% respectively) and also showed a trend of increased growth and yield. Mane *et al.* (2009) demonstrated that continuous contour trench (0.5 m x 0.6 m) was the best soil-conservation practice for cashew on lands with 7 to 8% slope.

Hence, the barren land even in steep slope can be effectively utilized for cashew cultivation using proper soil and water conservation techniques.

2) Supplementary irrigation

In addition to the above soil and water conservation measures, if protective irrigation is taken up during peak summer season, it will have a good impact in increasing yield further. Irrigation would be needed to meet the water deficit particularly during the non monsoon period. In this context it would be relevant to note that cashew requires some dry periods for flowering and fruit setting. Hence irrigation should be avoided during the post monsoon period between August to December. Irrigation should not be given before or at the time of flowering as it would bad to vegetative growth rather than fruiting.

Studies on the effect of black polyethylene mulch and supplementary irrigation on fruit retention of cashew nut carried out at Central Experimental Station, Wakawali, Dr. B. S. Konkan Krishi Vidyapeeth, Dapoli, Maharashtra revealed that black polyethylene mulch with irrigation at the rate of 60 l/tree once in fortnight resulted in maximum fruit retention of 66.15%. This treatment was significantly superior to other treatments of irrigation alone, polyethylene mulch alone and control. Similar results were obtained from the irrigation experiment on 13 and 14 years old trees for two years at National Research Centre for Cashew, Shantigodu, Puttur (Karnataka). Studies showed that irrigating once in I5 days @ of 200 l/tree from November to March resulted in two fold increase in yield over control (No irrigation). However, two years average data revealed that significant difference was not observed in yield for treatments of irrigation of 200 l/tree once in 15 days from November to March (10 irrigations) and January to March (5 irrigations). Based on the result of this experiment has been recommended that irrigation be given from January to March (5 irrigations) once in 15 days at the rate of 200 litres/tree to achieve 50-100 per cent increase in yield. In West Coast, it has also experienced that frequently rain extends up to November and December (North East Monsoon) resulting soil moisture levels going up to 15 and 10% by weight, respectively compared to lowest soil moisture from January to March ranging from 8.82 to 9.30%. The field capacity and permanent wilting point in this region for sandy loam type of soil is 20% and 7%, respectively. Compared to soil moisture level in March, relatively higher moisture level is available in November and December, which is congenial for cashew flowering. Soon after the flowering at the time of nut set and nut development in January to March the crop undergoes moisture stress. It is during this period the crop requires protective irrigation thereby doubling the yield due to increased nut retention (Samuel and Singh, 2000).

3) Micro (drip) irrigation in cashew

Use of micro irrigation system in cashew growing areas, where generally land in undulating, is recommended. It has been found at NRCC that drip irrigation at the rate of 80 l/tree once in four days from the second fortnight of December to the end of March, coinciding with the flowering season, resulted in significantly higher yields as compared to lower levels of irrigation or without irrigation. The increase in cumulative yield due to irrigation has been observed to be about 165 per cent of that of control plot receiving no irrigation. Total irrigation water added over the period from December to March is 2400 litres/tree. This works out almost 20 litres of water/day from December to March. Thus, irrigation helps to wet about 8% of the given ground area and 56% of the effective root zone. The advantages of irrigation have been observed in increasing the yield through increased tree canopy area, flowering laterals/tree, bisexual flower production, nut retention, number of nuts produced per tree and increased weight of nuts (Samuel and Singh, 2000). Studies conducted at Plasticulture Development Centre (PDC), at Navsari Agricultural University (erstwhile Gujarat Agricultural University) Navsari indicated that drip irrigation at 0.4 Potential Evapo-transpiration recorded maximum yield in terms of number of nuts per tree (535 per tree) and weight per tree (3.36 kg/tree) (Samuel and Singh, 2000).

Fig. 39: Drip irrigation in cashew

For a crop spacing of 8m x 8m, the total plant population would be 156 plants/ha. The required amount of water is delivered to the root zone of each plant with the help of drippers through the main and lateral pipe networks. The indicative cost details for setting up micro irrigation unit for cashew for a 0.40 ha plot and one ha plot is reported by Samuel and Singh (2000) is given in Table 32.

Table 32: Cost estimate for drip irrigation in cashew (8m x 8m)

SI. Item No.	Unit	Approx. unitcost	0.40 Ha.		1.00 Ha.	
			Qty.	Amt.	Qty.	Amt.
A. Variable Item						
1. PVC Pipe 50 MM	Meter	29.00	96	2784.0	156	4524.0
2. LDPE lateral 12 mm O.D	Meter	3.90	535	2087.0	1310	5109.0
3. Dripper 4 / 8 LPH	No.	2.00	260	520.0	680	1320.0
4. Micro tube 6 MM	Meter	1.60	240	384.0	475	760.0
5. Control valve	No.	500.00	1	500.0	1	500.0
6. Flush valve	No.	100.00	1	100.0	1	100.0
7. Accessories	Set.		1	600.0	1	200.0
Sub-total A				6975.0		13113.0
B. Fixed Items						
8. Screen filter up to 15 MVHr.	No.	1800.00	1	1800.0	1	1800.0
9. By pass assembly	No.	1100.00	1	1100.0	1	1100.0
Sub-total B				2900.0		2900.0
Total				9875.0		16013.0
C. Fertigation (Optional)						
Fertigation unit	No.	9000.00	1	9000.0	1	9000.0

□□□□

CHAPTER – 15

Intercropping and Covercropping

Intercropping

Canopy of cashew to cover the entire area in about 8 to 10 years when grown under normal spacing of 8m x 8m in fertile soil. However, in marginal lands, it takes 10 to 12 years for the canopy to cover entire area. Cashew when grown in hilly regions there is a possibility of soil erosion and weed growth during initial years of planting (Yadukumar and Nayak, 2000). Intercropping is the only available best options to prevent such erosions and conserve soil and moisture in hilly areas and to obtain sustainable income.

Recently, establishment of large plantations and adoption of systematic package of practices and intercropping is practiced to realize additional income in the initial years of cashew cultivations. When the cashew canopy starts touching the neighbouring trees possibility of intercropping is ruled out. Mixed cropping with other tree species is not widely accepted and even not advisable (Anon., 1997), though there are a few instances of growing casuarina as mixed crop in Andhra Pradesh and Orissa states (Yadukumar and Nayak, 2000).

Intercropping has received very little attention. The main objective of intercropping is to obtain some returns from the cashew plantations during the initial years of establishment till cashew trees come to flowering and start bearing economically. When once the canopy of cashew trees covers the area, it lives no scope for intercropping because of the dense nature and complete shading of interspaces. Further, the heavy leaf fall in cashew is not conducive for any field crop. Intercropping of vegetables in the initial growth stage of orchard during kharif season is recommended. Suitable intercrops such as cucurbitaceous crops like bitter gourd, snake gourd, ridge gourd, cucumber, pumpkin, okra, groundnut, turmeric, pineapple and tuber crops can also be grown as intercrop.

At Bapatla, cluster bean was found most economical intercrop in *kharif* season and black gram was remunerative in *rabi*. Groundnut was found economical intercrop at Jhargram and Vridhachalam, whereas bitter gourd was highly profitable at Vengurla.

Fig. 40: Intercropping cashew with pineapple

Intercrops like cowpea, French bean, cluster bean, rice bean, red bean, mung bean, soybean and groundnut could be grown along with cashew to get additional profit (Gupta, 1999). Maize and groundnut can be grown successfully as intercrops in newly-planted and two year-old cashew orchards (Abeysinghe *et al.*, 2003). Intercropping cashew with pineapple (Fig. 40), turmeric or elephant foot yam under normal planting density system during the first five years increased net benefit from cashew gardens (Yadukumar *et al.*, 2003). Weed-suppression was best in plots carrying cashew / cassava and cashew / plantation / cassava mixtures with 50-60% reduction in frequency of weeding per annum (Adeyemi, 1998).

The most viable and profitable intercrops with cashew in interior west coast regions are pineapple and ginger. Pineapple can be grown as an intercrop with cashew profitably for the initial 6 to 7 years (Mandal *et al.*, 1991; Yadukumar *et al.*, 1994). Both cashew and pineapple can be planted simultaneously in trenches in double row system in the interspaces across the slope. In the second and third year of planting majority of the yield can be realized and replanting can be done in the fourth year. Growing pineapple in the trenches across has additional advantage of conserving soil and moisture thereby improving the cashew yield (Yadukumar and Nayak, 2000).

A result from intercropping trial of cashew with pineapple conducted at NRCC has indicated that higher yield of cashew to the tune of 1.5 times could be realized compared to cashew as a monocrop. Further, it has been shown that net profit of Rs. 1,05,831/ha could be realized from pineapple alone in the first 7 years (Yadukumar and Nayak, 2000).

Ginger can also be grown as an intercrop in the initial 3-4 years of cashew plantation in the interior West Coast region. Depending upon the locations intercrops could be chosen. In the west coast region fodder crops, legumes and vegetables such as bitter gourd can be grown during the Kharif season (Mandal *et al.*, 1991). In the northern Kerala and Dakshina Kannada districts of Karnataka, pineapple and ginger can be grown most profitably (Yadukumar and Nayak, 2000).

Covercropping

Leguminous cover crops such as *Pueraria javanica, Calapagonium muconoides* and *Centrosema pubescens,* enrich the soil with the plant nutrients, add organic matter, prevent soil erosion and conserve moisture. The seeds of these cover crops may be sown in the beginning of rainy season with a seed rate of 7 kg/ha or as per the recommendation for that area. In the interspaces of cashew orchard, bed size of 30 cm × 30 cm are prepared in sloppy degraded soils by loosening the soil and mixing little quantity of compost. The seeds of the cover crops are sown in these beds and covered with a thin layer of soil. In order to get good germination, the seeds are soaked in water for 6 hours before sowing. Seed soaked with appropriate strain cultured enhance germination and yield. Different types of bio-culture are available commercially in the market.

□□□□

CHAPTER – 16

Interculture Operation

In the orchard, several cultural operations like weeding, mulching, etc. are very important. Similarly protection from fire is also essential. The points are discussed here.

Weed Management

The Luxuriant weed growth is a common sight in neglected old cashew gardens. Weeds are not only harmful by competing for the nutrients and moisture but also cause inconvenience to take up effective plant protection measures and harvesting of fallen nuts; thus resulting in loss of yield. Therefore, sanitation of cashew orchard is very essential, it also minimise the incidence of cashew stem and root borer (CSRB). Careful weeding-clearing the area within 2 m radius of the trunk is essential. Clearing of weed growth is to be carried out twice in a year. Weeds may compete for nutrients and moisture with cashew. The weeding cycles are generally confined to the pre-monsoon and post-monsoon periods to coincide with the fertilizer application.

Weeds should also be controlled between the rows. This can be done by inter-cropping with certain crops for 3/4 years before the tree canopies close in. In addition a legume can be grown to both control weeds and fix nitrogen for the trees. This improves the soil and makes the nutrient available.

Weed management trials have shown that two weedings first during June-July and second one at November-December are effective in keeping the weed growth under check, particularly November-December weeding helps in keeping the plots clean to enable easy harvesting of fallen nuts. Application of weedicides 2-3 times starting from July at monthly intervals and mulching around the runk were also found effective in reducing weed growth.

Alternatively, wherever labour is very costly, weedicides may be applied to control weed growth. Weedicides may also be applied after slashing, well in advance of the rainy season if the under growth is too dense. It is suggested to apply 2-4,D at the rate of 4 ml/litre of water followed by Grammoxone at the rate of 5ml/litre of water. Approximately, 400-500 litres of spray are required to cover one hectare area. The spraying is repeated in the post monsoon season, if the weed load is heavy.

Mulching

Mulching the basins of plants prevents soil erosion and also conserves soil moisture. Mulching with organic matter prevents weed growth and reduces surface evaporation during summer and also regulates the soil temperature. The green matter obtained during weeding may also be utilized for mulching the cashew orchards at the base of individual trees soon after the fertilizer application.

In low rainfall areas, mulching around the base of trees helps in the control of weeds, retention of moisture and modulation of soil temperature, especially in the hot summer months. This becomes an essential operation as cashew is usually planted in very dry areas where other crops are seldom grown.

Protection of Cashew from Fire

Fire in bearing cashew tree plots during summer is common in most of the cashew growing areas that cause tremendous loss in yield to the farmers. Mostly in summer season, there is possibility of fire in cashew garden especially in the uncleaned cashew plots. To protect the cashew garden from fire during summer, it is advisable to clean the grasses, weeds and bushes near the fence area up to 15-18 m and also make a fire protective trench just near to the fences.

□□□□

CHAPTER – 17

Maturity, Harvesting, Yield and Post Harvest Handling

Maturity of Cashewnut

Cashew fruit has two parts namely cashew nut (actual fruit) and cashew apple (pseudo fruit). Development of these is not parallel. Cashew nut develops first and the pseudo fruit develops later. It takes 60 days for both the nut and the pseudo fruit to develop completely since anthesis. Development of cashew apple and nut has been studied by number of workers in different cashew varieties (Rao *et al.*, 1962; Augustin and Unnithan, 1981; Mohan Kumar *et. al.*, 1984; Antarkar and Joshi, 1987). According to the report, specific gravity could be considered as physical index of maturity of nut in cashew varieties of V-1, 2, 3, 4 and 5 (Antarkar and Joshi, 1987; Mohan Kumar *et al.*, 1984). In Thailand, the nut reached maximum size in 4 weeks after fruit set, hardened in the subsequent week and declined in size by 10 per cent at harvest (Sumrit Feungehan *et al.*, 1989). Augustin and Unnithan (1981) have reported that the growth of nut is faster during early stage of development than that of fruit. All these reports indicate that the cashew nut develops earlier than cashew apple. Changes in the composition of cashew nut and apple during development have been looked into by various workers. Hariharan *et al.* (1985) have reported decrease in amino acid content in kernel with increased maturity. Cashew kernel lipid, free amino acids and CNSL have been shown to increase with increased maturity. Similarly, cashew kernel phenols have been shown to decrease with maturity. Changes in the kernel sugars and starch were not uniform (Anon., 1998).

In general, the literature indicates that the cashew nut develops first up to 30 days since fertilization and loses moisture during subsequent 10 days. Cashew apple develops only after the development of cashew nut stops. The whole process of development of cashew nuts and apples takes 60 to 70 days. Proportion of the nut harvested in different months is given below in Table 33.

Table 33: Proportion of the nut harvested in different month
i) Commencement of flowering – October
ii) During of bisexual phase – 55 days

Sr. No.	Months	% Harvest
1.	January	20.70
2.	February	45.95
3.	March	20.95
4.	April	7.60
5.	May	4.80
Total		100.00

Harvesting of Nuts

Cashew puts forth indeterminate inflorescence which takes 60 to 120 days to complete flowering resulting in continuous fruit set and maturity within a panicle. Therefore to get fully matured nuts, only fallen nuts are to be collected from ground. This method of harvesting not only helps in getting fully matured quality nuts but also in getting higher yield by weight per tree and fetches better market price as compared to existing practice of harvesting of nuts from tree before they mature by shaking or any other means, wherein potential hermaphrodite flowers and immature fruits at different stages drops off resulting in loss of yield and quality.

Fully mature nuts possess brown colour. In view of pilferage of nuts, cultivators have a tendency of plucking the nuts before the pseudo fruit develops completely. This results in harvesting immature nuts which affects the quality of the kernels. Fully mature apples are never plucked from the tree with the help of a long pole with a *sickle* tied to it at one end. When the cashew apple is fully ripe, with a simple tap on the fruit, the fruit drops off. Shaking the branches for harvesting the nuts is practiced. Although, this results in dropping of fully mature nuts and also leads to dropping of immature nuts. As such this practice of harvesting nuts should be avoided. The best method of harvesting the nuts is to wait till it falls and collect the fallen nuts. This damages the pseudo fruit. If one intends to utilize cashew apple, one has to pluck the fruit from the tree without causing any damage to the fruit.

Harvesting and Yield

Though cashew starts flowering and yielding after third year of planting, the trees will be in full production by the tenth year and continue bearing further period of 40 years. Main harvesting season is February-May. Flowering starts in the month of November and continues up to February. Percentage of harvest and collection of nuts in February is 15 per cent, March- April, 75 per cent and little about 10 per cent in the month of May. This indicates that the major crop is obtained during March-April. Fruits may be allowed to fall on the ground and nuts collected periodically. Some farmers harvest the nuts just before they mature to avoid pilferage. It results in poor development along with poor quality of kernel. Only full mature nuts are to be picked after they fall off from the tree. In very dry climates where the topsoil remains dry overnight, nuts can be left under the trees for several weeks without their quality being affected. However, where humidity of the air or soil causes moisture and dew formation, the nuts should be reaped at least twice a week. Of course, it is not very economical, unless it is carried out on small holdings with relatively high labour intensity, as there will not be adequate numbers of nuts to harvest sufficient quantities to achieve collection levels of 50 kg per day (Ohler, 1979). Apples to be used for processing into products such as jam or juices should be picked from the tree before they fall naturally. On falling to the ground, apples may become damaged. Once damaged, the apples may ferment and deteriorate quite rapidly. Ripen the apple, the sweeter the taste. It is therefore recommended that the apple be picked as it is about to fall. At this stage the nut is fully-grown for about two weeks and is ripe and ready for harvest.

For best quality of nut, freshly fallen fruits are collected, the apples are removed and nuts are sun dried for 3-4 days to reduce moisture from 25 per cent to 9 per cent. Drying of nuts is very essential to prevent spoilage during subsequent storage. Average yield per tree increases from 2-5 kg at the ages 3-5 years to 5-10 kg at ages 6-10 years and to more than 10 kg at ages 12-15 years onwards.

The individual productive capacity of the common cashew tree varies greatly, with plants that produce from less than two to more than 25 kg of nuts per harvest depending on the plant population per unit area, age of the plant, ecological conditions, management of orchard and selection of varieties planted. The yield adversely may also affected by insect-pest damage or disease occurrence, unprecedented drought prevail and cold/heat wave prevail during flowering and fruit set. The weight of the fruit also varies greatly, ranging from 3 to 33 g, with the weight of the peduncle ranging from 20 to 200 g.

The minimum age for stable production is greater than 8 years, and normally between 12 to 14 years. However, the yields of individual isolated trees continue to increase as the canopy spread increase and a maximum of 50 kg/tree may not be uncommon (Mandal, 2007). Using high yielding varieties, proper spacing and good maintenance of the plantation, it is possible to get about 2.0 to 3.0 tonne nuts per hectare.

The first characteristic of importance to the farmer is the yield capacity of nuts, expressed in nut returns, so this seems to be the most important selection criterion. Yield should be expressed in mass and quality of the kernels, as these comprise by far the greatest part of the nut value (Ohler, 1979 and Roe, 1994). Ohler (1979) stated that high yielding trees normally have more than one mature nut per inflorescence. In India, it was found that the number of perfect flowers produced, it governed the yield of the cashew tree. Different environmental conditions affect yield of the same cashew strain differently (Behrens, 1996). Yield parameters (age, yielding, and weight of nut and percentage of kernel) for selected material in different countries have been presented in Table 34.

Table 34: Yield parameters (age, yield/kg, weight of nut and percentage of kernel) of different parts of India and abroad

Location	Age (years)	Mean yield (kg/tree)	Nut weight (g)	Kernel (%)	Source
India					
Andhra Pradesh	33-38	13.5	7 .5	27	Reddy *et al.* (1985)
Tamil Nadu	17	7.40	5	20	Rao (1989)
Karnataka	25	19	7	31	Nalini and Santhakerman (1994)
Kerala	7-14	17.14	7.3	26	Rao (1989)
Anakkayam	-	3.29	3.6	46	Nalini and Santhakerman (1994)
Ullal	11-20	14.68	7	30	Kumar and Hedge (1994)
Senegal	29-32	36.48	6.9	-	Ohmstedt (1991)
Australia	5	3.91	5.46		Behrens (1996)
Tanzania					
Lulindi	5	3.6	-	-	Mutter and Bigger (1962)
Nachingwea	3	2.6	4.9	-	Northwood (1966)
Mahajanga	5-15	6.5	5.66	-	Rakotovao (1999)
Brazil	3-14	17.50	6	-	Gondins (1973)

The improvement in production through the use of selected plant materials from existing cashew plantations might be possible (Table 34). Selection and breeding offers considerable opportunities for increasing the cashew productivity. Total yield of nuts per tree was influenced by several genetically determined factors, including the number of panicles produced, number of perfect flowers produced per panicle, average mass per nut, pest and disease resistance and extent of premature nut drop (Ohler, 1979; Wait and Jamieson, 1986).

□□□□

CHAPTER – 18

Post Harvest Storage

Introduction

Once the cashew fruits fallen, are collected, the nuts are to be separated from the cashew apple and sun dried for 2 to 3 days. This is very important in view of the quality of kernels obtained during subsequent processing. Raw nuts at the time of harvest will have moisture content ranging between 16 to 25 per cent; while sun dried nuts have moisture content of 8 to 10 per cent (Nagaraja, 2000). If the nuts are not dried and stored properly, it leads to fungal spoilage resulting in poor quality of kernels. Fungi such as *Gonatobotryum, Helminthosporium* sp., *Corynespora* sp., *Alteraaria* sp., *Verticillium* sp. and many species of *Aspergillus* have been isolated from stored cashew nuts (Joseph, 1981). In view of the occurrence of fungi, drying of the raw nuts before storage becomes absolutely essential. Maximum permissible moisture content of raw cashew nuts is 8.7 to 9.1% (Ohler, 1979). Okwelogu and Mackay (1969) have reported that high whole nut or kernel moisture percentages are closely associated with high shell moisture contents and for nuts whose moisture content is in equilibrium with the ambient relative humidity, the moisture content of whole nut and kernel could be predicted from that of shell with more than 99% accuracy. A higher kernel shell weight ratio has been shown to be inversely proportional to moisture content.

Effect of storage temperature and relative humidity has been studied by Okwelogu and Mackay (1969), and reported that nut stored at 27°C with relative humidity of 70% had a moisture content of 9.2%. Nuts exposed to higher relative humidity of more than 75% comes down heavily with mould infection within few weeks. Therefore, relative humidity of the storage godown plays an important role. Under high relative humidity, nuts pick up moisture from the atmosphere till they attain equilibrium.

It has been shown that storage of raw nuts with a moisture content of 5 to 6% for a period of 12 months at ambient temperature (25 to 30°C) did not affect either processing or biochemical quality (Nagaraja, 1996; Nagaraja and Giridhar

Prabhu, 1996). Insects such as *Corcyra cephalonicas, Oryzaephilus surinamensis* Fab and *Tribolium castaneum* L. have been shown to infest cashew kernels during storage at RH ranging between 40 and 90% (Singh and Pant, 1986). Kernel rejects of 2 to 3% obtained during processing of cashew by steam roasting method has been shown to be influenced by immature nuts and floaters (Nagaraja and Giridhar Prabhu, 1996). However, storage period up to 16 months, initial moisture content of raw nuts ranging from 5 to 14%; time delay up to six months in drying the freshly harvested nuts; grading the nuts before processing; processing with and without steam roasting and steam roasting, time varying from 5 to 20 min have been shown no effect on per cent kernel rejects obtained during processing (Nagaraja and Giridhar Prabhu, 1996). The studies conducted at NRC Cashew, Puttur (Karnataka) on effect of maturity on the processing quality have clearly indicated that shelling percentage, peeling outturn, per cent wholes recovered increased while per cent of kernel rejects decreased with maturity (Table 35) (Nagaraja, 2000). It has been noticed from the survey of processing factories for establishing database that insecticides / pesticides are being used during storage of raw nuts. Among 132 processing factories responded for the questionnaire, 59% and 3% of them have indicated the usage of insecticides and pesticides, respectively, during storage of raw cashewnuts (Nagaraja, 2000).

Table 35: Effect of maturity on processing quality

DA A	Shelling(%)	Peeling outturn(%)	Wholes recovered(%)	Kernel rejects(%)
40	12.3	3.2	87.5	68.0
45	23.1	17.1	80.9	28.8
50	28.7	-	-	Nil
55	27.9	27.1	74.8	Nil
Fully mature	29.9	24.6	84.5	1.8

Precaution During Post Harvest Storage of Raw Nuts

Cashew nuts can be stored for up to 9-12 months in good condition provided the following conditions.

- Harvesting of immature nuts with the help of bamboo sticks is wrong practice that deteriorates the size, weight and quality of kernel. Therefore, only fully mature, fallen nuts on ground are collected and heaped at one place and then the nuts are removed from apple immediately when fresh.
- Raw nuts should be cleaned, if extraneous materials are adhered.

- Dry nuts in the sun for 2-3 days. Use either concrete floor or tarpaulins as the base. The objective is to reduce the moisture content up to 9 % for long term storage. Do not over dry the nuts as this could lead to overheating and damage to the kernel from leaching of CNSL in the kernel.
- After drying, nuts should be placed in 300 gauge polythene bags and tie the upper end of bag with jute string in order to airtight it and then the bag is placed in gunny bag and sew it with smooth jute string. Such air tight bags are then kept in a warehouse in dry place and 3-4 inches above the ground in order avoid further spoilage due to contact of surface moisture especially during rainy season and also to reducing damage from pests.
- Pests in storage can cause damage to the crop. The shell of the cashew nut is hard and it checks to most pests in storage; however, the moisture content more than permitted level then there may be more chance of attack of store pest. Similarly, warehouse which is being used for dried nuts storage should be rodent proof. Proper precautions should be taken against rats.

□□□□

CHAPTER – 19

Marketing of Raw Nuts

Raw cashew nuts are a seasonal commodity and the trading season is from March to May. Growers usually supply the primary or village markets where small traders collect and supply the urban markets. The cashew trade is seldom handled by exclusive traders. Usually, those traders who collect other plantation products also trade in cashew. Due to the highly competitive nature of the cashew trade, growers have few marketing problems. When large quantities are collected by middlemen, the processors enter the marketing chain and make wholesale purchases. Grades and standards for cashew are yet to be introduced in India. Quality is generally determined by appearance and cutting tests that traders employ prior to purchase. The raw cashew nut market involves a large amount of capital where nearly 80 per cent of the produce is transacted within a period of 35 days. The current value of Indian production is estimated at around Rs. 10,000 million. This capital is made available by industry for procurement and processing operations.

There are no growers' cooperatives or organizations for cashew marketing. In Kerala, however, the government has been involved in the procurement process and supply to large-scale processors. This adversely affected the cashew trade and has now been replaced by a free market policy.

In addition to the local production of nearly 7 lakh tonnes, India also imports a considerable quantity of raw nuts from several African and South-east Asian countries to satisfy the national processing capacity of about 12-14 lakh tonnes established in the country.

Marketing of Raw Cashewnuts

Marketing of raw cashewnut in India has not yet been organized in systematic manner except in Goa where co-operative marketing society is procuring raw nuts to the little desired extent. A major portion of the produce is brought by itinerant merchants and the agents of the processing units. A number of wholesale merchants and the processing factories open their collecting centres in important cashew producing areas during the harvesting period. The petty dealers who buy the nuts from the growers also dispose the nuts in these collecting centres. Cashewnuts are brought for sale to the assembling markets largely by the itinerant merchants. In certain areas, the most resourceful processors contact the producers thus avoiding the commission agents' role and enjoy good bargaining power by providing credit facilities to the producers.

As there are a number of intermediaries operating the field between the primary producer and the processing unit. The different costs and margins in the total spread between the producer and the processing unit are quite significant and the producers share in the price paid by the processing units generally low.

Role of BAIF under 'WADI' Yojna

In Gujarat, with spread of developmental activities by Bhartiya Agro Industries Foundation (BAIF-Dhruva) working as a big NGO countrywide has implemented 'WADI' Yojna for the tribal farmers growing mango and cashew. The institute collect the raw cashew nuts at the reasonable price and then process through the co-operative society or by itself at various centres. This helps to uplift the socio-economic status of tribal growers along with providing training on knowledge sharing with experts.

□□□□

CHAPTER – 20

Nutritive Value of Cashew Kernel and Apple

Cashew Kernels

The role of balanced diet in contributing to a long and healthy life is undisputed. Diet related diseases are quite common even among the people of the affluent West. Unlike in developing countries, where many diseases are caused by deficiency in the quantity and / quality of diet. The diet related diseases among affluent populations result from excessive consumption of undesirable foods. Over consumption of foods of animal origin like meat and eggs are implicated in the causation of atherosclerotic heart diseases. Annual food consumption has also been suggested as possible aetiological factor in some cancers. After years of research, scientists are finding health benefits in a largely plant based diet emphasising foods low in saturated fats. On the basis of the awareness that excess animal food consumption is harmful, recent dietary guidelines all over the world recommend a reduction in the intake of meat and meat products and a liberal increase in consumption of fruits and nuts. Like meat and eggs, nuts are abundant in proteins, the body building material essential for growth and maintenance of tissues. Nuts are also a concentrated source of energy, since they contain liberal amounts of fat.

As all other tree nuts, cashewnuts are excellent dietary supplements in the human diet; protein is present in abundance and that too of a good quality. Cashewnut provides a rich blend of minerals and many water soluble vitamins. A regular serving of cashewnut will supplement the micro nutrient intake of individuals. The fat content in cashewnut makes it an energy rich food and the fatty acid profile is ideal for human consumption.

A significant feature in favour of cashew is its delightful taste and that it can be incorporated into a wide range of cuisine. Plain cashew kernels are adaptable to any form of cooking or cuisine. It is extensively used in Indian and Chinese food and moderately in other countries.

Cashew kernel contains proteins (21%), carbohydrates (22%), fat (47%), minerals and vitamins. Cashew kernel proteins contain all the essential amino acids and are comparable with other nuts like almond. Cashew kernel proteins are rich in acidic amino acids (38.78%). The major basic aminoacids, leucine and arginine are present to an extent of 22.23%. Cashew kernel proteins are comparable with casein in terms of protein efficiency ratio (PER). Protein efficiency ratio of cashew kernel proteins is 3.2 which are comparable with that of milk protein casein. Cashew kernels do not contain any anti-nutritional factors. Proteise inhibitors such as trypsin inhibitor have not been detected. Cashew kernels contain 47% fat which is quite rich in unsaturated fatty acids. The major fatty acids present in cashew kernel fat is oleic acid (73.4%) followed by linoleic acid (11.9%). Stearic acid is the major saturated fatty acid present (11.9%). Ratio of unsaturated to saturated fatty acid is 5.9. Cashew kernels are free from cholesterol and contain sizeable quantity of mono unsaturated fatty acid (oleic acid) which is now believed to be as efficient as poly unsaturated fatty acid in lowering blood cholesterol. Besides these, cashew kernel contains about 1.3% dietary fibre. Cashew kernels provide more energy compared to animal food (147-272 κ cal/100 g) and fish (234 κ cal/100 g). Cashew kernels provide energy of 611 κ cal/100 g and this is very much equal to that of almond (612 κ cal/100 g).

Cashew kernel besides contains protein, fat and carbohydrates, it contains number of inorganic elements like sodium, potassium, calcium, magnesium, phosphorus, iron, copper, zinc, manganese, selenium and chlorine.

Table 36: Fatty acid composition of cashew kernels (%)

Saturated	Stearic	11.24
Mono unsaturated	Oleic	73.73
Poly unsaturated	Linoleic	7.67

Table 37: Amino acid composition of cashew kernel protein (%)

Glutamic acid	28.00
Leucine	11.93
Isoleucine	—
Alanine	3.18
Phenylalanine	4.35
Tyrosine	3.20
Arginine	10.30
Glycine	5.33
Histidine	1.81
Lysine	3.32
Methionine	1.30
Cystine	1.02
Threonine	2.78
Valine	4.53
Tryptophan	—
Aspartic acid	10.78
Proline	—
Serine	5.76

Table 38: Vitamin content (mg/100 g) of cashew kernels

Thiamine (B_1)	0.56
Niacin (pp)	3.68
Tocopherol (E)	210
Riboflavin (B_2)	Traces
Pyridoxine (B_6)	Traces
Axerophtol (A)	Traces
Vitamin D	Traces

Table 39: Mineral content of cashew kernels (%)

Ca	0.040
Fe	0.008
P	0.880
Cu	0.002
Na	0.005
Zn	0.004
K	0.570
Mn	0.002
Mg	0.280

(*Source*: Booklet-Nutritive value of cashew published by NRCC, Puttur, Karnataka)

Cashew kernels are rich in potassium and phosphorus. Potassium is known to be essential for upkeep of human kidney. Selenium present in cashew kernels could help in protecting against lung, liver, skin, brain and gastro-intestinal cancer.

Cashew kernel contains sizeable quantity of vitamin E, a naturally occurring anti oxidant (210 mg/100g) and few water soluble B vitamins such as thiamine, riboflavin, niacin, biotin, folic acid, vitamin B_6, B_{12} and pantothenic acid.

Comparison of Cashew with other tree nuts

Cashew kernel compare well with almond with respect to protein, carbohydrate, fibre and mineral matter. Compared to hazelnuts, walnuts and macadamia nuts, cashew kernel contains higher protein, carbohydrate and mineral matter. Fatty acid composition of cashew kernel oil compares well with that of almond and macadamia nuts. Compared to all other tree nuts, cashew kernel oil contains higher quantity of mono unsaturated fatty acids except hazelnut (Table 40, 41).

Table 40: Kernel composition of different tree nuts (%)

Constituents	Almond	Hazelnut	Walnut	Macadamia nuts	Cashew
Moisture	5.2	-	4.5	1.5-2.5	5
Protein	20.8	12.7	15.6	9.2	21.0
Fat (Ether Ext.)	59.9	60.9	64.5	78.2	47.0
Carbohydrate	10.5	17.7	11.0	10.0	22.0
Fibre	1.7	-	2.6	-	1.3
Mineral matter	2.9	-	1.89	-	2.4

Table 41: Fatty acid composition of kernel oil of different tree nuts (%) of total fatty acids

Constituents	Almond	Hazelnut	Walnut	Macadamia nuts	Cashew
Lauric (12:0)	-	-	-	0.62	-
Myristic (14:0)	0.2	2.2	-	0.75	-
Palmitic (16:0)	8.9	3.1	7.0	6.15	0.9
Palmitoleic (16:1)	-	-	-	19.11	-
Stearic (18:1)	4.0	1.6	3.0	1.64	11.24
Oleic(18:1)	62.5	88.1	30.0	67.24	73.73
Linoleic(18:2)	24.4	2.9	57.0	1.34	7.67
Linolenic(18:3)	-	-	2	-	-
Arachidic(20:0)	-	-	-	3.30	-
Lingoceric(24:0)	-	-	-	-	0.15
Unsaturated	86.9	91.0	89.0	87.69	81.4
Saturated	13.1	6.9	10.0	12.46	12.29
Mono unsaturated	6.63	13.19	8.9	7.04	6.62
Poly unsaturated	2.6	30.4	0.51	64.4	9.6

After years of research, scientists are finding health benefits in a largely plant based diet emphasising foods low in saturated fats. Tree nuts, like Cashews, Almonds, Pistachios, Walnut, Hazel nuts, Macadamia nuts and Pecans are low in saturated fats and packed with micro nutrients and substances known to help lower cholesterol, prevent cancer and prolong life.

On the basis of the awareness that excess animal food consumption is harmful, recent dietary guidelines all over the world recommend a reduction of intake of meat and meat products and a liberal increase in consumption of fruits and nuts. Like meat and eggs, nuts are abundant in proteins, the body building material essential for growth and maintenance of tissues. Nuts are also a concentrated source of energy, since they contain liberal amounts of fat (Table 42) (Varma, 2000).

From Table 43, it is reveals that the essential amino acids content of the cashewnut protein compares favourably with that of other nuts. In fact, cashewnut protein is richer in five essential amino acids when compared with rice protein, rated as one of superior vegetable proteins. However, when compared with animal proteins, cashewnut protein contains relatively lower amounts of lysine, threonine and methionine.

The limiting amino acid thus appears to be methionine, the sulphur containing amino acid. This in no way renders cashewnut undesirable since the major constituent of diet will provide adequate amounts of methionine. Including

Table 42: Nutrient values of selected animal foods and nuts

	Composting/100 g						
	Food	Edible portion	Water (%)	Protein (%)	Fat (%)	Carbohydrates (%)	Energy κ cals
Animal Food	Chicken egg	100	75.1	12.5	10.8	Trace	147
	Beef (Sirloin)	92	59.4	16.6	22.8	0	272
	Chicken, meat & skin	64	64.4	17.6	17.7	0	230
Fish	Herring (raw)	55	63.9	16.8	18.5	0	234
Nuts	Almond	100	4.2	21.1	55.8	6.9	612
	Cashew	100	2.4	20.5	50.9	18.8	611
	Peanut	100	1.9	24.5	53.0	7.1	602

Table 43: Essential amino acid content of selected foods

	Beef	Egg-chicken	Almond	Cashew	Peanut	Rice (raw)Milled
Approx. Total N g/100 g	3.6	2.13	3.33	3.39	4.05	1.09
Arginine	410	40	660	650	690	480
Histidine	200	150	140	130	140 .	130
Lysine	540	440	160	290	230	230
Tryptophan	70	90	50	110	60	80
Phenylalanine	260	360	300	270	310	280
Tyrosine	220	250	180	-	240	290
Methionine	160	210	100	90	60	150
Threonine	280	320	170	200	170	230
Leucine	510	520	480	510	400	500
Isoleucine	320	410	280	320	240	300
Valine	330	450	310	360	280	380

Note: Amino acid concentration expressed as mg/g of N

cashew in the daily diet (at least one serving a day) would ensure an excellent supply of protein, simultaneously providing liberal amount of energy. The replacement of animal foods brought about by substitution with cashew will enable a reduction in the risk of atherosclerotic heart disease (Varma, 2000).

Value Addition in Cashew

Commercial processing of cashew either by steam roasting or drum roasting results in kernels of different grades such as wholes, bits, pieces, etc. Cashew kernel baby bits are obtained to an extent 0.5 to 1 per cent when nuts are processed by steam roasting. Attempts have been made at National Research Centre for Cashew, Puttur, Karnataka to develop value added products from the cashew kennel baby bits. Milk and spread could be prepared from cashew kernel baby bits. Cocoa flavoured and sweetened milk could be reconstituted from the cashew kernel paste prepared and storage of this paste up to 34 days at <0°C had no effect on organoleptic quality and composition of the milk prepared (Table 44, 45).

Table 44: Protein and sugar content of milk prepared from kernel paste during storage at <0°C.

Storage (period days)	A (mg/5 ml)		B (mg/5 ml)		C (mg/5 ml)		D (mg/5 ml)	
	Protein	Sugar	Protein	Sugar	Protein	Sugar	Protein	Sugar
0	20.3	11.8	22.0	10.6	22.4	20.2	23.3	21.0
7	20.2	15.9	20.5	15.1	21.2	31.7	23.8	31.5
14	20.3	18.2	24.7	19.5	24.4	21.9	24.7	21.9
25	23.3	18.0	23.8	18.9	25.5	30.8	23.1	34.4
34	23.5	5.3	24.9	5.7	21.2	8.5	20.0	10.1

Values are mean of three individual estimations

A - Paste prepared from cashew baby bits alone.

B - Paste prepared from cashew baby bits with sodium benzoate as a preservative.

C - Paste prepared from cashew baby bits with cane sugar and sodium benzoate.

D - Paste prepared from cashew baby bits with cane sugar.

Table 45: Cumulative mean scores for colour, flavour and taste of sweetened and cocoa flavoured milk prepared from kernel paste during storage at <0°C.**

Storage period (days)	Mean cumulative Hedonic score			
	A	B	C	D
0	9.1	9.9	10.4	10.0
7	11.1	10.2	9.9	10.2
14	10.8	10.2	11.6	12.1
25	9.5	9.58	10.3	10.2
34	9.2	9.50	9.8	10.4

Values are mean of three individual estimations

A - Paste prepared from cashew baby bits alone.

B - Paste prepared from cashew baby bits with sodium benzoate as a preservative.

C - Paste prepared from cashew baby bits with cane sugar and sodium benzoate.

D - Paste prepared from cashew baby bits with cane sugar.

Cashew Apple

The pseudo fruit which otherwise called cashew apple is a juicy fibrous fruit which is quite nutritious. Cashew apple contains sugar, amino acids, tannins, ascorbic acid (Vitamin C) and fibre. Cashew apples are very rich in ascorbic acid (240 mg/100 g) which is almost six times that of citrus fruits (40 mg/100 g) and also known to be very rich source of vitamin C. Besides vitamin C, cashew apple contains free soluble sugars, most of which are reducing sugars. Cashew apple is rich in fibre and on a dry weight basis; the fibre content varies from 15 to 18 %. Phenols and flavonols present in cashew apple could serve as natural antioxidants which play a major role in destroying free radicals. Cashew apples are good source of Vitamin C and fibre. Consumption of cashew apple could help in overcoming the vitamin C deficiency and also constipation. Comparison of the nutritive value of cashew apple with other fruits is given in Table 46 and information of nutritive value of cashew kernel and cashew apple is given Table 47.

Table 46: Comparison of the nutritive value of cashew apple with other fruits (per 100g)

Name of fruit	Moisture (g)	Protein (g)	Fat (g)	Calcium (mg)	Iron (mg)	Carotene (µg)	Vitamin C(mg)
Cashew	86.1	0.2	0.1	0.2	0.4	23	261.5
Apple	84.6	0.2	0.5	10	0.6	0	1
Banana	70.1	1.2	0.3	17	0.36	78	7
Orange	87.6	0.7	0.2	26	0.32	1104	32
Mango	81.0	0.6	0.4	14	1.3	2743	16
Papaya	90.8	0.6	0.1	17	0.5	666	57

Table 47: Nutritive value of cashew kernel and cashew apple

Sr.No.	Important constituent	Cashew kernel	Cashew apple
1.	Moisture (%)	5.9	87.9
2.	Protein (%)	21.0	0.2
3.	Fat (%)	47.0	0.1
4.	Carbohydrate (%)	22.0	11.6
5.	Salt (%)	2.4	0.2
6.	Calcium (%)	0.55	0.01
7.	Phosphorous (%)	0.45	0.01
8.	Iron (mg/100 g)	5.0	0.2
9.	Vitamin C (mg/100 g)	630.0	261.5
10.	Riboflavin (mg/100 g)	190.0	0.2
11.	Carotene (I.U./100 g)	100.0	-

The nutritive value of cashew kernel and cashew apple indicates the importance of cashew in our diet (Table 47). However, the increasing price day by day of cashew kernel makes it difficult to reach poor's middle class families.

The importance of a varied and nutritionally sound diet is being increasingly recognized. An ideal diet should include liberal amounts of vegetable foods. Nuts have an important role in the daily diet of individuals irrespective of the culture and race. Two to three servings of nuts and fruits daily are recommended by most nutritional committees to ensure that people consume healthy diets.

As already discussed in this chapter, cashewnuts are excellent dietary supplements in the human diet. Protein is present in abundance and that too of a good quality. Cashewnut provides a rich blend of minerals and many water soluble vitamins. A regular serving of cashewnut will supplement the micro nutrient intake of individuals. The fat content in cashewnut makes it an energy rich food and the fatty acid profile is ideal for human consumption.

The high protein content and energy density of cashewnut makes it an ideal supplement in the diet of children, pregnant women, lactating mothers and convalescing subjects. Consumption of cashewnut primarily as a protein and energy source as an alternative to meat and meat products will reduce the risk of heart disease.

A significant feature in favour of cashew is its delightful taste and that it can be incorporated into a wide range of cuisines. Plain cashew kernels are adaptable to any form of cooking or cuisine. It is extensively used in Indian and Chinese food and moderately in other cuisines. It blends well with ice-cream, cookies, biscuits and confectioneries. Cashewnut is easily consumed in plain form and delectable while offered in roasted form whether in oil or dry roasted. It is also available in versions like honey roasted and other forms of coating like chocolate - rum coated etc. The incorporation of cashewnut in to the regular diets would enhance the nutritive qualities of the diet while improving palatability (Varma, 2000).

□□□□

CHAPTER – 21

Cashew Nut Processing

Brief History of Cashew Nut Processing Industry in India

As is known in the Cashew Trade Circles, cashewnut was not an indigenous product of India. A few hundred years back, the Portuguese brought it to Goa in India, from Brazil. The Portuguese also took it to their other colonies in Africa *viz*. Mozambique on the East Coast and Angola on the West Coast. In Mozambique, it spread all over the territory i.e. from Lourenco in the South to Ibo in the North and thence to British East Africa (now Tanzania) and Kenya. Though, now it is known as cashewnut, still in Mozambique and India, it is known by its original Portuguese name 'CAJU'.

For several years, it was a completely neglected crop in India. From Goa, it spread South-wards to Karwar Districts and thence to Kerala and Northwards to Ratnagiri District. The consumption was limited to the areas, where it was produced. Till the year 1915 or so, the people in Goa/Ratnagiri and Sindhudurg used to roast the raw cashew-nuts in open fire, which produced many scorched kernels. Scorching never mattered since the cashews were for home consumption only. There was also another method to take out the kernels and it was to dry the cashewnuts in the sun till they became reddish and then to strike the nuts with a handle. The kernels thus procured were with red skin and such kernels were called MOULA' and sold for years together at the rate of Annas (1/16 of Rupee before change over to decline currency) 3 to 4 per lb. The green raw cashew nuts were being opened by some pointed small sticks and they were being sold by the farmers at about 120 to 125 pieces per Anna and a delicious vegetable preparation called '*USAL*' was prepared by applying masala. Cashew Apples were totally neglected, except some good fruits being eaten by people of the area and the rest were being wasted or consumed by monkeys, bats, and birds. The same case was with the Cashew Nut Shell Liquid and the shells were used as fuel but afterwards, oil was extracted by country methods and mainly used for painting the country crafts to avoid occurrence of leaks in the crafts, called 'CHOPDAN' by the mariners of Ratnagiri/Goa. The above-mentioned state of affairs continued for a very long time (Joshi, 1980).

In the early period of 1920th Century, M/s. Torneys of Vengurla and M/s. Mahadev Bapu Prabhu, Zantye of Malvan (Sindhudurg District of Maharashtra) started to take interest in this commodity from business point of view and they started the industry at Vengurla and Malvan, respectively and small parcels began to come to Bombay for safe. In the initial stage, the parcels used to come in natural condition with the red skin, as it was presumed that the skin will enable the inside kernel to avoid deterioration for some time. But as there was no air-tight packing, the goods used to perish shortly. Anyway, Bombay began to take some interest and this encouraged the factories to continue the trade (Joshi, 1982).

In the year 1921 or 1922, the well-known Dr. Goheen, who was looking after the Vengurla Mission Hospital, which was a branch of the famous Miraj Mission Hospital, was presented with a parcel of cashew kernels by his friend in Vengurla and he took delight in the taste of cashews. He, therefore, procured some more parcels of peeled cashew kernels and despatched them to his friends in New York. Dr. Goheen was informed that the taste was very good and could find a market in the USA. This story was told by his son late Mr. Ridiard Goheen, who was In-charge of Goa Office during 1937/38, of the Vitapac Corporation, the Indian Branch of M/s. Becker-Bennett-Day, the old name of the present General Foods Corporation. In the year 1923 or 1924, Mr. Dajisaheb Torney of Vengurla ventured to send his first consignment of cashew kernels to New York. The steamers used to take 45/50 days to reach New York. In the absence of proper packing the consignment did not stand the hazards of transit resulting in the goods being infested with weevils. In this connection, Mr. Rajaram Torney, son of late Shri Dajisaheb Torney had been to New York and he was perhaps the first Indian to go to New York in connection with the Cashew Trade. No shipment was thereafter sent to New York till about 1928. During the period, Bombay trade had somewhat developed and attention was paid to local trade both by Torneys and Zantye, Some improvements were made in the packing and the goods were able to remain in good condition for about 3/4 months. M/s. Laxmidas Shamji and M/s. Narsi Mulji (both Kachchh – Gujarati) were the leading merchants in Bombay and they used to send small parcels to North Indian markets, where cashew kernels were not known and consumption was very limited till about 1945/46. Afterwards, Vithaldas Haribhai (Gujarati) took interest in the sale of cashew kernels. They are still one of the leading merchants in the line in Bombay (Joshi, 1982).

Mr. S.G. Nervekar, who was in charge of Cashew Trade of Narsi Mulji, negotiated with Zantyes of Malvan to start foreign trade and joined them afterwards in the year 1925 or so, In the meantime, M/s. T. M. Duche & Sons and M/s. Zenobia Company started correspondence with Zantyes through

M/s. Jacques Missir, New York about Cashew business. Zenobias sent their first representative *viz.* Mr. Abdullah Hougaz (accompanied by his sister Miss Helen Hougaz) to India during the year 1928. The negotiations materialised and The Second Consignment of Cashew Kernels in airtight packing was sent to New York during the year 1928 or 1929 by M/s. Mahadev Bapu Prabhu Zantye, in three grades *viz.*, Cashew Kernels White Wholes (Price Br. Sh. 125/- per Cwt) White Pieces (Price Sh. 90/-) and Wholes Natural i.e. with skins. The packing was in airtight tins of 25 lbs and such two tin to a wooden case. The consignment fared nicely. With success marking the export of this second consignment, enquiries began to pour in from New York and M/s. Becker-Bennett-Day (now General Foods Corporation) started taking keen interest. They sent two representatives to India, to encourage the existing factories and start the industry on bigger scale. One of them was the Famous Lady, Miss Josephine Hougaz, who was joined by her brother Mr. Abdullah Hougaz and started the branch of Becker-Bennett-Day in Goa (Joshi, 1982) under the name and style of 'Vitapack Corporation'. They looked after the factories at Goa, Malvan and Vengurla. Zantyes and Torneys had established factories in Goa by 1930, while a new comer M/s. Shantaram Govind Kamat came on the stage during 1930 or 1931. They were also in the sphere of Goa Office of Vitapack (Joshi, 1982).

At the same time Quilon in Kerala was found to be a suitable place for Cashew Industry. A couple of factories were already functioning there, one being the India Nut Factory of Saminathaa Becker-Bennett-Day sent Mr. Lindsey Johnson there as their resident-Representative at Quilon. During 1930-31 two more cashew factories were established at Vengurla *viz.* N. M. Teli and Gopal Pandurang Shirodkar but their trade was limited to Indian markets only till about 1948, Torneys started a factory at Vetapalem on the East Coast. Vetapalem is still a small centre of Cashew Industry, the production being sent to Indian markets (Joshi, 1982).

During 1931, M/s. Pukhade Brothers, Commission Agents at Bombay started sending Cashew Kernels consignments of Zantyes to London and Liverpool to the care of M/s. William H. Mackerrow and M/s. Landauer & Co. of 39, Eastcheap, London E.C.3. During the years 1932 to 1935, the Napack Corporation, introduced a new method of packing cashew kernels in airtight (an atmosphere of carbon-dioxide gas, called 'Vita-pack Packing'). The prices also were changed from the British Sterling to Indian Rupees. The cashew kernels were divided into so many grades viz. W 210, W 240, W 280. W 320, W 400, W 450, W 500, SW, and also Butts, Splits, etc. Regular parcels began to go to USA in big quantities and small consignments to European countries mentioned above. In the year 1937, Zantyes introduced Cashews to Rotterdam and Amsterdam and booked their orders of 25 C/s each as trial business on the Diwalee-Day as Muhurt-business. During these years, Quilon and Mangalore had also developed their cashew business to a great extent while Goa/Ratnagiri lagged behind. Many cashew factories were established in Quilon and vicinity and prominent among them were those of Peirce Leslie and Musaliars. Peirce Leslie had, in the meantime, started extracting Cashew Shell Liquid and to export the same in small quantities and Musaliars followed suit (Joshi, 1982).

During the World War II, and especially after the tragedy of Dunkirk, the US business circles in India feared about immediate invasion by Germany over England and they started to wind up or curtail their business activities in India. Accordingly, Mr. Phillip W. Swan, the then President of the General Foods & Sales Corporation, personally came to India during 1940 and closed the offices of Vitapack Corporation. Shortly afterwards, US herself was involved in War and all cargo steamers were commandeered by US Government. Cashew shipments came to a stop by about the end of 1941. Due to war conditions, it was also risky to send goods to other countries and foreign trade of Cashews came to a complete stop. Stocks accumulated in India and the Bombay and other important markets were dumped with cashews and sold at ridiculously low rates. Some Arab parties had purchased Cashew Kernels from Goan parties at Rs. 30/- per case of W 320 and took them to the Middle East by crafts. Imports of Raw Cashewnuts from Portuguese to East Africa were also stopped, though during 1943 Dar-e- Salaam used to send some consignments by crafts (Joshi, 1982).

By the end of 1943, the US Government allowed imports of Cashew Kernels subject however, to the condition that the parties exporting Cashews should supply to the US Govt. 1 lb. of Cashew Shell Liquid, (which was a strategic war material at the time) for every lb. of cashew kernels. Peirce Leslie and Musaliars had already started extraction of Cashew Shell Liquid by machinery and Zantyes started extracting the oil by country method, which was somewhat inferior but suitable to meet the US requirements. Some other Quilon factories also had started extraction of Cashew Shell Liquid and from 1944, the export of Cashew Kernels and Cashew Shell Liquid was resumed in full swing in spite of U.S. being still in war with Japan in the Far East. The prices improved. Seeds from both the Africa's were available in plenty and by the year 1947/48, the exports to US amounted to about eight lakhs of cases. Further developments were known to the Cashew-trade circles in India and elsewhere.

During the late thirties, many dry fruit brokers and Importers in New York started taking keen interest in Cashew Trade and besides M/s. Duche & Sons; M/s. Braun Importing Co., Zalooms etc. came forward as leading importers. M/s. Mitchel Beck Company became the leading brokers and controlled the major portion of the business. After India's Independence, Indian Government started taking keen interest in this Dollar-earning commodity and new markets were created in other countries. Government of India is still taking keen interest in promoting exports of Cashew kernels. U.S.S.R. is now the second important market for Cashew Kernels. However, it will be interesting to note in this connection that during 1950 or 1951, Zantyes had written to Russian Consulate in Bombay making a proposal for Cashew business and that a negative reply that 'this product will NOT find a market in Russia' was received by them (Joshi, 1982).

Cashew Processing

India has been a pioneer in cashew processing and registered as the first country to hit the world market with cashew kernels. Cashews are cultivated mainly in Kerala, Maharashtra and Goa. Cashew processing is a well established activity in Kerala and Goa, but it is yet to pick up in Maharashtra. Kokan region of Maharashtra is famous for cashews and Alphonso mangoes. It is also attracting number of tourists due to its vast coast line and scenic beauty. Konkan railway has made this region easily accessible. Most of the cashew processing is undertaken manually and reportedly there are few mechanised factories. There are some colleges in Ratnagiri imparting vocational training in cashew processing. Thus, cashew processing has good potential in the region.

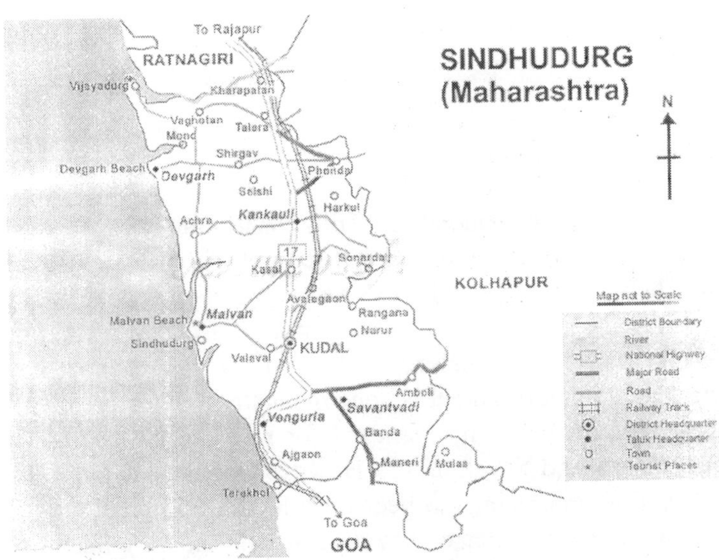

Fig. 41: Map of Sindhudurg District of Maharashtra

Cashew is a unique crop which provides nutritious food for health, row material for industry, economic income for rural masses and above all a good amount of foreign exchange to the national exchequer.

Currently cashew is grown both in the East and West Coast of India and provides employment to over 5 lakh workers in 3650 processing factories in India. Today, the Indian Cashew Industry demands 14-15 lakh tones of raw nuts to meet the present installed capacity. This means, the current level of production in India (7.0 lakh tones) is only half the requirement which needs to be increased at least by two times. However, to run our processing units reasonably, economically and make them viable, our country imports raw cashewnuts from African countries.

The status of Indian cashew industry is presented in Table 48.

Table 48: Cashew Industrial Scenario

State	Processing Units (Nos.)	Processing capacity ('000 MT)
Kerala	432	700
Karnataka	266	65
Goa	45	21
Maharashtra	2200*	20
Tamil Nadu	417	565
Andhra Pradesh	175	95
Orissa	60	11
West Bengal	30	08
Chhattisgarh	03	-
NE States	22	15
Total	3650	1500

**/750 Small scale cottage industries.*

Even today, the quality consideration of raw cashew nut is very poor. The nuts dropped off from the trees are collected, dried and sold to processors for kernel extraction. There is no variety wise or size wise grading is done before selling the raw cashew nuts. There is no proper non-destructive method has been developed to assess the moisture content of the raw cashew nuts. After processing, the grading of kernel is done manually to fetch premium price in the market. Moreover, quality standard for raw cashew has not been yet standardized (Saroj and Nayak, 2013). However, methods to judge the quality of raw nuts before processing has been described for different tests are being adopted to judge the qualities of raw nuts as suggested by Salam and Peter (2010) is as follows. Cutting test, floating test, count test and test for moisture content enable us to judge the quality of raw nuts.

1) Cutting test

This test is a physical method of quality evaluation. It involves cutting samples of raw nuts longitudinally and physical examination of the kernels on their fillage and quality. Normally 1 kg of raw nut will be subjected to cutting test, to examine the extent of white kernels, damaged kernels, discoloured ones and the kernel out turn. In factories, raw nut price is fixed based on cutting test. The recovery of good quality kernel should not be less than 25% to qualify the nut as good quality.

Kernel recovery is low, if the nut sample contains more dried immature nuts. Fungal infection, thrips damage, heat stress etc. cause drying up of immature. Panicles and tender nuts dry up due to tea mosquito bug at early stages of development and they may not contain good quality kernel. Therefore, such nuts should be separated from good quality ones. Depending on the extent of pest infestation and stage of maturity, the damage to kernel differs. Good quality nuts will contain white kernels and it will be full of kernels. Bad quality nuts will contain no or deteriorated kernels.

2) Floating test

This test is another method commonly use for quality evaluation. Dense nuts usually contain good quality kernels and they sink in water. Poor quality nuts with poor kernel development or damaged kernel may float while dipping in water. The percentage of floats in raw nuts in a sample indicates the quality of raw nut. If the number of floaters in a sample is more than 24 per 100 nuts, such lots may be treated as bad quality.

3) Count test

This refers to the number of nuts per kg of raw nut. It may vary from 100 to 300 nuts/kg. More the countless will be the kernel size.

4) Moisture content

The allowable level of moisture content in raw nuts is 8 to 10%. Raw nuts having moisture above 10 per cent may indicate signs of low quality.

Harvesting immature nuts, improper drying, improper storage of dried nuts, prolonged storage (beyond 12 months), exposure to moisture etc. are some of the causes of quality deterioration in raw nuts.

According to Prabhu (2001), the following quality parameters are to be considered while importing raw cashew nut.

(a) Count of raw cashew nuts/kg: not less than 180 nuts per kg

(b) Number of floaters/100 nuts: not more than 24

(c) Percentage of bad nuts in floaters: not more than 15%

(d) Total kernel yield: not less than 25%

Drying Raw Nuts

Drying seeks to reduce the moisture content to facilitate storage without rapid deterioration. Drying is done for 2-3 days only, moisture loss at this stage ranges between 3 to 10 per cent depending on the time of harvest. However, excess drying at high temperature (above 35^{0}C) may lead to reduced kernel quality. At the time of harvesting, the raw nuts may contain 12 to 16% of moisture. Since the raw nut processing is an activity spread over different months, it is necessary to store the nuts over 6 to 12 months before processing. To prevent damage of kernels during storage, the moisture content of the raw nut is to be kept below certain critical level. Sun-drying of raw nuts for 2-3 days may bring down the moisture content to safer level (9-12%). Well-dried nuts when pressed with hand produce a shrilling noise indicating that they can be safely stored.

Storage of Raw Nuts

To minimise damage, it is necessary to store nuts in moisture proof store houses. Usually dried raw nuts are packed in gunny bags, which can hold 40-80 kg nuts (Fig. 43). It is not advisable to store the gunny bags on the floor as it may absorb moisture from the soil and cause damage. It is desirable to keep the bags on racks keeping a minimum distance of 30 cm from the ground and walls of the store. A single layer of sand bags are used to spread on the floor, to avoid direct contact with ground surface. The store should be dry and free from insects, birds, squirrels, rats etc. Storage in well ventilated warehouse is also a pre-requisite for good quality kernels.

Fig. 42: Drying of raw nut

Fig. 43: Filling raw nut in bags after drying

Cashewnut Processing

Stages in Cashew Processing

Cashewnut processing covers all the stages from the drying of raw cashew nuts to the packing of processed kernel.

Roasting/Steaming

Roasting or steaming is employed to facilitate the removal of the shell in the subsequent process.

Roasting

Roasting can be done in two ways:

i. Drum roasting

This is one of the oldest and more widely used methods. The nuts are fed into a rotating red hot drum which will ignite the shell maintaining its temperature because of the burning of the shell liquid. The drum is kept in rotation for 3-4 minutes and the roasted nuts are discharged from the lower end of the drum and immediately covered by ash after sprayed with little water, so as to absorb the oil on the surface. This facilitates the removal of the remaining oil on the shell. It is one of the cheapest available methods though cashewnut shell liquid (CNSL) recovery is not possible and is lost by burning.

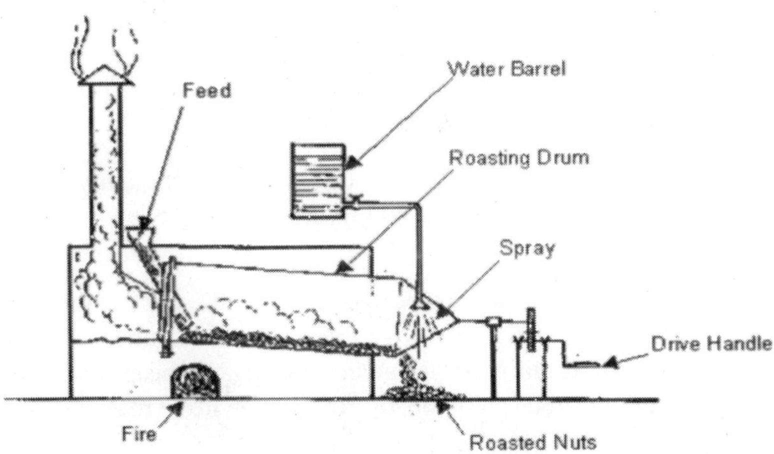

Fig. 44: Drum roaster fired from a furnace

Fig. 45: After Drum roasting, nuts are shelled manually by using mallets hammer may be of wooden.

ii. Oil bath roasting

In this method, conditioned nuts are passed through CNSL bath heated to 170-200^{0}C by conveyer buckets for 1-2 minutes during which period the shell gets heated rupturing the wall and releasing the oil into the bath. The oil is recovered by continuous over flow arrangement. The roasted nuts are centrifuged to remove adhering oil, cooled and shelled by hand and leg operated shelling machines. The kernel with the adhering testa is scooped out using a sharp needle. However, this method is generally not being used now.

iii. Steam bath roasting

Steaming is an alternative to roasting the cashew nuts. Well dried raw nuts are steam cooked at about 120-140 lbs/sq inch pressure. Appearance of white fumes and sweet smell during steaming gives good quality kernels. Excess boiling or steaming may lead to yellow coloured kernels and reduced quality. Shell oil can be extracted in later stages by crushing. The nuts are shelled by hand and leg operated shelling machines.

A general comparison of the above three methods would show that the oil bath method and steam roasting systems require more initial investment and higher maintenance costs: the drum roasting method being the cheapest. From the point of view of environmental pollution, the safest is the steam roasting method.

Decortification: Removal of the shell.

Shelling: Roasted (drum or oil roasted) nuts are shelled by workers using wooden mallets.

Cutting: Steamed nuts are cut by workers with hand and leg operated cutting machines called as cashewnut cutters.

Fig. 46: Boiler for roasting of raw nuts by steaming

Shelling/Cutting

Cashew nuts after roasting and cooling are to be shelled to remove kernels. One has to be very careful while shelling the nuts. Hands are to be protected from Cashew Nut Shell Liquid (CNSL) which is highly corrosive. Hand gloves should be used while shelling. For the same reason, it is advisable to dust the nuts with wood ash. Commercial processing units use foot operated shell cutters a mechanical device for shelling, which consists of a pair of blade (knives) shaped in the counter of half a nut which could be operated by foot. The blades cut through the shell all around the nut, leaving the kernel untouched. After shelling the kernels and shell, pieces are separated manually. The nuts have to be grouped into various sizes, each size matching a pair or blades of appropriate size.

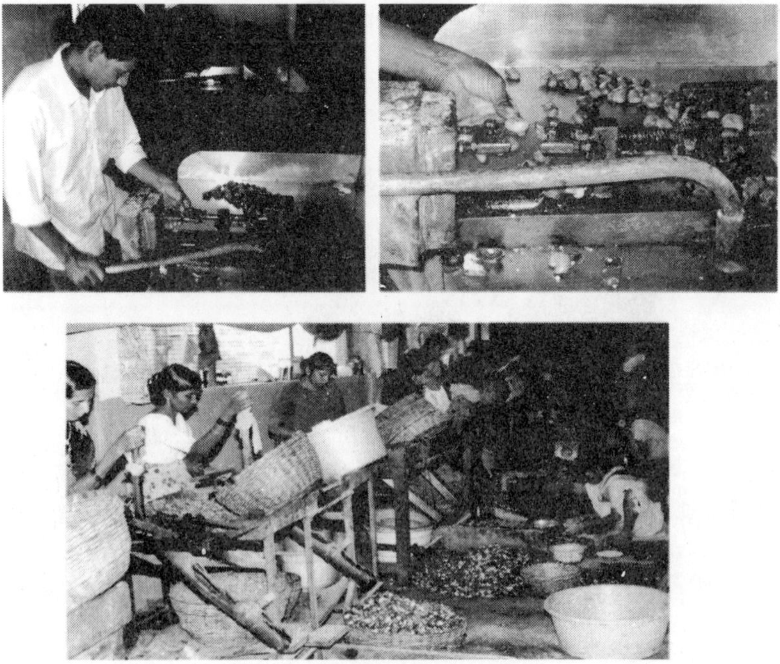

Fig. 47: Cutting of roasted cashew nut on shell cutters machine

Drying and Cooling of Shelled Kernels

The shelled kernels are dried in "*Borma*" (oven) drier at 80 to 90^{0}C. The process taken about 6 to 12 hours depending on the kernels and the types of Borma used. The old methods such as the 8 feet, 12 feet and the thattu Borma use direct application of heat generated using cashew shell as fuel. In Blower Borma hot oil is blown in to the chambers in which the kernels are kept. An Electric Borma is the recent arrival. Though the older method causes greater scorching, they are still widely used because of the higher investment and maintenance costs associated with recent innovations.

Drying makes the kernels harder, the moisture level being 5 to 6 per cent. This is done by drying the kernels in hot chambers at 70-80^0 C in perforated trays for about 6-8 hours. Uniform drying could be achieved with a cross flow drier using forced hot air circulation through the kernel layers. In order to ensure uniform drying, the position of the trays has to be changed frequently, as scorching may occur at hotter places. Excess drying of kernels leads to kernels becoming very brittle resulting in higher breakage. After drying, the kernels are kept in the moist chamber for 24 hours which facilitates easy removal of testa (peeling) and minimizes broken kernels. Proper drying and cooling is necessary to maintain the white colour of the kernels and to reduce excessive breakage in the subsequent processes.

Fig. 48: Dryer (outer view)

Fig. 49: Dryer (inner view)

Peeling

Fig. 50: Peeling of kernels (Removing testa)

This process involves the removal of testa (thin coat) from the kernel. Peeling is done using a sharp knife or bamboo piece. Care has to be taken while removing the testa. If kernels are scorched, more it results in poor quality kernels and afterward the initial grading as wholes and broken and by colour is done. The peeling worker has to be experienced and skilled if breakages are to be kept to a minimum. From 100 kg raw nuts, about 22-24 kg whole kernels can be obtained and the broken kernels will be about 3-4 kg.

Grading

Kernels undergo a final grading by hand and/or sieve (mesh) at standard specifications. Based on colour and appearance cashew kernels are classified as white scorched and dessert kernels.

Manual grading of kernels is generally adopted in cashew processing. In the International Market, bold whole kernels fetch premium price. The grading standards developed in India refer to white whole (undamaged) kernels and indicate the number of kernels per lb of weight. The largest kernels come in the grade W-180 and the smallest of the seven grades is W-500. Generally, Brazilian kernels have a relatively high proportion of large wholes, with the extra attraction of the 180's grade, known as Special Large Wholes. The white whole kernels are priced according to size. Further classification refers to broken kernels, butts, splits, pieces, small pieces and whether kernels are white or scorched.

Based on the shape, size and colour of the kernel, cashew kernels are graded into white or scorched wholes, pieces, splits, butts etc. The Government of India Act prescribes 33 different grades of cashew kernels of which only 26 grades are commercially available and exported. W-320 is the most popular among cashew kernels and also the most available, worldwide. Scorched wholes are another grade of cashew kernels. They are prepared by subjecting the kernels to a longer duration of roasting, whereby they attain a slight brown colour. They, however, have all the other characteristics of white kernels including nutritional qualities. Butts, splits and pieces are priced low and are ideal for cooking, preparation of sweets and savoury snacks and so on.

Whole : White wholes/ Scorched whole/ Dessert whole

Brokens : White brokens/Scorched brokens/ Dessert brokens

Fig. 51: Manual grading of kernels

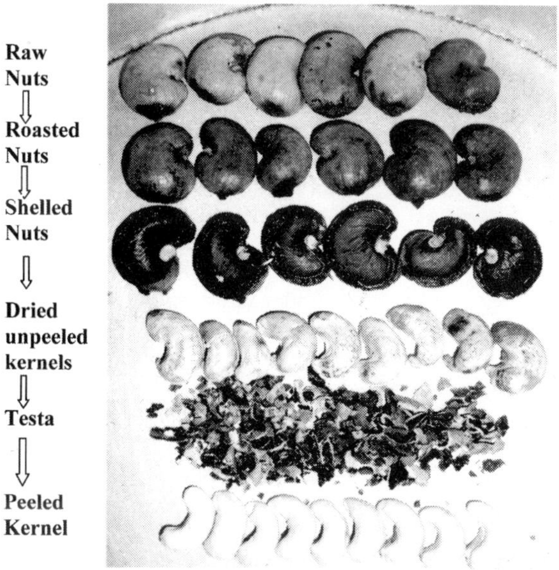

Fig. 52: Flow diagram of cashew nut processing (up to peeling)

The detailed gradation in cashew is given below.

Table 49 (A): Cashew kernels white wholes

Sr. No.	Grade Designation	Trade name	Colour/ characteristics	Count per 454 g size description	Max. moisture (%)
1.	W-180	White Whole	White/pale Ivory light ash characteristic shape	170 to 180	5
2.	W-210	White Whole	White/pale Ivory light ash characteristic	200 to 210	5
3.	W-240	White Whole	White/pale Ivory light ash characteristic	220 to 240	5
4.	W-320	White Whole	White/pale Ivory light ash characteristic	300 to 320	5
5.	W-450	White Whole	White/pale Ivory light ash characteristic	400 to 450	5
6.	W-500	White Whole	White/pale Ivory light ash characteristic	450 to 500	5

Table 49 (B): Cashew kernel scorched whole

Sr. No	Grade designation	Trade name	Colour/ characteristics	Count per 454 g size description	Max. moisture (%)
7.	SW 180	Scorched whole	Kernels may be scorched/slightly darkened due to overheating while roasting or drying in Borma)	170 to180	5
8.	SW-210	Scorched whole	Kernels may be scorched/ slightly darkened due to overheating while roasting or drying in Borma	200 to 210	5
9.	SW-240	Scorched whole	Kernels may be scorched/slightly darkened due to overheating while roasting or drying in Borma	220 to 240	5
10.	SW-320	Scorched whole	Kernels may be scorched/slightly darkened due to overheating while roasting or drying in Borma	300 to 320	5
11.	SW-450	Scorched whole	Kernels may be scorched/slightly darkened due to overheating while roasting or drying in Borma	400 to 450	5
12.	SW-500	Scorched whole	Kernels may be scorched/slightly darkened due to overheating while roasting or drying in Borma	450 to 500	5

Table 49 (C): Cashew kernel dessert whole

Sr. No.	Grade designation	Trade name	Colour/ characteristics	Count per 454 g size description	Max. moisture %
13.	SSW	Scorched whole secibs	Kernels may be scorched in alone shrivelled is coloured and light blue	N.A.	5
14.	DW	Dessert whole	Kernels may be deep brown, deep blue coloured	N.A.	5

Table 49 (D): Cashew kernel white pieces

Sr. No.	Grade designation	Trade name	Colour/ characteristics	Count per 454 g size description	Max. moisture %
15.	SSW	Scorched whole secibs	Kernels may be scorched in alone shriveled is coloured and light blue	N.A.	5
16.	DW	Dessert whole	Kernels may be deep brown, deep blue coloured	N.A.	5

Table 49 (E): Cashew kernel scorched pieces

Sr. No.	Grade designation	Trade name	Colour/ characteristics	Count per 454 g size description	Max. moisture %
17.	SPS	Scorched pieces second	Kernels may be over scorched immature shriveled discoloured and light blue.	Kernel broken in to pieces but not passing through 4 mesh sieve	5%
18.	DP	Dessert pieces	Kernels may be deep scorched, deep brown, deep blue discoloured and black spotted	Kernel broken in to pieces but not passing through 4 mesh sieve	5%

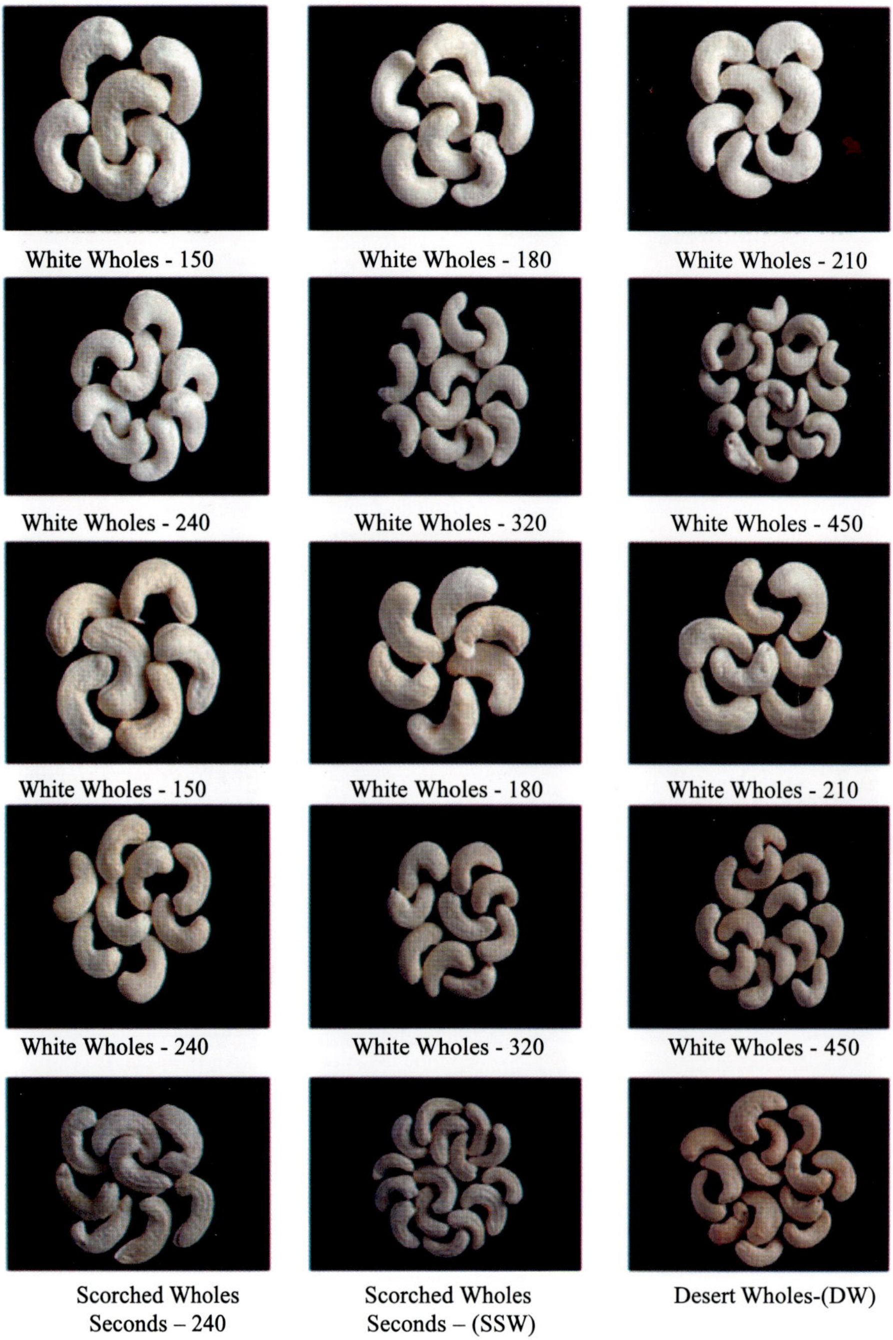

Fig. 53: Grade specification in cashew

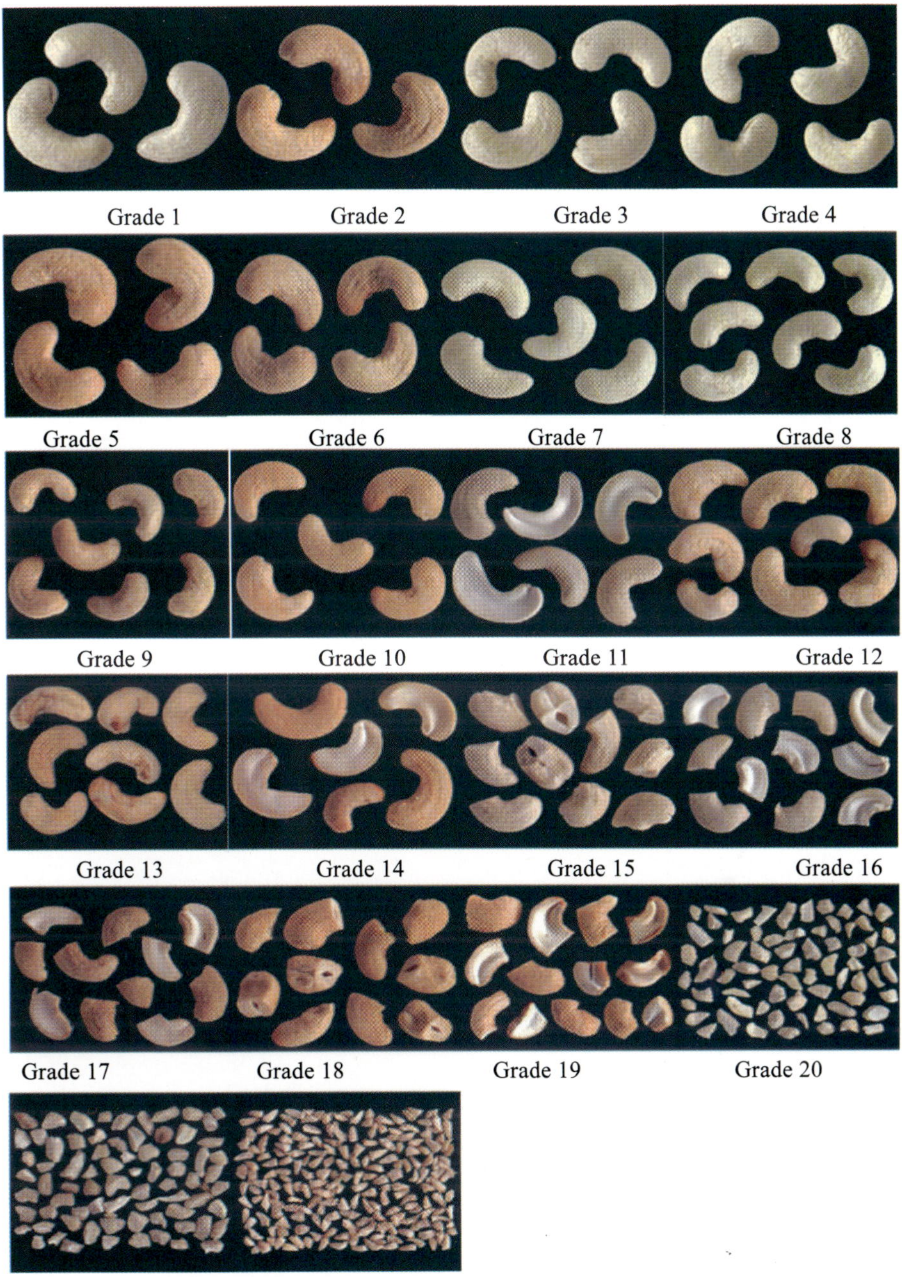

Fig. 54: Grade specification (Source: CEPC, Cochin)

Packing

The international trade of cashews was developed and started in India during the early 1920's. The shelled cashew kernels were sun dried, sorted and the un-peeled kernels were packed in wooden cases lined with newspapers and shipped to U.S.A. and U.K. Due to long transit time and inadequate shipping facilities, there was a constant problem of spoilage of the product due to insect infestation. In order to overcome these problems, the Indian industries along with the American Scientists (as U.S.A. was the main buyer at that time as well) developed a method of packaging cashew kernels which protected the product quality. The developed packaging system cantered around usage of a tinplate container of 18 litres capacity, commonly known as kerosene oil (KO) tins. Two such tinplate containers each containing 25 lbs. of cashew kernels were packed in an outer wooden case (Narayanan and Dordi, 2000).

For the first time cashew kernels were shipped in tinplate containers flushed with carbon-dioxide gas from Mangalore in 1925. Later a process called "Vita" packaging was developed in which air was first removed from the filled tinplate containers to create a vacuum and then an inert gas like carbon-di-oxide was flushed in and the containers were hermetically sealed. This type of packaging helped to retain the keeping quality of the product for a longer time. During the early 1970's the wooden cases were replaced by corrugated fibre board (CFB) boxes (Narayanan and Dordi, 2000).

Traditional packaging system

Till the mid 1990's the Indian cashew exporters used tinplate containers (KO type) of 18 litres capacity which could accommodate 11.34 kg (25 lbs.) of cashew kernels. The 3-piece tinplate containers were soldered i.e. the top and bottom were joined to the body by a double seam and then fully soldered. The solder used for seaming was 50 per cent lead and 5O per cent tin.

The traditional vita packaging was under threat as 14 States in the U.S.A. banned the use of lead in packaging of food products due to health and disposal problems. This was followed by many other states in U.S.A. and other countries. At about the same time when the ban on lead in some states in U.S.A. came into force, importers in U.S.A., Europe and Japan, looked for a new generation flexible packaging system which would retain the product quality, be eco-friendly and easy to handle (Narayanan and Dordi, 2000).

Commercial adoption in kernel packing

Currently a large number of exporters have partially or completely substituted the traditional tinplate packaging system by the recommended new generation alternate bag-in-box system. The alternate system has been successfully adopted and is promoted by the Government of India/CEPC by providing incentives to the cashew exporters to install the packaging machinery for the new system. Initially, some of the machines were procured from overseas sources. However, similar machines are now manufactured within the country and have been installed by a significant number of exporters (Narayanan and Dordi, 2000) and main feature of the new generation alternate packaging system are as follows.

Main features of new generation alternate packaging system

- The recommended packaging materials for the new packaging system are available globally.
- Adoption of the new packaging system does not involve major changes in the traditional filling system.
- Packaging machinery is available for vacuumising/nitrogen flushing and sealing the packages.
- The new alternate packages are lighter in weight as compared to traditional tinplate containers.
- The alternate packages are collapsible when empty and therefore storage requires less space.
- In the alternate packaging system, the materials used are eco-friendly and recyclable.
- The problem of lead solder has been overcome, so also the problem of sharp edges, making the system operator friendly.
- Economically beneficial by 10-15% compared to traditional packaging.

The international market demands the kernels be packed in new, clean, dry, leak proof and lead free containers with an airtight (hermetic) seal and without internal paper liners. It is also stipulated that the outer container should be of a new cardboard, free of infestation and visual mold, and sealed without staples unless otherwise specified. All cartons should be clearly marked on each end with the following.

1. Name of the product and trade name or brand name, if any;
2. Name and address of produces or packages;
3. Grade;
4. Net weight;

5. Country origin;
6. Buyers name and marks;
7. Destination;
8. Other marks agreed to by buyer and Sheller.

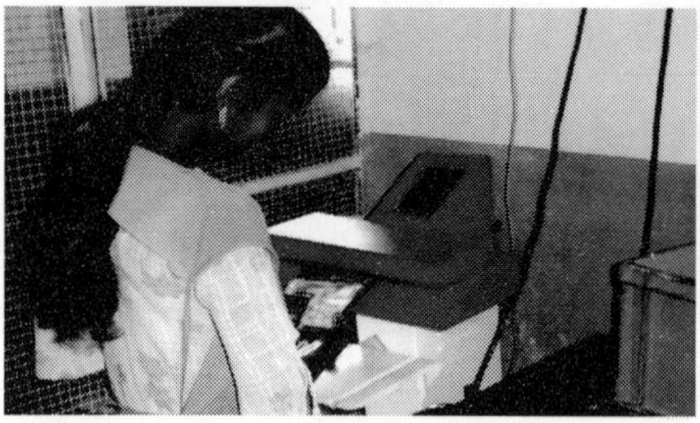

Fig. 55: Packaging of cashew kernel

Fig. 56: Recent air tight cashew kernel packaging technique

Processing at Orchard Level

The processing of cashew in the orchard is mainly confines to removal of raw nuts from cashew apple and drying. Harvest only fully matured nuts. Usually, nuts are picked when they fall off from the tree. Best quality nuts are obtained, where fresh fallen fruits are collected. The apples are removed and the nuts are sun dried for 2 - 3 days to reduce moisture from 25 per cent to below 9 per cent. It is very essential to dry the nuts after harvest to prevent spoilage during subsequent storage. This helps the kernel to retain their quality, particular the flavour. However, if cashew apples are used for processing, it is better to harvest them from the tree without damaging the apples. The mature nuts will sink in water while the immature/unfilled ones will float. This test could be used to find out whether the nuts are mature or not.

Further, the compiete process starting from collection of raw nuts to the final processing of kernels is shown in diagramme (Fig. 57). This is a general overview of process along with several processing options, and will be useful to understand the general method adopted in cashew industry.

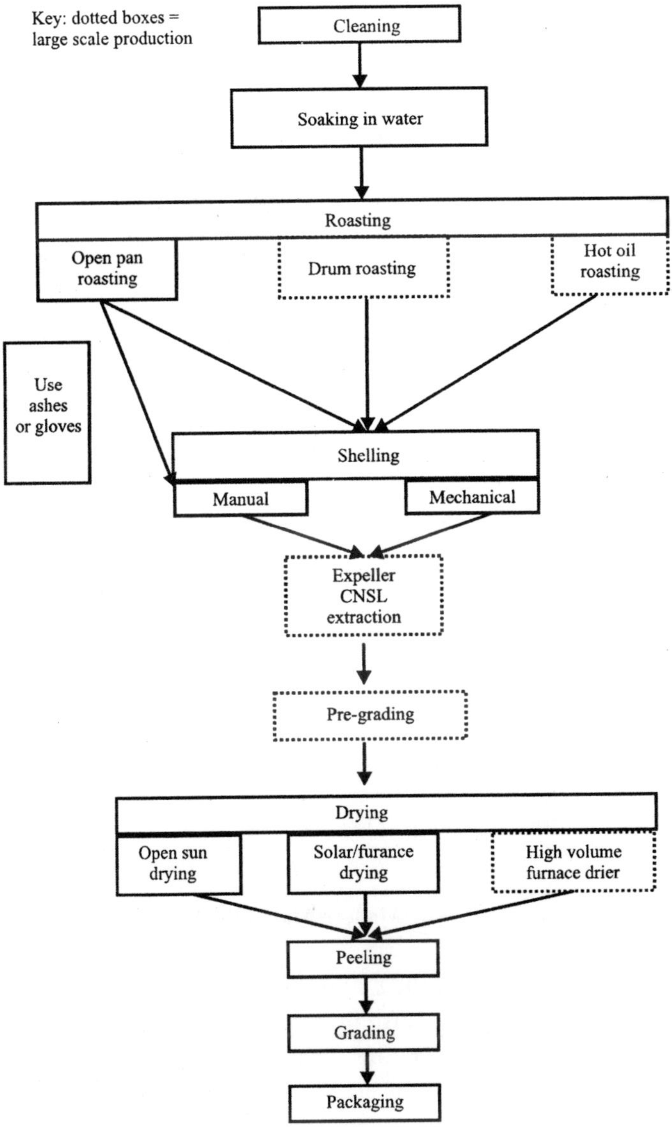

Fig. 57: A diagram showing a general overview of the process and the various processing options.

Cashew Nut Shell Liquid (CNSL)

CNSL is a versatile by-product of the cashew industry. A mesocarp layer of shell, inside which is a soft honey comb structure containing a dark reddish brown viscous liquid. It is called Cashew Nut Shell Liquid or CNSL.

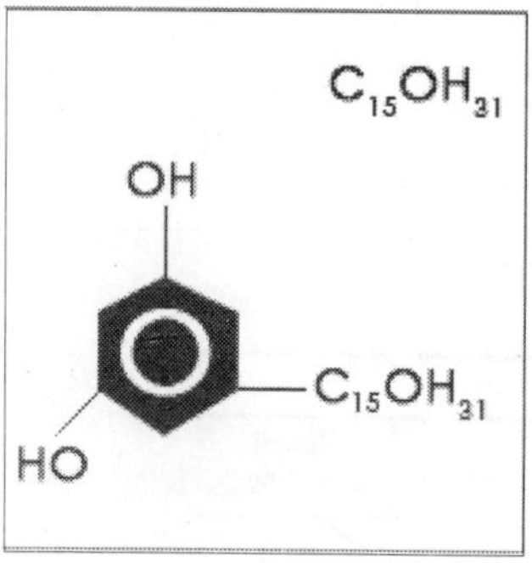

58: Chemical bonding of CNSL

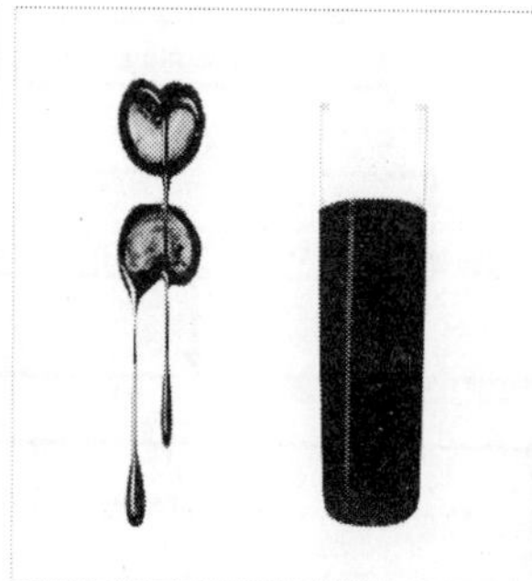

Fig. 59: Cashew Nut Shell Liquid

Fig. 60: Cardanol

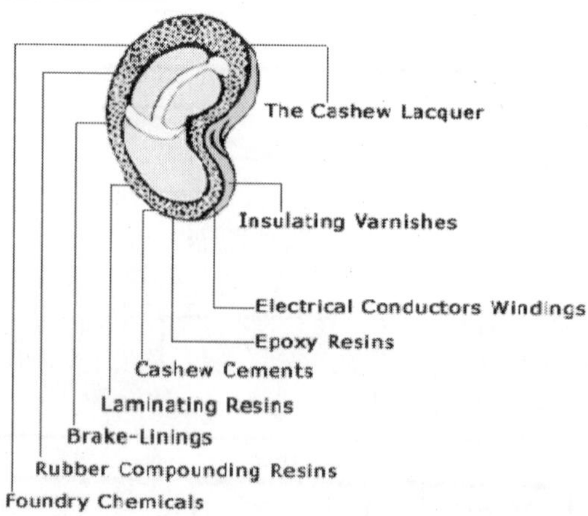

Fig. 61: CNSL-Cost Effective Modern Materials (*Source* Anon., 2010b)

CSNL used for manufacturing of CNSL Resin, cardanol or card phenol or cashew phenol, cashew friction dust, anti corrosive hiring (lining) chemicals, paints, varnishes, enamels, insecticides and fungicides, cashew lacquers, bakelite, electrical conductress, cashew cements, core oil, red oxide, wood, fuel, speciality chemicals, laundry chemicals and many other Industries.

There is considerable scope for its utilization in the development of drugs, antioxidants, fungicides, etc. The main factors that are affecting the utilization of CNSL are its dark colour, variable quality and above all its high price.

Uses of Cashew Nut Shell Liquid (CNSL)

Cashew Nut Shell Liquid (CNSL) is a very important industrial raw material which has got multidimensional uses. Considerable research work has been done on the utilization of CNSL in India and abroad. The cashew nut shell contains a viscous and dark liquid, known as cashew nut shell liquid (CNSL), which is extremely caustic. It is contained in the thin honeycomb structure between the soft outer skin of the nut and the harder inner shell. The CNSL content of the raw nut varies between 20 and 25 per cent. Cashew nut shell liquid (CNSL) is an important and versatile industrial raw material. There are more than 200 patents for its industrial application, in particular, its use as raw material for phenolic resins and friction powder for the automotive industry (brake linings and clutch disks). In drum-brake lining compounds, cashew resins are used as fillers and may also be used as binders. In disc pads, the role of cashew resin is restricted to the use of friction dust as filler. The advantage of the cashew resins compared with synthetic phenolic resins is that they are more economical and produce a softer material, which gives a quieter braking action (CTCS, 1993). CNSL is also used in mouldings, acid-resistant paints, foundry resins, varnishes, enamels and black lacquers for decorating vases and as insecticides and fungicides. In tropical medicine, CNSL has been used in treating leprosy, elephantiasis, psoriasis, ringworm, warts and corns. Like cashew nuts, CNSL also has an excellent international market and its imports have reached almost US $ 10 million annually, corresponding to the sale of the raw liquid. However, the exporting country would earn much more foreign currency if manufactured products were exported. After extracting the CNSL, the cashew nut shells can be burned to provide heat for the decorticating operation or can be used in the manufacture of agglomerates. Together with the testa, it may be used either in the manufacture of dyestuff or to provide durability to hammocks and fishing lines.

About 90% of the CNSL imported by the USA, the UK and Japan was used as friction dust for drum brake linings and clutch facings in motor cars. It is expensive compared with asbestos but the superior friction modifying properties make it economic. Since disc brakes replaced break shoes in motor cars, the use of CNSL has decreased (Wilson, 1975). Still then, a huge variety of other products can be made from CNSL (Table 50).

Table 50: Some uses of CNSL

Use	Authors
Wood protection against insects (raw CNSL)	Wolcott (1944)
Binders in particle boards	Dhamaney *et al.* (1979)
Special cements, resins for modification of rubbers, surface coatings, insulating varnishes and plasticisers	Evans (1955)
Germicides, fungicides, insecticides and photographic developers	Ramaiah (1976)
Lacquer	Rudeco (1989)
Medicines (including cancer treatment)	Duke (1989), Kubo *et al.* (1993), Muroi *et al.* (1993)

The cashew shell contains a viscous, balsam-like substance known as cashew oil or cashew nut shell liquid (CNSL). It has caustic properties and when heated, gives off pungent and choking fumes (Duke, 1989). The CNSL is a highly toxic fluid, about 90% of which is comprised of anacardic acid. The remaining 10% consists of cardol and is mainly responsible of the activity of heliquid (Cornelius, 1966). When in contact with the skin, the liquid may cause swelling, rubefaction, vesication and even acute dermatitis. Therefore, cashew nuts must be cleaned to remove the cardol and then roasted to remove the toxins before the kernels are ready for consumption. The CNSL also occurs in other parts of the cashew tree (Table 51). Hammonds (1977) reported CNSL level in different parts of the cashew tree (in ppm) is given below

Table 51: CNSL level in different parts of the cashew tree (in ppm)

	Root	Wood	Leaves	Bark	Apples	Kernels
CNSL	75	25	250	85	60	35

Several special treatments before roasting can bring the rate up to 90-95 %, and with solvents 100% can be achieved, but these methods are expensive and their use depends on the CNSL price (Ohler, 1966). Any sales add direly to the profit of the factory. If CNSL is not sold, it can be burnt or it has to be discarded thus creating additional costs. Natural CNSL consists of anacardic acid, cardol, cardanol and 2-methylcardol in various compositions (Table 52).

Table 52: Composition of cashew nut shell liquid

Author	Anacardicacid (%)	Cardol (%)	Cardanol (%)	2-methyl cardol (%)
Cornelius (1966)	90	10	-	-
Hammonds (1977)	82	13.8	1.6	2.6
Tyman (1978)	74.1-77.4	15.0-20.1	1.2-9.2	1.7-2.6
Ohler (1979)	90	10	-	-
Tyman (1980)	80	15	small amount	small amount

During the commercial extraction (hot oil bath) the liquid undergoes decarboxilation; the anacardic acid is converted into cardanol and polymer. Therefore commercial CNSL consists mainly of cardanol (60- 65%), polymer (20-25%), cardol (10-12%) and a small amount of anacardic acid (Ramaiah 1976; Tyman, 1980).

Chaudhari *et al.* (2012) studied the physical properties of cashew nut shell liquid at Dapoli, Dist. Ratnagiri (Maharashtra) and they concluded that, the specific gravity of the Crude CNSL extracted by screw press method is 0.98. The specific gravity of the heat treated CNSL extracted by screw press is 0.96.The Viscosity of the Crude CNSL extracted by screw press is 57.43 cP. The viscosity of the heated CNSL extracted by screw press is 28.96 cP and that for the CNSL extracted by hot oil bath method is 37.69 cP.

Polymerization Characteristics of CNSL

CNSL can be polymerised by a variety of methods:

- Addition polymerisation through the side chain double bonds using cationic initiators such as sulphuric acid, diethylsulphate etc.
- Condensation polymerisation through the phenolic ring with aldehydic compounds.
- Polymerisation after chemical modification to introduce speciality properties.
- Oxidative polymerization.
- Various combinations of the above.

Industrial Significance of CNSL

- Low cost phenol.
- Versatility in polymerisation and chemical modification.
- Possibilities for development of high performance polymers.
- Property advantage over phenolics in certain applications such as impact resistance, flexibility, faster heat dissipation etc.

Reactivity

CNSL undergoes all the conventional reactions of phenols. Cardanol differs from phenol only in the C_{15} side chain. It undergoes the well known formaldehyde condensation reaction of phenols that gives rise to phenolic polymers. Moreover, it can be polymerised through unsaturation in the side chain although the bulky nature of the side chain restricts the molecular weight attainable to oligomers. One of the significant advantages of the cardanol is its amenability to chemical modification to effect desirable structural changes so as to get specific properties for making tailor-made polymers of high value. Thus, structural changes could be effected at the hydroxyl group, on the aromatic ring and on the side chain.

Cardanol

Distillation of CNSL under reduced pressure gives cardanol. The residue will be rich in cardanol and is generally known as residol, which is conveniently used in the preparation of friction dust for brake linings, and also in rubber compounding formulations (The process know-how for the production of cardanol and conversion of residol into friction dust is available with RRL, Thiruvanathapuram).

Table 53: ISI Specification of the untreated CNSL IS : 840(1964)

Specific Gravity/30° C	0.950 - 0.970
Viscosity at 30° C cps, Max.	5 50.0
Moisture % by weight, Max.	1.0
Matter Insoluble in Toluene % by weight, Max.	1.08
Loss in Weight on heating, % by Weight, Max.	2.0
Ash % by weight, Max.	1.0
Iodine Value, Min	1.0
1. Wij's method	250
2. Catalytic method	375
Polymerisation Time	
a) Time in minutes, Max.	4
b) Viscosity at 30° C, cps, Min	30
c) Viscosity after acid washing at 30° C cps, Min	200
Colour shall be not deeper than dark brown when viewed by transmitted light.	

(*Source*: Anon., 2010b)

Specification for Treated Cashewnut Shell Liquid

The cashewnut shell liquid as extracted has a strong vesicant dermatitis action. Before this liquid is utilized for preparation of resins, it requires treatment to get rid of metallic impurities as well as traces of sulphur compounds. The liquid thus treated is known as "Treated Cashewnut Shell Liquid ".

The specification of treated cashewnut shell liquid is as follows:

Table 54: Specification for treated cashewnut shell liquid

Specific Gravity at 25° C / 24° C	0.955 - 0.975
Viscosity at 25° C (max.)	800 cps
Iodine Value (min.)	240.00
Ash (max)	1%
Moisture (max.)	0.5%
Acid Value (max.)	14

(*Source*: Anon., 2010b)

Specification for Cold Pressure Cashewnut Shell Liquid

Cashewnut shell liquid is also produced by the 'Cold Pressed' method in solvent extraction plant. The specification of the liquid is as follows:

Table 55: Specification for cold pressure cashewnut shell liquid

Specific Gravity at 26° C	0.9668-1.0131
Refractive Index at 41.5° C	1.5158
Saponification number	106-119
Iodine number	270-296
Acid number	94-107

Chemical Structures of Anacardic acid, Cardanol and Cardol

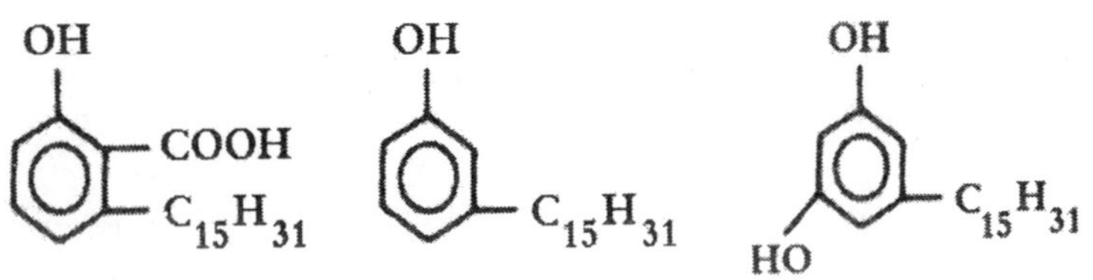

Anacardic Acid　Cardanol　Cardol

Market Potential

Demand and supply

Cashews are high value dry fruits with retail price ranging from Rs. 300/- to Rs. 625/- per kg. Their shelf-life is 4 to 6 months if processed properly or else, they develop fungus or taste bitter. Kernels are used in many sweet preparations, certain *farsan* items, dessert preparations and ice-creams. They are also used as table enrichers in some exclusive restaurants and star hotels. Due to their high price, their regular domestic use is limited to few elite families.

Marketing strategy

Market for cashews is gradually increasing whereas its supply is limited. Cashew plants require a very special climate and hence they are grown in Kokan region of Maharashtra, nearby state of Goa and Kerala. Thus, demand for cashews is increasing and there are fluctuations in prices according to the supplies.

Processing cost for 100 kg seed and feasibility of home scale processing given in following Table 56 and Table 57, respectively.

Table 56: Processing cost for 100 kg seed

Particular	Cost (`)
Total cost (seed, cost, labour, Packing, machinery etc.)	5462.00
Total Receipt (seed sale)	7230.00
Profit	1768.00
Profit /kg	17.68

Table 57: Feasibility of home scale processing

Number of Cashew trees (average)	100
Raw cashewnut production @ 8kg/tree (Average)	800
Receipt from Raw Cashewnut @ Rs. 45/kg	` 36,000
Shelling % (KPR)	25%
Receipt after processing (Wholes : Broken ratio is 4: 1)	` 59,200
Receipt through cashew shell @ Rs. 2.5 /kg	` 1,375
Total income	` 60,575
Labour cost	` 7600
Electricity charges	` 4000
Depreciation	` 6500
Total cost	` 54,100
New value addition	` 13,425
Unit weight benefit of raw material	` 75.72
Net returns/ kg of raw material	` 8.09

Import of cashew nuts in India

India does not produce sufficient quantities of raw nuts required by the processing units and resorts to import from African countries and South East Asian countries. Country-wise import of cashew nuts in India indicates that during 2007-08, the major portion of cashew nuts to India came from Côte d'Ivoire (34%), Guinea Bissau (16%), Tanzania (13%), Benin (9%), Indonesia (6.7%), Ghana (5%) and Mozambique (4.37%). The share of African countries in India's cashew nut imports during 2007-08 was 93 per cent. Cashew nuts are imported in the country for the purpose of re-export of processed kernels since India has a labour cost advantage in this commodity. High cost involved in processing cashew is the major reason for African countries to export raw cashew nuts. Moreover, consumption of cashew kernel in these countries is also low. Using a commonly accepted conversion ratio of raw nuts to kernel at 4.54, the percentage share of imports in the total quantity of nuts processed for exports was worked out. Results indicated that imports which accounted for 36.5 percent of the total quantity of raw nuts processed for exports during 1990-91, had increased to 61.6 per cent (2000-01) and 116 per cent of the latter during 2007-08, implying that kernels processed from imported raw nuts were also currently being consumed in the domestic market. Import of raw cashew nuts increased from 82639 MT in 1990-91 to 605970 MT in 2007-08, an increase of 633 per cent. The annual growth rate in the volume of cashew import was 11.32 per cent for the period 1990-91 to 2007-08. A comparison between the domestic production and import of raw cashew nuts indicates that the dependence of the Indian cashew industry on the imported cashew nuts has increased over the years. During 2007-08, as against the domestic production of 665000 MT, the volume imported was 605970 MT.

Cashew Kernel Products

Cashew kernel products and its preparation reported by Vaidehi (2000) are as follows:

1) Kaju Katli / Katri

Kaju katli is an exotic dessert of Indian origin. This dish has a long shelf life and is prepared on all important festivals and family functions. Kaju katri is an Indian sweet similar to a *burfi*. It is made from cashews, sugar, cardamom powder and ghee. Typically cut into rhombus shape and covered with edible silver foil, varakh. Boxes of kaju *katlis* are exchanged on festive occasions as a token of love and appreciation. Festivals like Diwali, Ganesh Chaturthy etc. are incomplete without preparing *kaju katli.*

Fig. 62: *Kaju Katli*

Method

Dissolve sugar in three fourth cup of water and bring it to boil. Add saffron and cook till it forms syrup of three thread consistency. Melt ghee and add to this syrup. Add cashewnut powder and mix thoroughly. Set aside to cool slightly. Knead it well and roll into half centimetre thick rectangle. Spread silver varq over it and cut into diamond shape of approximately one and half inch a side.

2) Cashew nut vegetable korma

Ingredients

100 g cashew nuts

100 g carrot (chopped)

100 g peas

100 g French beans

100 g paneer (cottage cheese)

2 tomatoes

2 onions sliced

¼ teaspoon turmeric powder

1 teaspoon red chilli powder

1 teaspoon coriander-cumin powder

1 teaspoon ginger-garlic paste

1 clove cardamom

Pinch of powdered cinnamon

1 lemon

Salt to taste

1 bunch coriander leaves (chopped)

Ghee or oil for frying

Method

Boil the carrots, peas and French beans for two to three minutes. Slice the onions and fry in ghee or oil. Add the ginger and garlic paste and fry. Add tomatoes, turmeric powder, chilli powder, coriander-cumin powder, and salt and stir together. Add the paneer pieces, boiled vegetables and cashew nuts. Mix all the ingredients and cook for five minutes. Remove from the heat; add the juice of a lemon to taste and coriander leaves to garnish. Serve with hot chapattis.

3) Green mango with cashew nuts and raisins

Ingredients

100 g cashew nuts

3 mangoes (green, medium size)

100 g raisins

6 g garlic flakes

2.5 cm (1 inch) ginger piece

Vinegar (add to taste)

8 chillies (red)

Sugar (add to taste)

Salt to taste

Method

Peel the mangoes and grate them. Grind the garlic, ginger and red chillies to a fine paste with a little vinegar. Melt the sugar, salt and vinegar over a low heat. Add the chilli paste and heat until it dissolves. Add the grated mango, raisins and nuts and cook over a medium flame until thickened. Hot-fill into clean sterile jars.

4) Cashew nut mushroom curry

Ingredients

75 g cashew nuts (green, tender)

100 g mushrooms

2 chillies (red)

Turmeric powder

1 teaspoon coriander

1g (a pinch) cumin

½ lime

2 onions

2 chillies (green)

10 g ginger

4 cloves garlic

½ coconut

1 teaspoon oil

1 sprig curry leaves

1 pinch mustard seeds

Salt to taste

Method

Clean the mushrooms and cashew nuts in hot water. Grate the coconut and extract the coconut milk from three-quarters of the coconut. Boil the cashew nuts and mushrooms. Grind the remaining coconut, red chillies, coriander, turmeric and cumin into a fine paste. Slice the onion, ginger, garlic and slit the green chillies. Heat the oil and add the mustard and curry leaves. When the seeds start to pop, add the ground spices and sliced ingredients. Cook until the onions are soft. Add the cooked cashew nuts and mushrooms, lime juice and salt. Simmer for 10 minutes. Add the coconut milk and simmer for 5 minutes. Season and remove from the heat.

5) Cashew nut banana cake

Ingredients

1 cup cashew nuts (grated or milled)

1¾ cup flour

11 cups sugar

1 mashed ripe banana

¼ cup butter

4 teaspoons sour milk

2 eggs

1 teaspoon soda

1 teaspoon baking powder

½ teaspoon salt

Method

Cream the butter and sugar together, add the banana and beat well. Add the soda, dissolved in the sour milk, and beat the eggs well before adding to the mixture. Mix in the flour, baking powder and salt. Add cashew nuts and bake in a moderate oven for about 40 minutes.

CHAPTER – 22

Cashew Apple Processing

Cashew apple, a highly potential nutritional source, is almost entirely wasted now. In Goa, cashew apple juice is fermented and distilled for production of alcohol. Cashew apple is juicy and swollen. It is a rich reserve of vitamins and minerals. The cashew apple is the peduncle of the nut. Normally the apple is 8 to 10 fold large by weight compared to cashew nut, at ripe stage. Cashew apples can be of different size and shape. The ripe apple is very juicy and fibrous with thin skin. When fully ripe, it gets detached from the pedicel and falls down. In Brazil, cashew is a popular fresh fruit and dwarf types are being evaluated to fulfil the market requirements of fresh cashew apple in terms of red colour, high sugar, low tannin etc.

Research studies have clearly revealed that the cashew apple, weighing about 8-10 times that of the nut, is an equally valuable produce from the crop, if it is economically exploited. However cashew apple is almost completely wasted in India, without any commercial exploitation, like in many other parts of the world. The production of cashew apple in India is estimated to be around 60 lakh tonnes per annum. Cashew apple is highly nutritious and comparable with many other tropical fruits. Large number of technologies has been developed at various research stations in India for the economic utilization of cashew apple by processing it in to various value added products. In spite of this, it is quite unfortunate that the country is wasting such an excellent fruit causing economic loss both to the farmers and the country.

Cashew apple grows steadily until nut maturation and thereafter it grows very rapidly and gets ripen (Pratt and Mendoza, 1980). Removal of nuts initiates rapid growth and early ripening of cashew apple. Thamburaj *et al.* (1980) reported that under Periyakulam conditions, fruit maturity in cashew was completed in 49 days. Augustin and Unnithan (1981) reported that the increase in dry weight in cashew apple was slow up to 36 days from fruit set and there after it was rapid in the early stages, growth rate of cashew nut was faster than the apple. Chattopadhyay *et al.* (1983) reported that the growth patterns of cashew apple and kernel were different. Initially the nuts grow much faster than the apples, but in later stages, apple sizes increases more rapidly and outgrow the nut. Both pericarp and kernel attain maximum size in about 30 days. Kumar *et al.* (1984) analyzed the physical parameters of growth of fruit in cashew. They found that optimum fruit development occurred 70 days after fruit set.

Composition of Cashew Apple

The ripe apple is very juicy, spongy, somewhat fibrous, having a unique smell, and has a very thin skin that gets easily bruised. Cashew apple is highly nutritious and is a valuable source of sugars, minerals and vitamins. The moisture content of cashew apple ranges from 85 to 89%. Cashew apple contains 0.2% protein, 0.1% fat, 11.6% carbohydrates, 0.9% crude fibre, 261 mg minerals/100 g, vitamins like thiamin - 0.02 mg/100 g, riboflavin- 0.5 mg/100 g, nicotinic acid - 0.4 mg/100 g and vitamin A - 39 IU (Augustin, 2001).

The major problem in cashew apple utilization for product development is the presence of tannins which is responsible for the astringency, seasonal availability and short shelf life. However, tannins can be removed by precipitating with gelatine. Clarified cashew apple juice (after precipitation of tannin) could be used for preparation of RTS, Syrups, Squash, Jams and Candy. Cashew apples are fermented to produce an alcoholic beverage called "Feni" in Goa.

Cashew apple juice contains sugar, protein, acid and large amounts of vitamin C. The juice also contains an astringent principle, polyphenolic in nature. The juice though one of the richest sources of vitamin C is not acceptable as a soft drink on account of its high astringency. The juice is a good media for alcoholic fermentation, as it contains all the nutrients for growth of yeast. The cashew wine contains about 4 per cent alcohol but has astringency. Brandy - distillate of wine - does not have the astringency but possesses the exotic flavour of cashew apple.

Composition of cashew apples compared with other tropical fruits (per 100 g fresh weight) is given in the following Table 58. The cashew apple is comparable with several other fruits in the content of most of the nutrients but superior in vitamin C.

Table 58: Composition of cashew apples compared with other tropical fruits (per 100 g fresh weight)

	Moisture	Carbohydrates (g)	Calcium (mg)	Phosphorus (mg)	Iron (g)	Ascorbic acid (mg)
Cashew	88	12	10	10	0.2	262
Guava	80	13	15	27	1.0	200
Mango	87	11	14	10	0.4	30
Orange	86	10	40	24	0.3	55
Papaya	90	7	11	9	0.4	50

The astringency of cashew apple is determined to a large extent by the tannin content, a phenolic compound, and its content varies from 0.06 to 0.22 g per 100 g. The pH of the apples varies between 4.1 and 4.7 and total sugars from 6.7 to 10.5 %. Ripe apples contain 0.76 to 1.17 % pectin.

Nutritive value of varieties of cashew apple shows in the following Table 59.

Uses of Cashew Apple

Several products from cashew apple, including products used for minor ailments, had been in vogue traditionally in households in cashew growing areas, some of which are still used. However it is not commercially exploited anywhere in the country except in Goa where it is used for the preparation of cashew fenny. However, many institutions in India, particularly Kerala Agricultural University, have developed technologies for the preparation of various nutritious and refreshing products from cashew apple. Cashew Research Station, Madakkathara of Kerala Agricultural University is profitably running a commercial unit for the production of five cashew apple products. This success story is to be replicated elsewhere in major cashew growing areas which will give additional revenue to fanners and can generate considerable employment for unemployed youth and women (Mathew *et al.*, 2010).

Fresh Consumption

Though there are innate disadvantages from the man's consumption point of view, cashew apple is widely eaten raw as fresh fruit. Either whole apples are consumed or they are cut into small pieces, mixed with table salt and eaten.

Cashew apple is a non-climacteric fruit, which must be harvested ripe, if it is to be consumed as a fresh fruit. The changes associated to ripening do not happen if cashew is detached from the tree and it will never reach good eating quality. Quality for fresh consumption is related to low astringency and acidity, sweetness, firmness, size and pear shape. The market requirements for

Table 59: Nutritive value of cashew apple powder varieties per 100 g on dry matter basis (with moisture value furnished)

Varieties Vitamin C	Moisture Tannin (%)	Protein (g)	Fat (g)	Energy (Kcal)	Crude fibre (g)	Dietary fibre ADF (%)	 NDF (%)	Ash (g)	Calcium (mg)	Phosphorus	
Ullal-1	8.5	8.4	3.4	332	6.4	7.3	28.5	8.3	220	160200	0.45
Ullal-2	8.8	8.1	3.3	304	4.5	8.0	20.0	5.6	240	151260	0.40
Ullal-3	8.7	7.8	2.8	314	4.3	11.9	23.8	8.7	235	154170	0.40
Ulial-4	8.4	8.4	3.6	300	5.2	9.7	21.8	2.0	218	148168	0.38
NRCC-2	8.1	7.0	3.6	324	4.8	10.9	25.7	7.3	228	150196	0.41
Vengurla-4	8.3	7.3	3.6	312	4.4	9.1	22.5	4.3	173	130220	0.46
Vengurla-5	8.4	8.5	2.8	310	4.5	9.8	21.9	5.3	100	127260	0.31
Bulk(Juice)	8.8	8.3	5.0	301	5.1	9.3	23.4	5.2	180	130172	0.51
Bulk (Without juice)	8.8	9.7	3.0	297	8.6	12.0	28.9	3.0	160	110165	0.27

Source: Vaidehi (2000)

appearance are to be taken into consideration. Cashew apple must be cosmetically perfect without misshaping or physical signs of injury to be sold as fresh fruit.

After harvesting, cashews must be placed in single layer into boxes with openings for ventilation and lined with form rubber. Upon arrival to packing house, fruits are washed, dried, selected, graded, packed, palletized, pre-cooled and stored under refrigerated conditions.

The main constraints affecting quality of cashew apple for fresh consumption are lack of knowledge on pre- harvest handling requirement, improper harvest indices and harvest boxes, long exposition to sun after harvest, delay between harvest and pre- cooling, in effective pre-cooling and package, inappropriate storage temperature and inadequate temperature management during transit, storage and retail (Mathew *et al.*, 2010).

Beverages

i) Fresh apple beverages

Several nutritious and refreshing beverages like clarified and cloudy juice, juice concentrate, squash and syrup can be made from the unfermented juice of cashew apple by adding varying concentrations of sugar, citric acid and preservative. Kerala Agricultural University has standardized the technique for the preparation of juice, syrup and ready to serve drink. The Cashew Research Station, Madakkathara is manufacturing cashew apple syrup and a ready to drink form "cashew apple drink" on a commercial scale and selling through the sales outlets of the university. Sri Lankan Cashew Corporation has standardized an RTS beverage named 'Cashola' (Mathew *et al.*, 2010).

Cashew apple provides a number of by-products of which the most available one in Brazil is the concentrated juice which ranks first in sales among the tropical juices. It has contributed to human nutrition particularly the poor people, by supplying low cost Vitamin C. Vitamin C content in cashew apple juice averages to about 200 mg/100 mg juice, which is nearly four times higher than other traditional fruit juices such as orange. Cajuda, cajuuina, cajuvita and cajuaperativo are different cashew drinks used in South Brazil. Cashew apple juice can also be blended with other fruit juices like lime, pineapple, orange, grape, watermelon and apple juices to produce mixed or composite beverages of greater appeal. Isolated studies indicated the possibility of carbonating the juice which is found to have better shelf life and preparation of spiced beverages (Mathew *et al.*, 2010).

ii) Fermented beverages

KAU and CFTRI (Mysore) have standardized technologies for utilization of cashew apple for the manufacture of the fermented products like wine, vinegar, liquor and alcohol. Cashew apple vinegar can be prepared by alcoholic and subsequent acetic fermentation of juice, which is perhaps the oldest known fermentation product. Cashew liquor is not made by blending of spirits, as done in case of foreign liquor, but distilled exclusively from the pure juice of cashew apple without addition of any extraneous matter. The steps involved in the production of cashew liquor are:

1. Extraction of cashew apple juice and detanning
2. Fermentation
3. Distillation
4. Ageing process in wooden casks and
5. Removal of astringent smell.

Use of good ripe apples is very important, as unripe or overripe apples would affect the quality of final product. One litre of 60- 62% ethyl alcohol can be obtained from eight litres of cashew apple juice. Kerala Agricultural University has standardized the method of producing liquor four different grades of cashew flavour viz., (1) Strong flavour (2) mild flavour (3) free from cashew flavour and (4) blended flavour (Augustine, 1987).

Goa is the only place in India where cashew apples are utilized widely for the preparation of the liquor, 'fenny', by distillation mostly through crude country methods on cottage industry basis in almost all plantations. On modern scientific lines, the cashew apple juice could be extracted from fruits, filtered, detanned, pasteurized, cooled, inoculated with a strain of pure yeast for fermentation, raked and distilled by modem techniques.

Cashew wine is a product of fermentation of hexose sugar of cashew apple juice by intact yeast cells to form ethyl alcohol and carbon dioxide. Kerala Agricultural University has developed methods for producing four grades of wine such as soft, medium, hard and sweet, based on the alcohol percentage and sweetness. Fermentation, filtration and ageing are done on detanned cashew apple juice. All grades of wine preparation, except soft wine, involve one more step of adding sugar. Sensory evaluation showed that the performance was in the order of sweet, medium, hard and soft wines. Wine can also be distilled to produce brandy (Mathew *et al.*, 2010).

Sri Lankan Cashew Corporation has standardized a method for production of semi sweet wine named Cazholeena by adding wine yeast in the dry form to single strength fresh pulp for fermentation, which contains 11.4% alcohol and 2% sugar and free from volatile acids and turbidity. Cashew apple wine can be blended with grape, pineapple and banana wines for better acceptability. Cashew apple wine can be mixed with fresh juices of orange, pineapple, tomato, grape and cashew apple as well as tender coconut water to produce wine coolers to serve as good health drink as they contain both wine with its medicinal properties and fruit juices with high amount of nutrients and minerals (Mathew *et al.*, 2010).

iii) Pulp products

The most important pulp product of cashew is jam. It can be prepared by boiling the cashew fruit pulp with a sufficient quality of sugar and a pinch of citric acid to a reasonably thick consistency, firm enough to hold fruit tissues in position. Mixed fruit jam can also be prepared by mixing cashew apple pulp with equal quantity of banana pulp or pineapple pulp. The Madakkathara Centre is commercially producing Cashew apple-Mango Mixed jam named Cashewman (Mathew *et al.*, 2010).

Fruit bar having 80° Brix can be prepared by heating layers of fruit pulp mixed with pectin, sugar, glucose and potassium metabisulphate to 90° C and drying to 15% moisture. Different layers of cashew apple paste mixed with 1% citric acid are sun dried and cut into required size after placing one on top of the other to form leather. The layers, after smearing sugar syrup and pressed together, can be eaten like fruit wafers. The cashew apple pulp, cooked in to thick jam like typical sweet (doce) and cooked pulp formed into balls and coated with sugar (caju cristalizado), are used in Brazil for serving as dessert (Mathew *et al.*, 2010).

iv) Confectioneries

Cashew apple impregnated with cane sugar and subsequently drained and dried is called a candied fruit. One kilogram of cashew apple on processing gives 745 g candies (Mathew *et al.*, 2010). The syrup left over from the candying process can be used for sweetening chutneys, in vinegar making or for candying another batch of fruits. Cashew apple can also be utilized for the preparation of tutty fruity. One kilogram of cashew apple on processing gives 715 g tutty fruity. The whole fruit can also be processed in to nutritious toffee, a feasible dessert item with extended shelf life (Mathew *et al.*, 2010). Vaidehi (1994) observed that toffee could provide 7.5 g of protein and 442 k calories per 100 g.

Frozen desserts and dairy confectionery items could be prepared from cashew apple juice by optimization of juice concentration and spray drying. The only constraint here is the large capital investment required for spray drier equipment. Different dairy products involving cashew apple have been prepared on a laboratory scale by the University of Agricultural Sciences, Bangalore (Mathew *et al.*, 2010).

Dehydrated cashew apple products can be prepared from dehydrated powder. Clarified juice is prepared from steam blanched, sulphur dioxide treated fruits and spray dried for preparation of cashew apple powder with juice. The pulp or the residue of apple can also be dried, powdered and sieved for use as cashew apple powder without juice. 10 to 30% dehydrated cashew apple powder can be used in various value added products like wheat *laddu*, *masala* biscuits, sweet and *masala* doughnuts, sponge cake, steamed *kabadu*, tomato cashew apple powder soup, powder *koftas*, chocolates, sweet and hot bread products and cashew apple blended chocolates. Nutri-Cashew, a ready mix have already been prepared using cashew apple powder for the elderly as high fibre fruit (drink) float mix for instant use (Mathew *et al.*, 2010).

10% to 15% clear and cool cashew juice mixed with skim milk powder can be spray dried for the production of cashew milk powder and can be utilized for the preparation of products like milk shakes, ice creams and ice candy. A ready-to- serve beverage mix, fruit-milk/*lassi* mix has been prepared from clarified juice by homogenization, spray drying and mixing with milk/ *lassi* powder (Vaidehi, 1994).

Culinary preparations

Pickle can be prepared from sliced raw green fruit using chilli powder, gingelly oil, fenugreek powder, asafoetida, turmeric powder, garlic, mustard powder, a pinch of sodium benzoate and salt to taste. Madakkathara Centre is commercially producing and marketing Cashew apple pickle. Chutney can be prepared from sliced cashew apple using sugar, onion, ginger, spices like cumin seed, pepper, cardamom, cinnamon and coriander powder, salt and vinegar. Dried pulp prepared from semi- boiled apples is preserved for off- season and used for culinary uses particularly for the preparation of chutney.

Culinary preparations like aviyal, mixed vegetable curry and bugi (stuffed preparation) are popular among traditional Kerala farmers. Mature, but unripe fruits, popularly known as "*karakka*" in Onattukara region of Kerala are used in their preparations after washing in starch solution or butter milk. Coconut oil is usually added to the curries for reducing tannin content. Cashew apple juice along with tender coconut toddy is used by several farmers for making "*apparrf*, a Kerala dish made from fermented rice flour (Mathew *et al.*, 2010).

Cashew apples are ground into a paste along with rice, folded in banana or teak leaves and steam cooked. This preparation, locally known as "Madith Deedi Adye" in Tulu language and "Panege" in Konkani language is consumed as breakfast in villages (Anon., 2004).

Canned products

Cashew apple is peeled after treating with boiling NaOH solution followed by subsequent treatment in boiling solution of 0.2 N H_2S0_4. The treated fruits can be steamed and hot sugar syrup can be poured over the fruits for preparation of canned apple. Canned curried vegetables from raw green fruit of cashew in combination with potatoes (1:1) or potatoes and tomatoes (2:1) with or without tamarind are also reported (Mathew *et al.*, 2010).

Medicinal uses

Cashew apple preparations have been extensively used traditionally for several ailments. Cashew apple is used as a curative against scurvy and stomach ailments like dysentery and diarrhoea. It is used as a tonic to mothers in confinement. It is a medicine for women after parturition.

Cashew apple juice, without removal of tannin, is prescribed as a remedy for sore throat and chronic dysentery in Cuba and Brazil. Fresh or distilled, it is a potent diuretic, possessing anti-scorbutic properties, and is useful for kidney troubles, and in advanced cases of cholera. It is given for uterine complaints and dropsy. The brandy is applied to relieve the pain of rheumatism and neuralgia (CSIR, 1985). The cashew fenny is used to cure various ailments of infants and aged. Cashew apple liquor is used for medicinal purposes for aliments like worms, sickness, cold, body ache, fever or flue, toothache, fresh wounds and cuts, cramps due to chilling weather, muscular pain, irregular movement of bowels, low blood pressure, loss of sleep for aged people and cholera.

Cashew apple juice kept in sunlight for a fortnight can be preserved for 2 to 3 years, which would have effect in treating fever and diarrhoea of both human beings and domestic animals. It is believed that cashew apple juice induces sleepiness when given along with medicines for patients affected by fever, thus helping in fast recovery (Mathew *et al.*, 2010).

Use in Nutraceutical Extractions

Cashew apple contains ascorbic acid, fibre, carotenoid pigments, minerals and host of other chemicals, which are of significance to human health. Cashew apple powder lipids are rich in unsaturated fatty acids, the major ones detected being palmitoleic and oleic acids. Cashew apple is rich in fibre and on a dry weight basis the fibre content varies from 15 to 18%. Phenol and flavonol present in cashew apple could serve as natural antioxidants which play a major role in destroying free radicals (Anon, 2005). Vitamin C, an anti oxidant present in cashew apple was analyzed and it showed variation from 40.1 to 177.8 micro g/100g (Mathew *et al.*, 2010).

A valuable by-product that can be obtained from cashew apple waste is pectin. Pectin is used in manufacturing jams, jellies, marmalades, preserves *etc.* It is useful as thickening, texturising and emulsifying agent and finds numerous applications in pharmaceutical preparations, cosmetics etc. Pectin has been isolated from cashew apple powder, the yield of which varies from 1.6 to 2.03%. The cashew apple pomace or the fruit waste has been identified as the ideal medium for pectinase enzyme production for *Aspergillus foetidus* 115 through solid state fermentation (Venkatesh, 2003).

Agricultural uses

When bulk quantities of cashew apple is utilized for the manufacture of soft drinks or fermented beverages on a commercial scale, considerable amount of cashew apple residue is obtained as waste. Nutrient status of cashew apple residue on dry weight basis is: total ash 1.6%, total tannin 5.2%, ether extractives 4.6%, calcium 20.6 mg/ 100g, phosphorous 152.7 mg/100g, proteins 8.8%, crude fibre 8.4% and iron 35.0 mg/ 100g (Mathew *et al.*, 2010).

Vermicompost

The highly perishable and seasonal fruit waste can be converted to value added products with good manurial value without creating problems for disposal. Apple residue could be effectively utilized for the production of vermi compost of 1.69% N, 0.44% P and 0.58% K using *Eudrilus euginae*. The pH of the compost from cashew apple is 8.9 and could be used as a good ameliorant for acidic soils of Kerala (Mathew *et al.*, 2010).

Fig. 63: Vermicompost from cashew apple

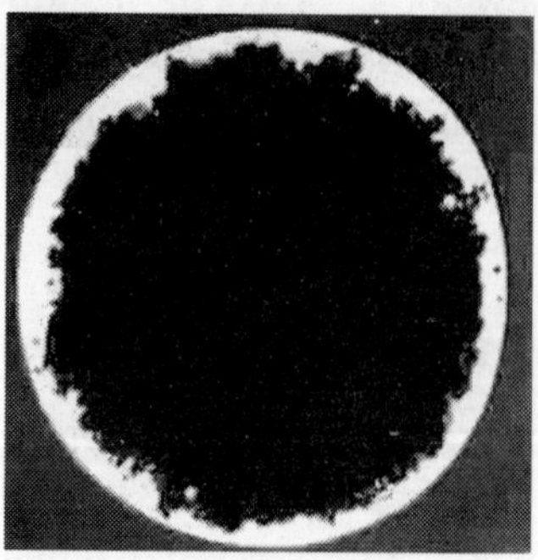

Fig. 64: Compost from cashew apple

Compost

When bulk quantities of cashew apple is utilised for the manufacture of soft drinks or fermented beverages on a commercial scale, considerable amount of cashew apple residue is obtained as waste.

Nutrient status on dry weight basis is :

Total ash 1.6 %

Total tannin 5.2%

Calcium 20.6 mg/100g

Phosphorous 152.7 mg/100g

Proteins 8.8%

Crude fibre 8.4% and iron 35.0 mg/100g.

Ideal medium for pectinase enzyme production for *Aspergillus foetidus* 115 (Venkatesh, 2003).

Animal Feeds

The ripened cashew apple or its residue could be utilized for the preparation of cattle feed, pig feed and poultry feed. Cashew apple is a promising feed source for dairy cows in Vietnam. Cashew apple or its residue could be preserved for long term use as cow feed by anaerobic ensiling with poultry litter. Cashew peel (7.6% protein, 12.3 % fat and 59.2% carbohydrate) is a good poultry feed (Mathew *et al.*, 2010).

Apples and effluent from apple processing plants can be used a yeast to enrich livestock feed stuffs. The yeast *Candida utilis* (NCIM 3336) grows well on cashew apple juice medium to produce useful biomass than a strain of *Saccharomyces cereviseae* occurring naturally on the fruits.

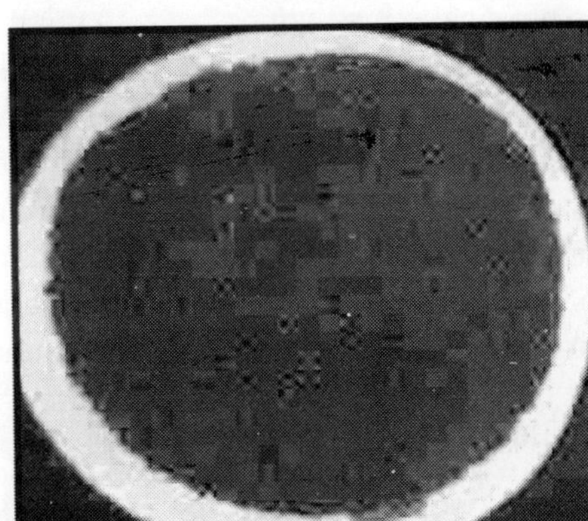

Fig. 65: Animal feed from cashew apple

Apples are also dried and preserved as cattle feed for rainy season. Daily feeding of 3-4 kg fresh apples along with normal feed to cow is found useful. However cattle will have stomach problem causing diarrhoea, when fed excess quantity of cashew apple.

Pest Management

Cashew apple extract has been found to be an effective insecticide against red palm weevil (*Rhynchophorus ferruginous* Olive) in coconut (Krishna Kumar *et al.*, 2003). Cashew apple and gum extract, in combined form or alone, acts as an effective repellent against leaf feeding pests of vegetables. The cashew apple is dried and powdered into meal which can be used as bait for catching crustaceans.

Energy Production and Industrial Uses

Fuel

There is a great scope to utilize cashew apple for production of alcohol to be used as a bio fuel. The residue, after extracting juice for fenny preparation, is used as fuel in liquor industry in Goa.

Biogas

Ripened fruits can be used as raw material for biogas plant.

Tannin extraction

Cashew peel can be used for the extraction of tannin which is useful in leather industry (Johnson, 1977).

Cashew Apple Processing Issues

The seasonal production of cashew apple is one of the greatest handicaps for the processing industry, along with its astringent and acrid principles. In addition, the utility of cashew apple is limited because of its high susceptibility to physical injury, which leads to microbial spoilage by yeast and fungi during harvest, transportation and storage. More than 63 per cent of cashew apple collected at ripe stage exhibits moderate to heavy damage. The storability of cashew apple is thus very poor and complete spoilage can occur within hours after harvest. Fragmented and scattered nature of cashew plantations also creates problem in collection and utilization of cashew apple in some cases. The system of collection of cashew nuts from fallen fruits after considerable delay also limits the availability of quality cashew apple for processing purposes

Suitability of Cashew Apples for Processing

Cashew apple suited for processing should have medium to large fruit size with more than 70% juice containing more than 11% sugar and 0.39- 0.42% acidity. Though the yellow coloured apple is preferred in some regions because of its sweetness, more or less sweet or astringent apples have been found in all colour groups.

Kumar and Aravindakshan (1985) listed K-27-1, BLA 139-1, BLA-1 and Sawantwadi as the varieties suitable for processing. According to Kutty (2000), varieties H-1593, H-1600, V-4, K-19-1, Madakkattara-2, Dhana, Kanaka and M 26/2 have desirable attributes with reference to quality parameters. Amrutha, Anakkayam-1, Dharasree and V-5 are identified as suitable varieties for preparation of candy and tutty fruity.

Pre-requirement for Processing

1. Harvest at full ripe stage and separation of the nut with minimum damage to the apple
   - Crisp, firm, tight and full colour developed apples
   - 44 to 46th days after fruit set, physical and chemical properties are optimum, and best time for collecting apples without spoilage.
   - To be collected every day before it falls to the ground
   - If left ungathered, rotting of cashew apples takes place
   - Once damaged, the apples may ferment and deteriorate rapidly.
2. Sorting to select the best quality ones
3. Washing

For removal of tannin, whole cashew apples are boiled in 2% salt solution for 15-30 minutes followed by washing in water.

Juice

Clarification of juice is carried out by using polyvinyl pyrrolidone (PVP), however, use of powdered sago or starch solution is the cheap and easy method.

Commercial Cashew Apple Products

Fenny is the most famous commercial product from cashew apple produced in Goa. It is liquor produced from cashew apple juice by fermentation and distillation.

An FPO licensed cashew apple processing unit has been established at Cashew Research Station, Madakkattara, Thrissur under Kerala Agricultural University during 1997 for the manufacture of unfermented cashew apple products. It is the first ever unit established in India for cashew apple processing. The unit is producing cashew apple syrup, cashew drink, mixed cashew apple-mango jam, cashew apple pickle and cashew apple candy. The first ever cashew apple processing unit in private sector has been recently established at Iritty, Kannur, Kerala under the trade name 'TOMCO PRODUCTS" and they are marketing cashew apple syrup

The details of commercial cashew apple products from Kerala Agricultural University are given below:

1. Cashew apple syrup and drink

Selected cashew apples are cleaned thoroughly, juice extracted and clarifying agent, preservative and citric acid are added immediately. The clarified juice is siphoned out and material for the preparation of syrup and drink. Sugar and citric acid rifted juice in required quantity to produce syrup and drink as per demand.

The nutrients, Vitamin C and riboflavin, which are high in cashew apple, are preserved in these beverages also. Cashew apple syrup contains 276 mg Vitamin C and drink contains 140 mg vitamin C /10Og. These are natural products and price is comparatively less. Taste is better if served chilled. Syrup has a storage life of one year.

Cashew apple drink is an RTS (Ready-To-Serve) beverage. Drink is marketed both in glass bottles and in attractive food grade pouches. Pasteurized drink in glass bottles has a storage life of three months under ambient storage conditions.

2. Cashew apple-mango mixed jam

The ripe apples are collected from the farm, selected, cleaned and soaked in salt solution for three days to remove tannin. Apples are again washed in water, cooked, made into pulp and is mixed with equal quantity of mango pulp. Pulp is mixed with sugar and citric acid to prepare jam. Vitamin C content of the product is 18 g/ 100g. It is marketed under the trade name Cashewman mixed jam.

3. Cashew candy

It is a sweet product and quality apples with good shape are selected for candy preparation. As in jam preparation, tannin is removed from apples, cooked, pierced using fork and dipped in sugar solution. Concentration of sugar solution is gradually increased so as to reach 70° brix. After two weeks of soaking, sugar solution is drained out and candy is dried in shade. It takes about 2-3 weeks for making the final product About 745 g candy can be obtained from one kilogram of cashew apple. Vitamin c content of the product is 28.4 mg/100g.

4. Cashew pickle

Mature but unripe cashew apples are collected directly from plant carefully without disturbing the flowers and tender nuts. After cleaning, the fruits are cut into small pieces and astringency is removed by immersing in salt water. After removing from salt water, it is again washed and pickle is prepared using oil, chilli powder, fenugreek powder, turmeric powder, ginger and garlic paste.

Cashew apple products and its preparation reported by Vaidehi (2000) are as follows:

Cashew apple fruit chutney

Ingredients

3 kg cashew apple slices

3 kg sugar

168 g salt

84 g onion (peeled and finely chopped)

21 g cumin powder

21 g black pepper

21 g cardamom powder

42 g green ginger (peeled and grated)

7 g cinnamon (coarsely powdered)

7 g coriander

42 ml acetic acid glacial (vinegar)

Methods

Type 1

The cashew apples are immersed in 2 % brine solution (common salt solution) for three days. After washing free of salt, the apples are steamed for five to seven minutes at 10 lb pressure. The steamed apples are then cooled and washed in water. Any bad portions are trimmed off before the apples are sliced into segments. The apple slices are cooked with the addition of salt, sugar, spices and the acetic acid. The ginger, cinnamon and coriander are tied loosely in a muslin cloth bag and added to the mixture. This bag is removed when the chutney is cooked.

Type 2

Soak cashew apple slices in 2 % brine solution for three days, remove from the brine and wash in clean water. Remove any undesirable portions and then slice the apples. Chutney is prepared as usual, using 1 kg sugar, 1 onion (large size), 30 g ginger, 1 teaspoon of cumin seed, pepper, cardamom, cinnamon and coriander powder, salt to taste and 20 ml glacial acetic acid for every 1 kg of fruit slice.

Tie all the powdered spices in a clean thin piece of cloth, (such as muslin). Prepare sugar syrup by adding an equal quantity of water to the sugar. Add the sliced apples, chopped onion, grated ginger, vinegar and salt to the sugar syrup. Drop the spice bag in when the mixture begins to boil. Boil the mixture until it is sufficiently thickened and hot-fill into clean, sterilised jars.

Cashew apple pickle

Ingredients

65 g cashew apple pieces

12 g salt

5 g chilli powder

5 g coriander

2 g fenugreek seeds

5 g mustard powder (fried and ground)

1 g turmeric powder

1 g acetic acid

1.5 - 2 g citric acid

5 g jaggery

5 g ginger oil

Method

Soak cashew apple pieces for 10 days in 12 % brine solution and then wash thoroughly in water. Heat the oil and add the spices. Pour the oil and spice mixture over the brined apple pieces, mix well and allow cooling. Add preservatives (acetic acid and citric acid) to the apple mixture cover with the remaining oil and store in a sterile container.

Sweet cashew pickle in oil

Ingredients

1 kg cashew apples

63 g salt

20 g chilli powder

1.5 g ginger paste

30 g chopped onion

3 g mustard seeds

3 g cumin

3 g cloves

1.5 g cardamom

3 g cinnamon

3 g garlic paste

125 ml vinegar (acetic acid)

125 g sugar

Edible oil (enough to immerse the pickles in the container)

Method

Soak cashew apple cubes in 12 % brine solution for 10 days and then thoroughly wash in clean water. Heat a small quantity of oil until it is bubbling and adds the mustard seeds, onion, garlic and ginger. When the mustard seeds stop popping, add the remaining spices and the fruit cubes and heat for a few minutes. Cool the mixture to room temperature, then leave to stand in the sun for three or four days. After this time, add the sugar and vinegar and stir into the mixture. Heat sufficient oil to immerse the pickle in the container, cool it and pour over the pickle mixture. If hot oil is added to the pickle, it will develop black colour of pickle. Seal the container and store at room temperature.

Recipes using dried cashew apple powder

Cashew apples are highly perishable; they cannot be stored for more than 24 hours. To resolve this problem, the pulp can be dried and ground into a powder that can be used in a range of products.

Preparation of cashew apple powder

Cashew apple pieces are dried (either with or without cashew apple juice) for 12 hours (with juice) or 4.5 hours (without juice) at 70°C. The dried pieces are ground and stored in airtight containers for use in a range of recipes.

Cashew apple sweet cookies

Ingredients

72 g wheat flour

8-16 g cashew apple powder

50 g ghee (clarified butter)

40 g sugar

5 g baking powder

Method

Cream the ghee or butter and sugar to form a light and fluffy cream. Sieve the wheat flour, cashew apple powder and baking powder into the mixture and mix to form stiff dough. Roll out the dough into a layer an eighth of an inch thick (3-4 mm), and cut out small biscuits. Bake at 180°C for 20 minutes until golden brown.

Cashew apple masala cookies

Ingredients

135 g wheat flour

15-30 g cashew apple powder

75 g ghee or butter

10 g sugar

5 g salt

5 g baking powder

45 g curds

20 g green *masala*

(Made from a mixture of coriander leaves, green chillies curry leaves and ginger in equal quantities)

Method

Sieve the wheat flour with the cashew apple powder, baking powder and salt twice to ensure that it is well mixed. Rub the ghee or butter into the flour until the mixture resembles breadcrumbs, then add the sugar and stir well. Add the green *masala* and curds and knead the mixture into soft dough. Roll out to form a thin sheet of one eighth of an inch thickness (3-4 mm) and cut with a fancy biscuit cutter. Bake in an oven at 180°C for 20 minutes until golden brown in colour.

Cashew apple porridge mix

Ingredients

70 g malted wheat flour

30-40 g cashew apple powder

15 g powdered sugar

Method

Sieve all the ingredients thoroughly to form a uniform mixture. Add 100 g porridge mix to 600 ml of water and cook, stirring continuously for about 5-10 minutes.

Cashew apple sweet doughnuts

Ingredients

90 g wheat flour
10 g cashew apple powder
1.5 g baking powder
3 g yeast
15 ml + 35 ml water
4 g milk powder
20 g sugar
20 g ghee or butter
1.5 g salt
Half an egg
A few drops of vanilla essence
Oil for frying

Method

Sieve the flour, salt, cashew apple powder and baking powder together twice. Gently rub the ghee or butter into the sieved powder mix. Then mix the yeast with 15 ml water and 1 tsp of the sugar and leave to stand for about 10 minutes until it froths. Add the rest of the water (warmed to about 35°C) to the yeast and mix. Add the sugar, lightly beaten egg and milk and beat until soft dough is formed. Knead lightly until smooth and rest for 30 minutes. Roll out to about one quarter of an inch (6-8 mm) thick and cut with a doughnut cutter. Rest it for 10 minutes. Fry the doughnuts in oil until golden brown in colour, and then roll in sugar powder.

Cashew apple masala doughnuts

Ingredients

90 g wheat flour

10 g cashew apple powder

½ teaspoon yeast

15+35 ml water

5 g (1 teaspoon) milk powder

5 g sugar

½ teaspoon salt

Half an egg

10 g ghee or butter

40 g onion (chopped)

6 green chillies

4 g curry leaves (chopped)

10 g coriander leaves

Oil for frying

Method

Sieve the flour, baking powder, cashew apple powder, milk powder and salt together twice. Dissolve the yeast in 15ml of lukewarm water with a pinch of sugar. Chop the onion, green chillies, curry leaves and coriander leaves. Dry them a little by keeping in an oven or frying for a short time in a pan. Cream the butter or ghee and the sugar together until smooth and then fold in the flour mixture. Add the yeast, the egg and the *masala* (mixed leaves and chilli) and leave to ferment for about 1 hour. Knead to a smooth dough add a little water and rest for 30 minutes. Roll out to a quarter inch (6-8 mm) in thickness. Cut with a doughnut cutter and rest for 10 minutes. Deep fry in oil until it becomes of golden brown colour.

Cashew apple sponge cake

Ingredients

90 g wheat flour

10 g cashew apple powder

¼ teaspoon baking powder

50 g ghee or butter

100 g powdered sugar

1 egg

½ teaspoon vanilla essence

½ teaspoon salt

Milk to mix

Method

Sieve the flour and cashew apple powder along with a pinch of salt and baking powder. Cream the ghee or butter and sugar until the texture is light and fluffy. Beat the egg thoroughly and add to the creamed mixture a little at a time. Mix well, and then gently fold in the flour. Add the milk to bring the mixture to a dropping consistency, spoon into a lined tin and bake at 150° C for 35-40 minutes.

Processing for cashew wine

- Cashew apples are cut into slices in order to ensure a rapid rate of juice extraction when they are crushed in the juice press. The fruit juice is sterilized in stainless steel pans at a temperature of 85°C in order to eliminate any wild yeast.
- The juice is filtered and treated with either sodium or potassium metabisulphite, to destroy or inhibit the growth of undesirable types of micro-organisms such as acetic acid bacteria, wild yeast and moulds.
- Wine yeast (*Saccharomyces cerevisiae* - var. ellipsoideus) should be added. Once the yeast has been added, the juice should be thoroughly stirred and allowed to ferment for about two weeks.
- The wine is separated from the sediment and clarified by mixing fining

agents, such as gelatine, pectin or casein, with the wine. Filtration is carried out with filter-aids such as fullers' earth. The filtered wine is transferred to wooden vats.

- The wine is pasteurized at 50 to 60° C. The temperature should be controlled, so that it does not rise exceed 70° C, since alcohol vaporizes at a temperature of 75 to 78°C. The wine is then stored in wooden vats and subjected to ageing.
- At least six months should be allowed for ageing. If necessary, the wine should be clarified again before bottling. During ageing and subsequent maturing in bottles, many reactions, including oxidation, occur.
- The formation of traces of esters and aldehydes, together with the tannin and acids already present, enhance the taste, aroma and preservative properties of the wine. The product is packaged in glass bottles with corks and should be kept out of direct sunlight.

Cashew wine is made in many countries throughout Asia and Latin America. It is a light yellow alcoholic drink, with an alcohol content of 6 to 12 per cent.

Flow diagram indicates the step-by-step process involving from selection of cashew apple to ageing in wooden vats.

Flow diagram

Selection
↓
Mature, sound cashew apples
↓
Slicing to increase extraction of juice
↓
Crushing
↓
Sterilization at 85°C
↓
Filtering
↓
Inoculation with *S. Cerevisiae*
↓
Fermentation two weeks
↓
Pasteurization 50°C to 60°C
↓
Ageing in wooden vats for 6 months

Economics of Cashew Apple Processing

Mini *et al.* (2006) have worked out the economics of processing cashew apple for syrup production. By processing one tonne of cashew apple, a net profit of Rs. 10368/- can be obtained. Considering that the average yield of nuts in India is 800 kg/ha, a production of 6.4 t/ha of cashew apple can be anticipated. A production of about 2 t/ha of good cashew apple can be ensured, taking 30% of the total production as fot for processing. Thus, the additional income from a hectare of cashew orchard from the processing of cashew apple worked out to Rs. 20736/-, if a farmer or farmers' groups can venture into this endeavour. The income can be further enhanced by processing cashew apple for high value products like alcohol and wine.

Processing of cashew apple is an economically viable enterprise in cashew growing tracts. Women Self Help Groups can very well take up this enterprise, thereby contributing to the cause of women empowerment. If legal permission is available for production of fermented products like liquor and wine, it can substantially enhance the income from cashew apple processing many folds (Mathew *et al.*, 2010).

Cashew Fenny

The Indian Made Foreign Liquors (IMFL) is made by blending of spirits whereas *fenny* is distilled exclusively from the pure fermented juice of either cashew apples or coconut toddy without addition of spirit. Konkan region of Maharashtra grows both cashew fruits as well as coconuts in plenty. Invariably, cashew apples are not further processed after separation of cashewnut. Since cashew apples are not available round the year, coconut toddy can be distilled during off-season. At present, *fenny* is produced only in Goa and it has become the speciality of this state. This product can be produced in other regions as well.

Fenny is made from pure juice without addition of spirit. Cashew apple is a valuable source of minerals, sugar and vitamins and coconut juice is also equally nutritious. For production of *fenny*, it is necessary to obtain licence from the state as well as the Central Govt. BIS has specified standards vide 5287:1978 and 3752/3:1967.

Demand and Supply

The state of Goa is famous not only for beaches but also for fenny. Fenny is now regularly exported to many countries. But somehow, production of fenny is confined only to Goa in spite of domestic as well as export market. Major feature of fenny is, it is distilled directly from fresh juice without blending of spirits. Cashew apples and coconut toddy have considerable nutrients.

Marketing strategy

Apart from export market, there is a very large untapped domestic market as well which needs to be tapped systematically. Proper positioning of the product, adequate marketing network and attractive publicity campaign are critical aspects.

Manufacturing process

Juice is extracted from cashew apples and coconut toddy and it is fermented till formation of a film floating over the juice. Time required for fermentation is 65-70 hours. Around 70 litres of arrack is distilled from 100 litres of fermented juice. Arrack and fermented juice are distilled in the ratio of 1:2 to obtain fenny. To obtain one litre of fenny around 30 kg of cashew apples or coconut toddy is needed. It is advisable to hire the services of an expert to achieve the desired quality.

The Process Flow Chart is as under

Juice Extraction
↓
Fermentation
↓
Distillation
↓
Packing
↓
Ethanol

Total demand for ethanol is expected to reach around 3 x 10^{11} in 2010 (Inoue *et al.*, 2010). However, due to increasing petroleum shortage, fermentation production of ethanol from renewable resources has received considerable attention (Ames, 2001). Nowadays, fermentation technology produces nearly 80% ethanol as clean fuel and for which *Saccharomyces cerevisiae* is considered to be of most important yeast strain because of its biotechnological application in the field of fermentation technology (Sreenath and Jeffries, 1996). In order to improve its substrate utilization range, several manipulations of the strains was carried without any success (Mobini-Dehkordi *et al.*, 2007). However, immobilized cell technology has been suggested as an effective mean for improving ethanol production (Yu *et al.*, 2007). Ethanol production by immobilized yeast cells has been extensively investigated during the last few decades. The most widely used immobilization methods are based on cell entrapment in gels, such as carrageen and Ca alginate (Godia *et al.*, 1987).

Fresh cashew apple

Nuts are removed and apple wash thoroughly with tap water

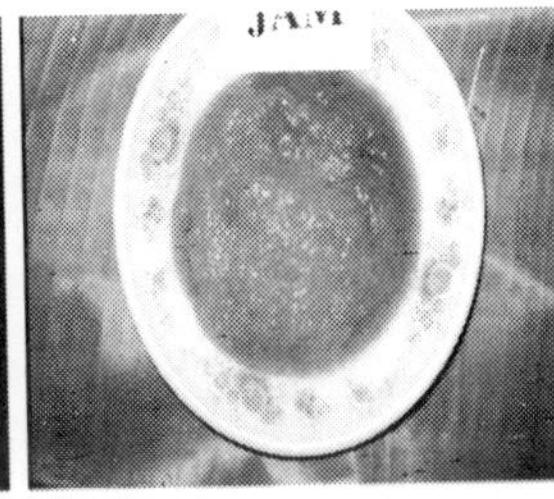

Cashew apple jam

Cashew apple Tutty fruity

Cashew apple Toffee

Cashew apple pickle

Cashew apple pickle

Traditional Konkani dish '*Panege*'

Canned apple

Various cashew apple product viz., Pickle, nectare (RTS), syrup and jam

Fig. 66: Various cashew products from fresh cashew apple

Among the different substrates cashew is considered to be an in expensive substrate for ethanol production (Rocha *et al.*, 2007). However the conditions optimum for ethanol production has not been studied extensively. Neelakandan and Usharani (2009) concluded that that cashew apple juice commercially for any industrial application and are poorly disposed could effectively be used for ethanol production through the process of immobilized yeast and fermentation.

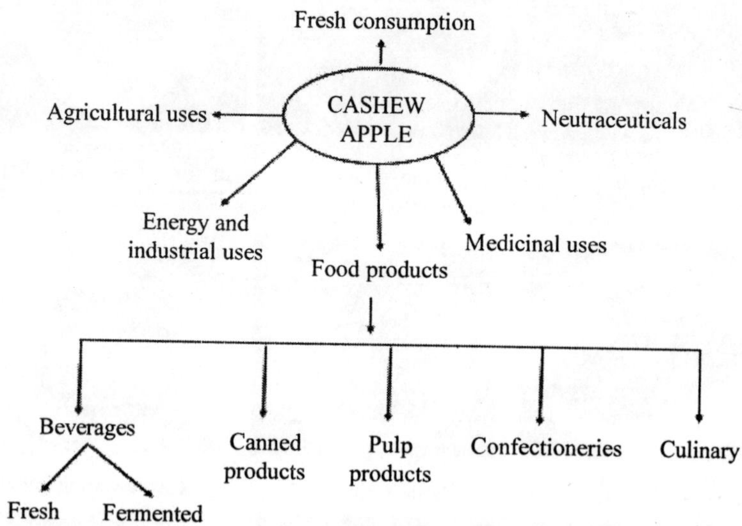

Fig. 67: Cashew apple uses

□□□□

CHAPTER – 23

Pests and Diseases of Cashew

Cashew (*Anacardium occidentale* L.) is one of the most important commercial crops of India. The crop is grown on commercial scale in coastal Maharashtra and other states of the country. The main reason for the low productivity is the pest incidence. The insect pests cause crop loss ranging from 30 to 40 per cent in cashew (Satapathy, 1993). Keeping in view, the strategy of increasing production, one of the important aspects is to develop technologies to save the crop from the pest incidences.

Cashew is attacked by around 180 species of insect and non-insect pests in India resulting in substantial yield loss (Sundararaju, 1993b). Godse *et al.* (2004) reported over 58 species of insects and pests on cashew crop in the Kokan region of Maharashtra region. The infestations of insect pests have been identified as one of the major factor responsible for low productivity. The most important pests that limit production are the cashew stem and borer (CSRB) and the tea mosquito bug (TMB). In addition, leaf and blossom webber, shoot tip caterpillar, and apple and nut borer cause damage. Recently, apple and nut borer and leaf miner are emerging as major pests and they also occasionally becoming serious in some orchards.

Tea Mosquito Bug *(Helopeltis antonii)*

Tea Mosquito Bug (TMB) damages the tender shoot, inflorescence and immature nuts. Both nymph and adult of TMB (*H. antonii*) suck the sap from tender shoot panicles, immature nuts and apples. Nymphs and adult suck the cell sap from tender shoots, panicles, nuts and immature apples resulting in development of water soaked lesions at the feeding point. Due to action of phytotoxin present in the saliva, the water soaked lesions turn pinkish brown and finally black. In case of severe infestation, trees show burnt appearance. The infested shoot dries within 24 hours.

The studies on the seasonal incidence at Regional Fruit Research Station, Vengurla indicated that the TMB outbreak initially noticed on vegetative flush during November-December and it became severe during the months of January-February. The severity coincided with emergence of panicles and reached in peak during February and further declined gradually during April-May as the temperature rises.

The adult bug (6-8 mm long) is reddish brown in colour with black head, red or black thorax and black and white abdomen. Tea mosquito lays eggs 24 hours after matings. The single female lays on an average 50 eggs singly inside the tissue of tender shoots/panicles. A pair of fine hair like chorionic processes projecting outside is indication of the presence of eggs inside the tissue. The eggs hatch within 6 to 7 days and the small size reddish colour nymphs come out. These nymphs look like small ants. The nymphal period ranges from 12 to 15 days. Total life cycle is completed within 20 to 25 days. The adult usually feeds during early hours (6-10 am) and makes about 150 feeding puncture per day.

During rainy season, tea mosquito population in not present in the field; but just after rain (October onwards), when the cashew trees are in their most active growth phase with the new flush and tender inflorescence, the condition becomes quite congenial for the multiplication and feeding; and during this period this pest rapidly becomes abundance. The pest population reaches their peaks invariably in January-February i.e., in the third month of commencement of the population build up when the trees are on full blossom.

Young trees are liable to suffer more in a longer period than older trees, because of the availability of succulent plant parts throughout the year. The population build up of the pest is negatively correlated with meteorological factors like minimum temperature, relative humidity, rainfall, etc. and positively correlated with sunshine (Pillai, *et al.*, 1979).

For the management of TMB, regular monitoring and early detection of the pest infestation and timely spray of insecticides are essential. A three spray schedule comprising the first spray of monocrotophos 15 ml/10 lit or Lambda cyhalothrin 6 ml/10 lit or Profenophos 10 ml/10 lit of water during flushing. The Second spray of Lamda cyhalothrin 6 ml/10 lit of water during flowering and Carbaryl 20 gm/10 lit or Lamda cyhalothrin 6 ml/10 lit or Profenophos 10 ml/10 lit of water during fruiting stage is recommended.

Table 60: Control Measures (Need based)

1st Spray	Profenophos (0.05%) or Lambda – Cyhalothrin (0.003%)	Vegetative flush
2nd Spray	Endosulphan (0.005%) or Lambda – Cyhalothrin (0.003%)	Flowering stage
3rd Spray	Dimethoate (0.05%) orPhosalone (0.07%) orCarbaryl (0.1%) or Lambda – Cyhalothrin (0.003%)	Fruiting stage

Hot spot areas of Tea Mosquito Bug (TMB) of cashew in India are given as below in Fig. 68.

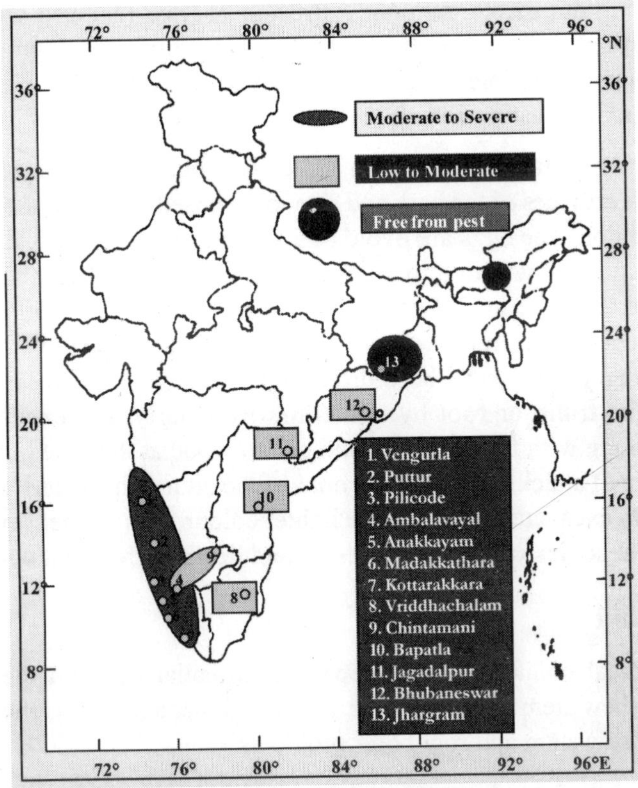

Fig. 68: Hot spot areas of Tea Mosquito Bug (TMB) in cashew (*Source*: Rao, 2010)

Resistant Varieties

Varietal screening conducted so far indicates that resistance is unlikely to be encountered. However, in one of the accessions (Goa 11/6) a phenological evasion has been noticed, which enables the accession to escape severe infestation (Sundararaju, 1998). Similarly there are also reports that hybrids H-3-17, H-1600, H-S-1, H-S-7, H-S-8 and H-15 showed moderate tolerance to this pest. It will be worthwhile to identify varieties whose flowering phase does not coincide with peak population of tea mosquito bug (Salam and Peter, 2010).

Cashew Stem and Root Borer (*Placaederus ferrugineus* L.)

The stem borer is an internal tissue borer pest. The borer is a hidden dreaded enemy of cashew tree as it is capable of killing the tree outright and the infestation by the pest is more severe in neglected plantations. Stem borer infestation could be identified by the presence of small holes in the collar region, oozing out of the gum and extrusion of frass (chewed up fibres and excreta) through holes. The affected trees show different degrees of foliar yellowing and during later stages; there will be shedding of leaves, drying of twigs and gradual death of the tree (Pillai, 1975; Pillai *et al.*, 1976).

The mated female beetle laid 100 to 125 eggs either singly or in batches of 3-8 deep in the crevices of the bark and stem or exposed root or in the soil near to the collar region. The eggs are ovoid (4.5 x 22 mm) or elliptical, dirty whitish in colour and hatch within 8 to 10 days. Grubs are whitish creamy in colour. Grub period ranges from 200 to 250 days. The pest requires 250 to 300 days for its one generation. Grubs feed on internal content of tree mainly bark portion. The emerging young grubs bore directly into the bark and feed on phloem tissues of the trunk or root by tunnelling irregularly. Presence of gummy substance along with faecal matter and bits of wood at the base of the plant is the symptom of attack. The older portions of the tunnels have dark colour frass while freshly excavated tunnels have lighter colour frass. Under severe attack, the canopy becomes yellow and dries up and death of the plant occurs.

Management

- Periodical monitoring of the crop is very essential to prevent the infestation by cashew stem and root borer. Neglected management is the root cause of cashew stem and root borer problems.
- Trees with symptoms of yellowing, dried and dead trees should be removed immediately along with roots by digging pit of 2 feet around the base of the trunk and disposed off. The cocoons present around the trunk and root system 1 to 2 feet below the soil surface should be destroyed mechanically.
- Tree base must be kept clean and weed free so as to observe the gum exudation which is the early symptom of borer attack.

- More than 50% of the bark circumference on the trunk/root should not be removed while extracting the grubs and / or cocoons.
- Extraction and killing of the grubs and cocoons by chiselling the affected portion of bark of trunk or root followed by swabbing and drenching of Chlorpyriphos 20 EC @ 10 ml/litre of water (total 5 lit. solution per tree) is recommended as corrective treatment for the management of cashew stem and root borer.
- Application of 10 ml Chlorpyriphos + 50 ml kerosene or 10 ml DDVP + 50 ml kerosene through entry hole (after removing of frass) with the help of pipe and plugging the hole by moist soil.

It is essential to adopt two important approaches, *viz*., a) phytosanitation, to achieve reduction of pest population in a given location and, b) saving the infested trees in the initial stages of infestation itself. To minimize CSRB infestation in the plantations, dead trees and trees beyond recovery (those with over 50% damaged bark circumference and/or showing yellowing of the canopy) should be uprooted and removed from the plantation immediately before and after the monsoon, since these serve as natural inoculum for multiplication and spread of the pest (Raviprasad *et al*., 2009; Sundararaju and Bakthavatsalam, 1990).

Newly-planted grafts should be trained to bear branches at a height of 0.75m to 1.00m from ground level for facilitating inspection and pest management techniques effectively. (Bhat *et al.,* 2010).

Leaf and Blossom Webber (*Lamida moncusalis* W.)

This is a major pest in Maharashtra and other cashew growing states, appearing regularly throughout the year particularly at the time of emergence of new flush. Generally, young trees suffer more. The adult moth lays eggs at the growing point. The caterpillar on emergence webs the terminal leaves as well as panicles and feeds inside, by scrapping the epidermal layers. The pest can be identified by such webs on the plants. As a result flowers fail to open and there will be no crop. It also feeds on tender, developing nuts and apples by scraping the epidermal layers, which drop prematurely.

Control measures

Spraying with Chlorpyriphos (2.0 ml/litre) and carbaryl (2 g/l) during flushing, flowering and nut formation stages, respectively are essential to control the pest.

Apple and Nut Borer (*Thylocoptila panrosema* M.)

This pest causes direct loss to the crop. It appears at the time of fruiting. The incidence of this pest occurs during the month of January onward when the nuts become pebble sized. The larvae bore into either apple or nut at the joint and feed inside. As a result, the development of kernel is arrested, the nuts get shrivelled, apples get hallowed and shrivelled and they drop prematurely. As the caterpillars reside inside the apple and nuts, they are out of reach for the chemicals. As such preventive measures are more important.

Control measures

All the foliage, floral and fruit and nut feeders can be kept under control by application of the insecticides as per the schedule mentioned above for TMB. Similarly, for effective management, collection and destruction of affected nuts and apples along with the caterpillars is necessary. Spraying with Carbaryl (0.1%) or Dimethoate (0.05%) or Dichlorvas 76%EC (DDVP) (0.05%) is effective to control the pest.

Flower Thrips (*Rhynchothrips raoensis, Scirtothrips dorsalis*)

They cause immature shedding of flowers. Scabs on floral branches, apples and nuts are affected. The incidence seems to be more severe on developing nuts and apples. Immature stages of this pest scrap the upper epidermis of nut and apple and suck the sap which results in development of corky patches on nuts and apples. Under severe infestation, breaking of apple and disfigurement of nuts is observed. Thrips causes about 30 % reduction in nut weight.

Control measures

Spraying with Phosalon (0.05%) or Dimethoate (0.05%) is effective treatment.

Leaf Miner (*Acrocerops syngramma* M.)

This pest appears regularly at the time of emergence of new flush in the months of June-July and October-November. The caterpillars mine into the young and developing leaves and cause blisters on the leaves and feed on the tissues causing development of serpentine tunnels on the leaves. Afterwards the infested leaves turn brown and dry subsequently.

Control measures

Spray Chlorpyriphos (2.0 ml/litre) and carbaryl (2 g/l) during flushing.

Shoot and Inflorescence Tip Borer (*Hileria haligramma* M.)

The pest appears throughout the year, particularly at the time of flowering and fruiting. The caterpillar feeds on leaves, tender shoots, inflorescence stalks, apples and nuts. At the time of new flush, the caterpillar folds the tender leaves from one margin and feeds inside. Then it enters the tender shoots by making a hole at the tip and feeds inside, causing drying up of twigs and panicles. At fruiting time, the larva enters the fruit at the joint of the apple and nut and feeds inside, causing premature fruit drop.

Control measures

Spray Chlorpyriphos (2.0 ml/litre) and carbaryl (2 g/l).

Mealy Bugs

Mealy bugs (*Planococcus lilacinus*, *Ferrisia virgata*, Pseudococcidae: Hemiptera) are soft-bodied, and covered with milky white coating on body. *Planococcus lilacinus* is found on tender shoots and fruits while *Ferrisea uirgata* infests the inflorescence resulting in shedding, and drying of flowers. Honeydew excreted by mealy bug attracts ants and sooty molds. Remove and destroy the affected parts.

Aphids

Nymphs and adults (*Toxoptera odinae*, Aphididae: Homoptera) are soft bodied, ovate and brown. Adults may be winged or wingless. Both nymphs and adults suck sap from tender shoots, inflorescence, apples and nuts. Honey dew excreted by aphids falls on leaves below on which sooty mold grows. Severe infestation results in shedding and drying of flowers, or distorted nuts and apples. Normally the population is kept low by the natural enemies. In population explosions, systemic insecticides like dimethoate, phosphamidon or methyl demeton may be sprayed.

Bark Eating Caterpillars

This (*lndarbela tetraonis*, *I. quadrinotata*, Metarbelidae: Lepidoptera) is a polyphagous insect, attacking a large number of fruit and forest trees. Moths emerge with the onset of monsoon. Larva continues damage throughout the year. Spherical single eggs are laid in cracks of bark by the creamy white moth having brown markings on forewings. Larva is brown to dark brown, sparsely hairy and about 5 cm long when fully grown. Caterpillar bores short distances through the trunk or branches.

At night, the larva comes out and feeds on bark of the tree. Small trees may succumb and large trees may fail to flush, if infestation is severe. Removal of the galleries plastered on the tree trunks, or pouring kerosene or dichlorvos during early stages of infestation reduces the damage.

Nut Crinkler (*Paradasynus rostratus* Distant)

In India, the coreid bug (Paradasynus sp.) is called as 'nut crinkler', which infests tender nuts. Causing them to shrivel, turn black and dry up (Pillai *et al.*, 1976). The bug thrusts its long stylets into the tender nuts to suck sap. Besides cashew, this pest also infests crops such as coconut, tamarind, guava, cocoa, passion fruits etc. The adult bugs lay eggs on leaf surface in groups of up to 52, equally spaced and arranged in regular rows of 5 to 6. The incubation period is 8-11 days. The nymphs emerge by pushing open the operculum. There are 5 nymph instars and nymph period ranges from 21-36 days with an average of 27 days.

The third round of spraying of carbaryl (2 g/l) at fruit set takes care of this pest.

Mites

Infestation occurs mainly in prolonged drought situations. Mites (*Oligonychus mangiferus*, Tetranychidae: Acarina) feed on leaves, remaining on the underside. Affected leaves loose, luster, become whitish and crispy. Normally chemical measures are not followed. If infestation is very serious, spray sulphur, dicofol or ethion.

Diseases of cashew

As compared to the number of pests attacking cashew and intensity of damage caused by them, the magnitude of the disease problem is rather comparatively low. Among few diseases attacking cashew, die-back and damping off of seedlings are found to be most destructive.

Damping off seedlings (*Fusarium* spp, *Phythium* spp, *Phytophthora palmivora* spp, *Cylindrocladium scoparium*)

The fungi attack either root or collar region of the tender seedlings. In case of *P. palmivora* attack, affected seedlings become pale, showing water soaked girdles of dark tissues around the stem and seedlings droop and die. In case of *C. scoparium* wilting and withering of seedlings and presence of necrotic lesions leading to rotting of underground parts occur. The disease occurs in the nursery stage, causing mortality of seedling.

Fig. 69: Adult Tea Mosquito Bug

Fig. 70: Tea Mosquito Bug damages tender shoot and immature nuts.

Fig. 71: Adult CSRB

Fig. 72: CSRB Grub infesting

Fig.73: Extraction and killing of the Grubs Cocoons

Fig.74: Infested tree cashew trunk grub & cocoons swabbing with Chlorpyriphos

Fig. 75: Life cycle of CSRB

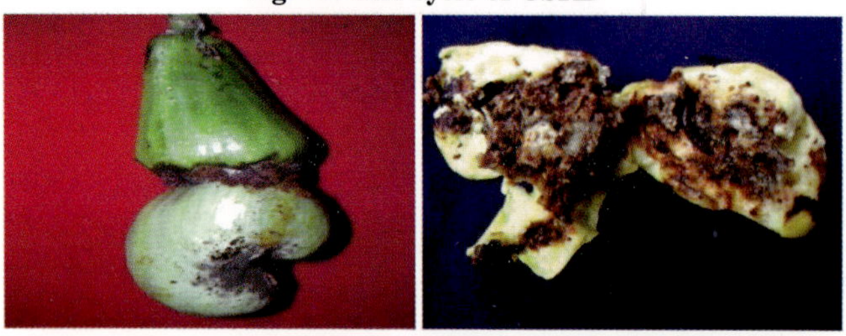

Fig. 76: Infestation of apple and nut borer

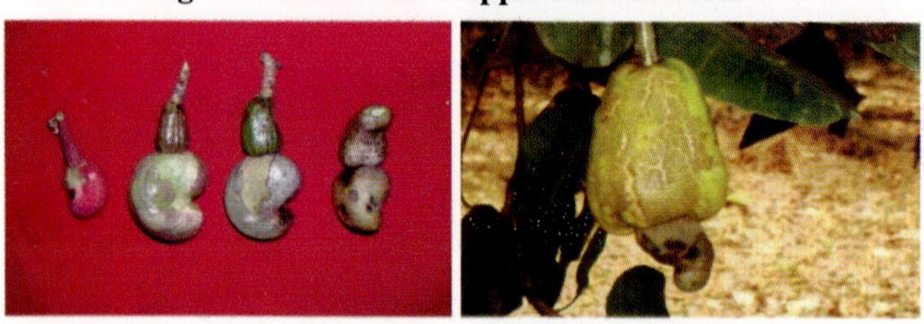

Fig. 77: Flower thrips scabs on tender nuts and apples

Fig. 78: Leaf minor infested tender leaves

Fig. 79: Mealy bug

Fig. 80: Aphid infested to the tender cashew nut

Control Measure

Drenching of seedlings/grafts at fortnightly intervals with Bordeaux mixture 1% during rainy season and cloudy weather conditions.

Incidence of Inflorescence Blight

The inflorescence blight in cashew resulting in heavy crop loss was thought to be caused by *Helopeltis antonii* in association with *Gloeosporium mangiferae and Phomosis qnacardii* (Anon., 1965); and hence combination spray of insecticides and fungicides were recommended for the control of both tea mosquito and the fungi. However, Nambiar *et al.* (1973) observed that infestation of tea mosquito alone was primarily responsible for the malady and the fungi associated with blighted inflorescence were only secondary saprophytic colonizers and hence insecticides alone need be applied.

Anthracnose (*Colletotrichum gloeosporiodes* Penzig.)

Reddish brown, shiny, water soaked lesions along with the resinous exudation on the affected plant parts. Affected inflorescences have a blackening of the floral peduncle and affected flowers turn black, wither and fall off. The lesions grow longitudinally, resulting in ultimate killing of shoots.

Affected tender leaves, becomes crumped, cover with tiny necrotic patches at the edges and the tip. Nuts and apples become shrivelled, decayed or dried out. The nut show small necrotic spots on their epicarp while apple turn black and mummified. The fungus enters the fruit through the floral stigma. The symptoms vary according to the organ attacked.

The disease developed under warm and humid conditions. It becomes virulent when heavy rainfall coincides with flowering season of cashew. Besides inflorescence, it also attacks tender leaves, twigs, immature nuts and apples. This disease appeared on epidemic form in Trichi (Tamil Nadu) in 1965 (Singh *et al.*, 1967). In Brazil, it caused severe economic loss to cashew (Agnoloni and Giuliani, 1977).

Control Measure

Regular plant sanitation, removal of affected leaves, spraying of Carbendazim 1g/l or 1% Bordeaux mixture as soon as the disease appears. It is advisable to repeat the spray to cure completely. Monitor for the disease particularly during rainy season and cloudy weather conditions which enhance the disease very fast.

Die-back or Pink Disease

This is a fungal disease caused by *Corticium salmonicolor* which occurs in the rainy season. The affected branches show white or pinkish growth on the bark. The fungus penetrates into the deep tissues and causes gradual death of the shoots. In advanced stages, the bark splits and peels off. The leaves turn yellow and fall off giving a barren appearance to a tree.

Control Measure

For effective control of the disease, prune all the affected branches below the spot of injection and destroy them. Protect the cut surface by applying Bordeaux paste. The disease can be prevented by giving prophylactic spray of 1% Bordeaux mixture twice a year, first in May-June before the onset of monsoon and second in October.

Inflorescence Blight

Inflorescence blight is characterized by the drying of floral branches with an estimation of crop loss as high as 30 per cent. The inflorescence blight is caused by attack on insect, *Helopeltis antonii* sign, (tea mosquito bug), fungi like *Gloeosporium mangiferae P. Henn and Phomosis anacardii. Earl and Fusarium* sp. have also been found associated (Anon., 1972) (Nambiar *et al.*, 1973). They further stated that the inflorescence blight is primarily caused by tea mosquito infestation and the fungi associated with it, are only secondary saprophytic colonizers which are not pathogenic. This is a major problem of cashew in India where insects play a major role and the saprophytic fungi associated are secondary colonies.

Leach (1973) opined that a toxic substance present in the saliva of tea mosquito is introduced through the stylets, which diffuses into the host tissues. The fungus enters the host slowly through the punctured tissues and colonized as a saprophytic agent.

Control Measure

Since this malady is primarily caused by tea mosquito infestation, spraying with insecticide alone may be stressed upon. As per the spraying scheduled given for control of tea mosquito may be follows at emergence of new flush and inflorescence, would help to bring down the insect population and thereby the incidence of inflorescence blight.

Leaf Blight

Three sprays of Carbendazim (0.1%), or Mancozeb (0.2%) or Bordeaux mixture (1.0 %) at one month interval are recommended for effective management of leaf blight disease of cashew. The first spray is given before the onset of monsoon (Anon., 2013).

Gummosis

This is caused by *Corticium salmonicolor, Diplodia natalensis* and *Ceratocystis* sp. Gummosis is characterized by exudation of a brownish liquid, which later turns black in colour. Main stem and branches are affected and longitudinal cracks can be seen on the affected branches from which gum exudes.

Control Measure

The affected portion may be chiselled out and applied with 1% Bordeaux paste.

Powdery Mildew

This disease is caused by *Oidium* sp. It affects the blossom under cloudy weather conditions and thereby affects the fruit set to a considerable extent.

Control Measure

Powdery mildew can be controlled by dusting with sulphur 2% (w.p.).

Sooty Mould

This is caused by the fungus, *Capnodium* spp. Black fungal growth appears on the leaf lamina and leaves dry up.

Control Measure

Spraying of 1 % Bordeaux mixture or 2% starch controls the disease.

Leaf Spots (*Pestalotia paconiae* Servazzi, *P. dichatae*)

The leaf spot fungus was identified as *P. paeoniae* which is the first record in Nigeria on Anacardium occidentale (Olunloyo, 1975). The minute brown circular spots on the upper surfaces of the leaves mostly near the tips are observed. These spots are enlarged and later coalesced in a leaf area of the lamina from the Up down. The affected cells turn brown and die. Old leaves are mostly affected and the symptoms become prominent on the peripheral leaves during the flushing period (September-December).

Control measure

Spraying of Bordeaux mixture 1%, Copper oxychloride 0.30% or Benlate 0.3% is effective to control this disease.

Yellow Leaf Spot

Microscopic examination did not revealed that presence of any fungi or bacteria on the affected tissue of the yellow spot leaves (Menon *et al.*, 1978) and hence no causal organism has so far been detected. Calton *et al.* (1961) reported that manganese and iron are suspected to be involved in the physiological process of unthrifty affected cashew trees under continuous drying of the soil, possibly interfering with macro-nutrients. Harishu Kumar (1981) stated that affected cashew leaves showed toxic levels of manganese and iron. The molybdenum deficiency has also been reported by some workers. The manganese-iron toxicity is reported to be combined with magnesium deficiency syndrome in the acid-soil complex. Yellow leaf spot has been observed in Maharashtra, Goa, South Kanara, Andhra Pradesh and Orissa.

The symptoms of yellow leaf spots are well manifested during the North East monsoon season (October-November), even though its appearance is observed throughout the year. During the dormant phase, the symptoms appear in most of the leaves ranging from mild to severe intensity. With first flushing, the symptoms are carried over from the older leaves of the lower branches to the newly matured leaves. The number of such spots may vary from one to many. These spots enlarge as the leaves mature and become bright yellow with indistinct and uneven margins. Such spots coalesce to form yellow blotches when the leaves turn completely yellow with big yellow spots, they fall off. It is reported that this malady destroys the leaf chlorophyll, thereby lowering the photosynthetic rate, resulting in low yield.

Control Measure

To reduce the Iron toxicity level of the acid-soil complex, liming may be necessary. The magnesium deficiency may be improved by the application of adequate magnesium sulphate or dolomite. It is reported that the affected leaves spread with ergostin reduced the leaf spot about 23% and improved the photosynthetic activity. The chlorophyll content of affected leaf was only 28% which are rose up to 65% after the treatment (from 3.6 mg/g to 8.3 mg/g of leaf), while the chlorophyll content of healthy leaf is estimated to 12.8 mg/g of leaf. Thus efficacy of ergostin spray to the yellow leaf spot affected leaves has been established (Mandal, 2007).

Fig. 81: Anthracnose of cashew

Fig. 82: Powdery mildew of cashew

CHAPTER – 24

Economics for Cashew Cultivation

Cashew is one major crop of Konkan region of Maharashtra and requires comparatively less expenditure for cultivation as compared with other crops. Under normal planting density (7m x 7m), about 200 plants/ha are occupied. If the cashew grower has its own land, then for first five year of orchard establishment and maintenance, total expenditure of Rs. 47447 is come up. Year wise expenditure (first five years) on cashew cultivation per hectare is presented in Table 61.

Table 61: First five years expenditure per hectare on cultivation of cashew (2000-01)

Year	Expenditure (Rs.)	Percentage
First	19582.00	41.27
Second	3946.80	8.32
Third	5094.60	10.74
Forth	7887.40	16.62
Fifth	10936.20	23.05
Total	47447.00	100.00

From Table 61, it is seen that, out of total expenditure on cashew cultivation in first five years, of which maximum expenditure (41 %) acquired in first year while, remaining 60% expenditure come up for next four years.

From full matured and well established cashew orchard with use of various inputs as per the recommendation of Dr. B.S. Konkan Krishi Vidyapeeth, Dapoli, then the total cost required for cashew cultivation per hectare and from that benefit is generated is given in Table 62.

Table 62: Cost of cultivation and production of cashew (for 200 tree/ha)

Sr. No.	Particular	Quantity	Rate (Rs.)	Expenditure (Rs.)
1.	Labour Male	120 days	120	14400.00
	Labour Female	82 days	120	9840.00
2.	FYM	20 CL	250	5000
3.	Fertilizers			
	a) N (1.0 kg/tree) (urea 2.17 kg/tree)	434 kg	6.10	2647.4
	b) P_2O_5 (250 g/tree) (SSP 1.56 kg/tree)	313 kg	7.40	2316.2
	c) K_2O (250 g/tree) (MOP 417 g/tree)	84 kg	16.80	1411.2
4.	Plant protection: propenophos/Lamda cyhalothrin/ Carbaryl (at flushing, flowering at fruit set; need base spray)	9 liter	350.00	3150.00
	Total expenditure on labours, manure, fertilizers and insecticide			**38764.80**
5.	Machinery and equipment depreciation			500.00
6.	Land cess (Shetsara)			50.00
7.	Running investment interest (13 %)			2512.00
8.	Stable investment interest (10 %)			500.00
9.	Land holding (1/6 th part of yield)			1333.33
10.	Amortization value			4852.00
11.	Maintenance charge (10 % of actual expenditure)			1933.00
	Total expenditure			**50445.13**
12.	Yield	20 quintal	8000.00	160000.00
13.	Net profit by reducing expenditure on labours, fertilizers and insecticide)			121253.2
14.	Net profit by deducting all expenditure			**109554.87**
15.	Expenditure for production of 1 kg nut			**31.52**
16.	C:B Ratio			**3.17**

Note: The economics of cashew can be changed to some extent depending upon the conditions.

It is seen from Table 62 that, the total expenditure for cultivation of cashew per hectare is Rs. 50445.13, out of which the actual expenditure on inputs is Rs. 38764.80, which is about 76.84 % of total expenditure. If we considered average 10 kg yield from a full mature cashew tree, then 20 quintal (2 t/ha) nut yield per hectare is obtained and from that we get total income (Present average cashew raw nut rate @ Rs. 80/kg) amounting to 1,60,000. If we reduce the total cost of expenditure of Rs. 50445.13, then net profit from one hectare of Rs. 109554.87 is obtained with C: B ratio of 1: 3.17.

□□□□

CHAPTER – 25

Constraints, Technologies Developed, Future and Development Strategies and Innovation in Horticulture

Some of the constraints in cashew production in the country which limits the productivity of cashew per unit area

- % Large area under cashew in the country is still covered with non-descript genetically inferior seedling progenies.
- % Cashew is a neglected plantation crop among the farmers and generally grown under neglected condition.
- % The land which is not suitable for other plantation crops only are available to cashew crop. Hence, the land which is available for cashew is usually eroded and degraded with poor soil fertility.
- % Farmers are not fully aware of the latest cashew production technologies due to inadequate transfer of technology.
- % Even if farmers are aware of the latest cashew production technologies, non adoption of recommended package of practices to the extent required.
- % Absence of compact and dwarf high yielding varieties, because of which high density planting cannot be adopted to realize the full potentiality of high density planting technology.

Technologies Developed for High Production

In India several techniques have been developed in cashew which is available for farmers' adoption. Hence, there is urgent need to transfer these technologies to actual farmer's field.

- **High yielding varieties of cashew:** Total 40 cashew varieties have been developed and released in the country for cultivation. Out of 40 varieties, 27 varieties are recommended for cultivation and 13 varieties have been withdrawn.
- **Multiplication of clonal materials:** Softwood grafting technique has been standardized and commercially accepted for multiplication. There are about 100 Regional Cashew Nurseries under public and private sector catering to the needs of graft requirement in addition to large number of small to medium cashew nurseries. About 50,000 ha of area are brought under cashew every year in the country by planting over 10 million cashew grafts at the rate of 200 plants/ha (Bhat *et al*., 2010).
- **High density planting:** High density planting 4m x 4 m (625 plants/ha) has shown to be better than 8 m X 8 m (156 plants/ha) resulting in a yield increase of more than 2.5 times over control in the initial ten years.
- **Canopy management:** Canopy management through pruning and training - Pruning and training operation were not followed in the cashew plantations. The plants become unmanageable due to erratic branching. The lower branches die due to shading effect resulting in less number of flowering and fruiting branches, severe pruning can be adopted to improve the yield. Limb pruning also resulting into higher yield. Pruning operation can be practiced after harvesting of the crop. However, preventive measures need to be taken to monitor each plant against stem borer.
- **Rejuvenation of senile cashew plantation by top working:** A Seedling origin plantation becomes unproductive and senile. Such trees can be rejuvenated by grafting with scion sticks of desired high yielding variety. Top worked trees start flowering and fruiting in very next year. Sept - October is the proper time for beheading for grafting. Precautions need to be taken must against attack of cashew stem and root borer.
- **Soil and Water conservation techniques:** Terrace with crescent bund and opening of catch-pits treatment is the best soil and water conservation technique for cashew garden grown in flat to medium slopes and serves the purpose in undulating soils. The available water during pre-monsoon and post-monsoon season can be effectively utilized for higher yields.
- **Vermicomposting of cashew biomass** - Cashew is a deciduous tree, which provides approximately 5.0 tonnes of cashew biomass per hectare in a well- established orchard. The biomass decomposes gradually thereby steadily releasing nutrient to the soil. Use of earthworm for production of vermicompost from cashew biomass is a low cost technology for

adoption in the cashew orchard. The vermicompost can be incorporated into the soil of cashew orchards for increasing production.

- **Irrigation, drip irrigation and fertigation:** Cashew is mostly cultivated as a rainfed crop. However, provision of protective irrigation from January to March increases the flowering intensity, fruiting and nut yield. Irrigating 80 litres of water / tree once in four days through drippers from December to March can also increase the yield.
- **Intercropping in cashew orchard:** Suitable intercrops such as black gram, green gram, horse gram, pigeon pea, groundnut, vegetables like cucumber, bottle gourd, pumpkin, tuber crops, fruit crops such as pine-apple, spices such as turmeric, ginger and pepper have been found suitable and profitable intercrops in cashew plantations. During early stages of cashew plantation the land utilization provide sanitation and economic return.
- **Management of insect pests** - The Cashew Stem and Root Borer (CSRB) is a dangerous pest of cashew and kill the plant slowly. Integrated pest management, phytosanitation and chemical insecticides are very effective in controlling the pest. Swabbing the pest infested portion after removal of different stages of pest occurring internally and drenching the root zone with chloropyriphos (0.2 %) as post extraction prophylaxis was effective in reducing pest infestation. Regular visit and inspection and clean cultivation will reduce the infestation. Recommended need based spray schedule can control tea mosquito bug and other pests. New molecules of insecticides I - cyhalothrin was found to be very effective in reducing the damage.
- **Utilization of cashew apple** - Annually about 40-50 lakh tones of cashew apple is produced and large quantity goes to waste. Cashew apple is a non-climacteric fruit, which must be harvested ripe for fresh consumption. Cashew apple can be utilized for preparation of juice, syrup, squash, candy, canned apple, chutney, pickles, jam, vinegar, wine and liquor. Cashew apple contains ascorbic acid, fibre, carotinoid pigments, minerals and most of other chemicals which are of great significance to human health. By-products such as pectin, animal feed and vermicompost can be prepared.
- **On farm level cashew nut processing:** Cashew growers normally sell their raw nuts to the traders who in turn sell the nuts to the cashew processors. The profit, which otherwise would have gone to middlemen, can be earned by the farmers themselves. The farmers can get better return from the cashew nut by processing themselves. Value addition can also

be made at farmer's level. The formation of Self Help Group (SHG) or Farmers Interest Group can promote marketing.

Transfer of Technology

The technology developed in the research station needs to be transferred to the farmers field through public extension agencies, private agencies, and electronic media. The technology is demonstrated by the scientists in the demonstration plots and through training and interactive discussion. The extension workers and NGOs should be trained on latest technologies to carry out the message to the farming community, Para agricultural workers and farmer promoters also be trained for technology transfer. Celebration of Annual Cashew Day also means to create confidence among the farmers on cashew production technology. The plant protection campaign during flushing, flowering and fruiting period will create awareness on occurrence of pest and their control. The technologies can be transferred through region based literatures on cashew production technology, pest and disease management. In the present context development of CD packages on production and protection technology will be helpful to the farmers. The popular technologies are spread through electronic media, Doordarshan, AIR, Video conferencing *etc.* Success stories of farmers will motivate other farmers. Exposure visit is means to reflect the various technologies in the minds of the farmers through interaction. Marketing strategies need to be developed for high return to the farming community (Lenka and Pattnaik, 2010).

Future Strategy

Strategies mentioned by Lenka and Pattnaik (2010) are as follows.

Crop Improvement

- Collection, conservation, evaluation and cataloguing of both exotic and indigenous germplasm accession with high yield and tolerant to biotic and abiotic stress.
- Molecular characterization of germplasm and varieties through DNA and Isozyme markers.
- Identification of dwarf and compact cashew varieties suitable for high density planting for better land use planning (LUP).
- Development of plant ideotype concept for cashew.
- Standardization of micro-propagation techniques for multiplication of cashew elite types / rootstocks.

- Genetic transformation studies and testing and evaluation of transgenics.

Crop Management

- Integrated plant nutrient management.
- Physiology of flowering and off-season flowering and studies on plant growth regulators useful for various horti-techniques, better fruit set and retention and higher final produce.
- Canopy management for rejuvenation of old and senile cashew plantations. High density planting system for increasing yield. Organic farming research including certification. Integrated cashew based farming system. Micro-irrigation studies.

There should be need based research study on farmers' problems along with critical examination with on farm trials on large scale and thereby finding to minimize the input costs and higher return which can boost up the socio-economic status of tribal farmers.

Post-harvest Technology

- Designing and development of value added products from lower graded kernels. Developing technologies for alternate use of by-products of cashew processing industry such as cashew kernel rejects cashew nut shell liquid, cashew shell cake, cashew kernel testa *etc.*
- Cashew apple utilization for production of industrial alcohol, bio-fuel, syrup, liquor on commercial scale.
- Development and refinement of on-farm processing machineries.
- Extraction of neutraceuticals from cashew apple pomace such as natural colours, flavour, pectin and nutritionally beneficial compounds.

Crop Protection

- Development of eco-friendly IPM strategies including entomo-pathogenic nematodes for control of major insect pests.
- Studies on kairomone and pheromones for effective and economic control of tea mosquito bug (TMB) and cashew stem and root borer (CSRB).
- Studies on pest complex in post harvest and pre-processing stages.
- Development of weather - pest disease models.
- Analysis of pesticide residues in cashew produce and cashew eco-systems to address concerns of food quality standards.

Developmental Strategies

Area expansion and transfer of technologies are important in increasing productivity and production of raw cashew nut in the country. The promotion of cashew in India has been taking place in private sector, Forest Corporation and state level corporations in different states. Plantations are mostly distributed in small and marginal sector with the holding size of < 1 ha (43 %), 1 - 2 ha (36 %), 2 - 5 ha (12 %) and > 5 ha (9 %). With the set in senility of about 36 % of the plantations, now available, it is not possible to attain self-sufficiency.

The following possible solutions and strategies to overcome cashew production constraintsare:

1. Development of compact and dwarf high yielding cashew variety suitable for high density plantation.
2. Massive replanting programme to replace senile and unthrifty orchards.
3. Development of cost effective cashew production technologies.
4. Increasing productivity by adopting improved cashew production technologies.
5. Enhancing the production of raw cashewnut to reduce the imports by expansion of cashew area both in traditional and non-traditional areas.
6. Development of cashew based integrated farming systems.
7. Effective pest control measures for tea mosquito bug (TMB) / cashew stem and root borers (CSRB).
8. Promotion of quality planting materials for area expansion. Ninety regional nurseries are established in the country and capable of producing 90 lakh grafts per year. Under National Horticulture Mission, small and mega nurseries are coming up to cover 50,000 ha clonal plantation per annum.
9. Organization of intensive thematic campaign and farmers' training programmes on cashew production technology.
10. Introduction of crop insurance policy for cashew.
11. Participation of cashew industry in promotional efforts.
12. Conservation of soil, soil moisture and water conservation techniques.
13. Development of better packaging and marketing strategies for domestic and international market. Promotion of SHG for marketing of cashew.

14. Provision of minimum support price for cashew.
15. Post harvest management including value added products and effective utilization of cashew apple.
16. Popularization on farm processing technology especially for small and medium size land holders for higher income.
17. Promotion of organic cashew cultivation with precision farming development concept.
18. Production forecast in relation to climate change.
19. Declaration of plantation crop status to cashew by all cashew growing states.

There is great demand for cashew kernel in the domestic market and as well as foreign market. The number of cashew processing units in various states is 3650 with a capacity to process more than 14-15 lakh tons of raw cashew nut. In the state of Maharashtra, out of total 2200 processing units, 700 small scale cashew processing units are operating under cottage industry. To achieve the target, the technologies developed at Research Stations need to be extended to the farmers' field. North Eastern hilly regions are suitable for growing organic cashew. Under National Horticulture Mission, there is target to extend 50,000 ha cashew area with clonal materials with suitable horticultural packages. Within 5 year, 2.5 lakh ha clonal cashew area will be available in the country. The varieties developed at National Research Centre for Cashew and All India Coordinated Research Projects on Cashew have developed varieties with a potentiality of 2 tonnes raw nut per hectare. The concentrated and coordinated efforts of developmental agencies and researchers will revolutionize the recent cashew industry (Lenka and Pattnaik, 2010).

□□□□

References

Abd-Allah, A.S.E. 2006. Effect of spraying some macro and micro nutrients on fruit set, yield and fruit quality of Washington Navel orange trees. *J. App. Sci. Res.* 2(11): 1059-1063.

Abeysinghe, D. C., Sangakkara, U. R. and Jayasekera, S. J. B. A. 2003. Intercropping of young cashew (*Anacardium occidentale* L.) and its effects on crop productivity and land utilization. *Trop. Agric. Res.* 15: 10-19.

Adams, B. R. 1975. Container production of cashew seedlings root stocks -seed germination in beds as an alternative to direct sowing. *Acta Hort.* 49: 99-107.

Adeyemi, A. A. 1998. Effects of intercropping on weed incidence in cashew (*Anacardium occidentale*) plantations. *Nigerian J. Tree Crop Res.* 2:83-94.

Agnoloni, M. and Giuliani, R. 1977. Cashew cultivation. Library of Tropical Agriculture, Ministry of Foreign Affairs. Instituto Agronomico Per L' oltremare, Florence. pp.168.

Aikpokpodion, P.E., Uloko, B. and Edibo, G. 2009. Nutrient dynamics in soil and cashew (*Anacardium occidentale* L.) leaf and kernel in Kogi State, Nigeria. *J Appl. Biosci.* 25:1573-1578.

Akinwale, T. O. 2000. Cashew apple juice: its use in fortifying the nutritional quality of some tropical fruits. *European Food Res. Technology.* 211:205-207.

Alexander, A. 1986. Optimum timing of foliar nutrient spray. In: Alexander, A. (Ed.) Foliar Fertilization. Martinus Nijhhoff Publisher Dordrecht, Netherland. pp. 44-60.

Aliyu, O. M. 2005. Application of tissue culture to cashew (*Anacardium occidentale* L.) breeding: An appraisal. *African Biotech. J.* 4(13):1485-1489.

Aliyu, O. M. and Awopetu, J. A. 2007. Assessment of genetic diversity in three populations of cashew (*Anacardium occidentale* L.) using protein-isoenzymeelectrophoretic analysis. *Genet. Resour. Crop Evolution.* 54: 1489-1497.

Ames, B. N. 2001. DNA damage from micronutrient deficiencies is likely to be a major cause of cancer. *Mut. Res.* 475:7-20.

Amiri, M. E., Fallahi, E. and Golchin, A. 2008. Influence of foliar and ground fertilization on yield, fruit quality, and soil, leaf and fruit mineral output. *Chinese Germplasm*, 2: 25-28.

Van Eijnatten, C.L.M. 1991. *Anacardium occidentale* L. In: Plant Resources ofSouth-East Asia No. 2: Edible Fruits and Nuts (Eds.) E. W.M. Verheij and R.E. Coronel. Pudoc-DLO, Wageningen, Netherlands. pp. 446.

Ananthakrishnan, G, Ravikumar, R. and Girija, S. 2004. Selection of efficient arbuscular mycorrhizal fungi in the rhizosphere of cashew and their application in the cashew nursery. *Sci. Hort.* 100:369-375

Ananthakrishnan, G, Ravikumar, R., Anand, R. R., Vengadesan, G and Ganapahi, A. 1999a. Induction of somatic embryogenesis from nucellus-derived callus of *Anacardium occidentale* L. *Sci. Hort.* 79:91-99.

Ananthakrishnan, G., Ravikumar, R., Anand, R. R., Vengadesan, G. and Ganapahi, A. 1999b. *In-vitro* plant regeneration from embryo axis of cashewnut. In: Kavi Kishor, R. B. (ed.) Plant Tissue Culture and Biotechnology - Emerging Trends. Universities Press (India), New Delhi. pp. 123-128.

Andrjanirina, G. 1990. Travaux de recherches sur I'amelioration de la production de I'anacarde des plantations industrielles de la SEM. FAMAMA. Madagascar. Doc. FOFIFA. pp.11.

Ankaiah, S. and Rao, P.V. 1983. Studies on the age of leaf, time of absorption and concentration of urea spray on cashew. *Indian Cashew J.* 15: 9-11.

Anonymous. 1965. *Annual Progress report of CCRS, Ullal.* pp. 45.

Anonymous. 1972. Annual News Bulletin for 1971-72, *Indian Forester*. 98: 715-781.

Anonymous. 1998. *NRCC Annual Report 1998.* National Research Centre for Cashew, Puttur, Kamataka. pp. 97-111.

Anonymous. 2002. Cashew production technology, *Technical Bulletin,* Published by DCR, Puttur.

Anonymous. 2004. *Indigenous Technology Knowledge (ITK) in Cashew*, Technical Bulletin published by NRCC for Cashew, Puttur, Karnataka.

Anonymous. 2005. *Cashew Nutritive Value*, Booklet published by NRCC for Cashew, Puttur, Karnataka.

Anonymous. 2006. *IFA Publications Manual: Cashew*. pp. 1-3. http://www.fertilizer.org/ifa/publicat/html/pubman/cashew.htm

Anonymous 2010 b. Cashewnut Shell Liquid. www.cashewindia.org

Anonymous 2011. Cashew Vision-2030. www.directordcrcaju@yahoo.co.in

Anonymous. 2012b. http://wwwdocnet.nic.in/cashewcocoa

Anonymous. 2012c. http://www.niam.res.in/pdfs/DDG Hort_lecture.pdf.

Anonymous. 2012d. http://www.ngoindia.com/baif/default.htm)

Anonymous. 2012a. *Indian Horticulture Database 2011*, National Horticulture Board, Gurgaon, New Delhi, pp.14.

Anonymous. 2013. *Research Recommendations- 2013*, Dr. Balasaheb Sawant Konkan Krishi Vidyapeeth, Dapoli-415 712, Dist. Ratnagiri, Maharashtra. http://www.dbskkv.org

Anonymous 2010. http://www.dbskkv.org/research/research_rsi_vengurla. html-132k.

Antarkar, M. V. and Joshi, G. D. 1987. Studies on physical indices of maturity of some cashew nut varieties. *The Cashew*. 1(4):180-21.

Archak, S., Gaikwad, A. B., Gautam, D., Rao, E.V.V.B., Swamy, K.R.M. and Karihaloo, J. L. 2002. Comparative assessment of DNA fingerprinting techniques (RAPD, ISSR and AFLP) for genetic analysis of cashew (*Anacardium occidentale* L.) accessions of India. *Genome*. 46:362-369.

Ascenso, J. C. 1986. *The Standard Encyclopedia of Horticulture* Vol. I. (17th edition), MacMillan, New York. pp. 1200.

Ascenso, J. C. 1986a. Potential of the cashew crop -1. *Agriculture International*. 38:324-327.

Ascenso, J. C. 1986b. Potential of the cashew crop -2. *Agriculture International*. 38:368-327.

Augustin, A. 1987. Cashew liquor and wine. *Indian Cashew J.* 18(2):19-21.

Augustin, A. 2001. Utilization of cashew apple. *In: Souvenir of World Cashew Congress-2001*, India, Cashew Export Promotion Council of India, Cochin. pp. 57.

Augustin, A. and Unnithan, V. K. G. 1981. An attempt on maturities of cashew apple. *Indian Cashew J.* 14(4):9-11.

Azam-Ali, S. H. and Judge, E. C. 2001. Small scale cashew nut processing. *A Report of Food and Agricultural Organisation of the United Nation*. pp. 86.

Azevedo, D.M.P., Crisóstomo, J. R., Almeida, F. C. G. and Rossetti, A. G. 1998. Estimates of genetic correlations and correlated responses to selection in cashew (*Anacardium occidentale* L.) *Genet. Mol. Biol. (online)*. 21(3). DOI: 10.1590/S1415–47571998000300019

Badhe, V. T. and Magar, S. S. 2004. Influence of different conservation measures on runoff, soil and nutrient loss under cashew nut in lateritic soils of south Konkan region. *Indian J. Soil Conser*. 32:143-147.

Badrinath, M. S., udhir, K. and Chikkaramappa, T. 1997. Soil fertility evaluation for cashew cultivation in coastal Karnataka soils. *The Cashew*. 11:22-23.

Baily, L. H. 1949. *Manual of Cultivated Plants*. MacMillan Co., New York (USA).

Baily, L. H. 1958. *The Standard Encyclopedia of Horticulture Vol. 1*, 17th Edition. MacMillan, New York.

Balasimha, D. and Yadukumar, N. 1993. Effect of plant density on phosynthesis in cashew. *Indian J. Plant Physiology*. 36(1):5-7.

Balasubramaniam, P. P. and Singh, H. P. 2000. Cashew Development in India: An integrated strategical approach. Singh, H. P., Balsubramanian, P. P. and Huballi, V..N. (Eds.) *Indian Cashew Industry*, Directorate of Cashew and Cocoa Development. Kochi-11. pp. 17-29.

Barros, L. M. 1988. Melhoramento. In: Lima VPMS (ed), A 459 *Braz. J. Plant Physiol.* 19(4): 449-461.

Barros, L. M. 1995. Botânica, origem e distribuição geográfica. In: Araújo JPP, Silva, V.V. (eds), Cajucultura: Modernas Técnicas de Produção, pp. 55-72. EMBRAPA/CNPAT, Fortaleza (Frutas do Brasil, 30).

Barros, L. M., Paiva, J. R. and Cavalcant, J. J. V. 1998. Cajueiro anão precoce. Melhoramento genético: Estratégias e perspectivas. *Biotecnologia*. 2: 18-21.

Barros, L.M., Araújo, F. E., Almeida, J. I. L. and Teixeira, L. M. S. 1984. A cultura do cajueiro anão. EPACE, Fortaleza (EPACE Documentos 3).

Beena, B., Salam, M. A. and Wahid, P. A. 1995. Root activity of cashew *(Anacardium occidentale* L.) varieties in relation to phenological phases. *J. Plantation Crops*. 23(1): 35-39.

Beena, Bhaskar 1992. Uptake Pattern of Major and Minor Nutrients in Selected Cashew Types. *M. Sc. Agri. Thesis*, Kerala Agricultural University. pp. 123.

Behrens, R. 1996. Cashew as an agroforestry crop: prospects and potentials. *Tropical Agriculture*. 9: 83.

Bezerra, F. C., Fragoso, H. A. and Hernandez, F.F.F. 1999. Avaliação do estado nutricional de cajueiro anão-precoce, clones CP-76 e CP-09. *Rev. Bras. Frut*. 21: 208-211.

Bezerra, M. A., Claudivan, F. de Lacerda, Enéas Gomes Filho, Carlos, E. B. de Abreu and José, T. P. 2007. Physiology of cashew plants grown under adverse conditions. *Braz. J. Plant Physiol*. 19(4): 449-461.

Bhaskaran, M. N. and Subrahmanyam, S. V. S. 2012. Water soluble fertilizers - NFCL Experience. *Indian J. Fert*. 8(3): 56-61.

Bhat, M. G, Kumaran, P. M. and Rao, E.V.V.B. 1998. Pollination technique in cashew. *The Cashew*. 22(4): 21-25.

Bhat, M. G, Nagaraja, K. V. and Rupa, T. R. 2010. Cashew research in India. *J. Hort. Sci*. 5(1): 1-16.

Bhat, M. G, Yadukumaran, N. and Gangadhar Nayak, M. 2007. *Cashew research achievement, technologies developed and development strategies* In: National Seminar on Research, Development and Marketing of Cashew 20th to 21st November, 2007. Souvenir and extended summaries. pp. 3-15.

Bird, S. M. and Gray, J. E. 2003. Signals from the cuticle affect epidermal cell differentiation. *New Phytology*. 157: 9-23.

Blaikie, S. J. and Chacko, E. K. 1998. Sap flow, leaf gas exchange and chlorophyll fluorescence of container-grown cashew (*Anacardium occidentale* L.) trees subjected to repeated cycles of soil drying. *Aust. J. Exp. Agric*. 38: 305-311.

Blaikie, S. J., Chacko, E. K. and Lu, P. 2001. Productivity and water relations of field-grown cashew: a comparison of sprinkler and drip irrigation. *Australian J. Exptl. Agri*. 41: 663-673.

Boggetti, B., Jasik, J. and Mantell, S. 1999. In vitro multiplication of cashew (*Anacardium occidentale* L.) using shoot node explants of glasshouse-raised plants. *Plant Cell Reports*. 18: 456-461.

Boma, F., Topper, C.P. and Anthony, J. 1998b. Evaluation of various sulphur formulations for the control of powdery mildew (*Anacardium anacardii* Noack) on cashew in Tanzania. In: Topper, C. P., Caligari, P.D.S., Kullaya, A.K., Shomari, S.H., Kasuga, L.J., Masawe, P.A.L. and Mpunami, A.A. (eds) Proceedings of the International Cashew and Coconut Conference. BioHybrids International, Reading, UK. pp. 228-235.

Boma, R., Topper, C.P. and Stathers, T. 1998a. Population dynamics of Helopeltis spp. on cashew in south-em Tanzania. In: Topper, CP., Caligari, P.D.S., Kullaya, AX, Shomari, S.H., Kasuga, L.J., Masawe, P.A.L. and Mpunami, A.A. (eds) Proceedings of the International Cashew and Coconut Conference. BioHybrids International, Reading, UK. pp. 185-187.

Bondada, B. R., Syvertsen, J. P. and Albrigo, L. G. 2001. Urea Nitrogen Uptake by Citrus Leaves. *Hort. Sci.* 36(6): 1061-1065.

Bose, T. K., Mitra, S. K., Farooki, S. K. and Sadhu, M. K. 1999. *Tropical Horticulture* Volume-1, Naya Prokash 206, Bindan Sarani, Kolkata, pp. 563-575.

Brown, P. H. 1999. Foliar Fertilization of Tree Crops. In: *Proceedings of 2nd Intl. Workshop on Foliar Fertilization*. Suwanarit, A., (Ed.), Bangkok. pp. 139-162.

Calton, W. E., Vail, J. W. and Padhve, U. P. 1961. Leaf composition of some tropical crops. *East Afr. Agric. & For. J.* (Kenya). 27(1):13-19.

Cardoso, J.E., Cavalcante, M.B., Calvacante, J. J. V. and Vieira, A.C.F. 1994. Identificacao de fontes de resistencia em clones de cajueiro anao-precoce as principals doencas foliares. Resumos de Congresso Brasileiro de Fruticultura, Sociedade Brasileira de Pruiicultura. *Salvador*. 13: 297-298.

Cardoza, V. and D'Souza. 2002. Induction, development, and germination of somatic embryos from nucellar tissue of cashew, *Anacardium occidentale* L. *Sci. Hort.* 93:367-372.

Carneiro, P. T., Fernandes, P. D., Gheyi, H. R. and Soares, F.A.L. 2002. Germinação e crescimento inicial de genótipos de cajueiro anão-precoce em condições de salinidade. *Rev. Bras. Eng. Agric. Amb.* 6:199-206.

Carlos, E.B. de Abreu, Prisco, J. T., Nogueira, Ana R.C., Bezerra, M. A., Lacerda, Claudivan F. de and Gomes-Filho, E. 2008. Physiological and biochemical changes occurring in dwarf-cashew seedlings subjected to salt stress. *Braz. J. Plant Physiol.* 20(2):105-118,

Cavalcante. A.A.M., Cavalcante. G., Rübensam, J.N., Picada, E., Gomes da Silva, J.C., Fonseca, M. and Henriques, J. A. 2003. Mutagenicity, antioxidant potential, and antimutagenic activity against hydrogen peroxide of cashew (*Anacardium occidentale*) apple juice and cajuina. *Environ. Mol. Mut.* 41: 360–369.

Chacko E. K., Baker, I. and Downton, J. 1990. Towards a sustainable cashew industry for Australia. *J. Australian Institute Agric. Sci.* 3(5):39-43.

Chadha K. L. 1991. Presidential Address. National Group Discussion of Cashew Res. Workers, held at CPCRI, Kasargod (Kerala), 30-31 August and 1st Sept., 1991.

Chattopadhyay, N. and Ghose, S.N. 1994. Studies on the effect of time and extent of pruning in increasing the yield of cashew. *J. Plantation Crop.* 22: 111-114.

Chattopadhyay, P. K., Pal, H., Roy, R. N., Sadhu, M. K. and Bose, T. K. 1983. Some aspects of developmental physiology of cashew *(Anacardium occidentale* L.). *Indian Agric.* 27(2): 149-154.

Chaudhari, A. P., Thakor, N. J., Mahale, D. M., Haldankar, P. M. and Sonawane, S. P. 2012. Physical Properties of Cashew Nut Shell Liquid. *Internatl. J. Application or Innovation in Engineering & Management* 1(2): 172-175.
Web Site: www.ijaiem.org

Chaves, M. M. 1991. Effects of water deficits on carbon assimilation. *J. Exp. Bot.* 42: 1-16.

Chhonker, V. S. and Singh, R. 1967. Effect of plant regulators on air layering in cashew nut (*Anacardium occidentale* L.). *Indian J. Hort.* 24(1-2): 26-30.

Chundawat. B. S. and Sen, N. L. 2002. *Principle of Fruit Culture.* Agrotech Publishing Academy, Udaipur- 313 002. pp. 117.

Conrad, H. H. 1944. Gifts of the Americas - The cashew. *Agric. in Americas.* 4(4): 79.

Cornelius, J. A. 1966. Cashew nut shell liquid and related materials. *Trop. Sci.* (U.K.) 8(2): 70-84.

Cook, J. A. and Boynton, D. 1952. Some factors affecting the absorption of urea by McIntosh apple leaves. *J. Expt. Bot.* 21: 102-111.

Crisóstomo, L.A., Rossetti, A.G., Pimentel, C.R.M., Barreto, P.D. and Lima, R.N. 2005. Produtividade, avaliação econômica, teores foliares de nitrogênio e potássio e atributos industriais de castanha em cajueiro anão-precoce adubado com doses crescentes de nitrogênio e potássio em cultivo sob sequeiro. In: Yamada T, Roberts TL (eds), Potássio na Agricultura, Potafos, Piracicaba. pp. 823-831.

Croxford, A.E., Robson, M. and Wilkinson, M. J. 2005. Characterization and PCR multiplexing of polymorphic microsatellite loci in cashew (*Anacardium occidentale* L.) and their cross-species utilization. *Mol. Ecol. Notes.* 6: 249-251.

Damodaran, V. K. 1966. The morphology and biology of the cashew flower *(Anacardium occidentale* L.). II Anthesis, dehiscence and receptivity of stigma, pollination, fruit set and fruit development. *Kerala Agric. Res. J.* 4(2): 78-84.

Damodaran, V. K., Abraham, J. and Alexander, K. M. 1965. The morphology and biology of the cashew flower I - Flowering habit, flowering season, morphology of the flower and sex ratio. *Kerala Agric. Res. J.* 3(1 & 2): 23-28.

Darlington, C. D. and Janaki Ammal, E. K. 1945. *Chromosome Atlas of Cultivated Plants.* Allen and Un win, London. pp. 397.

Das, S., Jha, T. B. and Jha, S. 1996. In vitro propagation of cashewnut. *Plant Cell Reports.* 15: 615-619.

De Castro, P. 1994. Summary of the study. In: *The World Cashew Economy,* (Eds.) A.M. Delogu and G. Haeuster. NOMISMA, L'Inchiostroblu, Bologna, Italy. pp. 11-12.

De Costa, C. 1578. Tratodo das progase Medicinas das Indias Orientals 1964 Edn. junta de investigacoes do uttramar.

Delogu, A. M. and Haeuster, G. 1994. The World Cashew Economy. NOMISMA, L'Inchiostroblu, Bologna, Italy. Goldblatt, P. (ed.)., 1984. Index to Plant Chromosome Numbers, *1979-1981. Missouri Bot. Garden.*

Deshmukh, M. T., Sawke, D. P., Borude, S. G., Hurni, H. and Tato, K. 1992. Effects of platform bench terraces on the growth and yield of mango and cashew grafts. *Erosion, conservation, and small scale farming report,* Department of Agricultural Engineering, Konkan Agriculture University, Ratnatgiri, India. pp. 477-482.

Dhanaraj, A.L., Rao, E.V.V.B., Swamy, K.R.M., Bhat, M.G., Prasad, D.T. and Sondur, S.N. 2002. Using RAPDs to assess the diversity in Indian Cashew (*Anacardium occidentale* L.) germplasm. *J. Hort. Sci. Biotechnol.* 77:41-47.

Dhanaraj, A.L., Theertha Prasad, D., Rao, E.V.V. B., Swamy, K.R.M. and Sondur, S.N. 2001. Use of RAPD markers in diversity analysis and germplasm management in cashew. In: Plant and Animal Genome IX Conference January 13-17, San Diego, California, USA. Poster: RAPD P307.

Dhamaney, C. P. 1979. Development of bamboo board based on cashew nut shell liquid. *Indian Pulp & Paper.* 22(1): 4.

Dixit, C. K., Yamadagni, R. and Jindal, R. C. 1979. Mineral composition of kinnow (a mandarin hybrid) leaves as affected by foliar application of zinc and iron. *Haryana J. Hort. Sci.* 8: 1-3.

Dixon, R. C. 2012. Foliar Fertilization Improves Nutrient Use Efficiency. *Fertilizer Tech. Bul.* 40: 22-23.

Doering, H. W. and Gericke, R. 1986. The efficiency of foliar fertilization in arid and semi arid regions. In: Foliar Fertilization, Eds: A. Alexander, Martinus Nijhhoff Publisher Dordrecht. pp. 96-125.

Dong, S., Neilsen, D., Neilsen, G. H. and Fuchigami, L. H. 2005. Foliar N application reduces soil NO_3 - N leaching loss in apple orchards. *Plant Soil*. 268: 357-366.

Dora, D. K. and Kabita Sethi 2013. Canopy Management in Cashew - A Lucrative Venture. *Canopy Management and High Density Planting in Subtropical Fruit Crops*, 2013, Eds. Singh, V. K. and Ravishankar, H., CISH, Lucknow. pp. 140-147.

Downing, A. J. 1869. *The fruits and fruit trees of America*. Wiley, New York.

Duke, A. J. 1989. CRC Handbook of Nuts. CRC Press Inc. Boca Raton Florida. pp. 164.

D'Silva, I. and D'Souza, L. 1992. In vitro propagation of *Anacardium occidentale* L. *Plant Cell, Tissue and Organ Culture*. 29:1-6.

Eichert, T. and Burkhardt, J. 2001. Quantification of stomatal uptake of ionic solutes using a new model system. *J. Expt. Bot.* 52: 771-781.

Eichert, T. T. and Goldbach, H. E. 2008. Equivalent pore radii of hydrophilic foliar uptake routes in stomatous and astomatous leaf surfaces - further evidence for a stomatal pathway. *Plant Physiol.* 132:491-502.

Eichert, T. T., Kurtz, A. and Goldbach, H. E. 2006. Investigations on the contribution of the stomatal pathway to foliar uptake. *Acta Hort*. 721: 307-312.

El-Otmani, M., Ait-Oubahou, A., Zahra, F. and Lovatt, C. J. 2002. Efficacy of foliar urea as an N source in substainable citrus production systems. *Acta Hort*. 594: 611-617.

Elsy, C. R., Namboodiri, K. M. N., Vidyadharan, K. K. and Achamma Ommen 1986. Role of pollen and pollinating agent in cashew yield. *Cashew Causerie*. 8(4): 3-4.

Evans, E. M. 1955. The application of cashew nut shell liquid resins. *Rubbers and Plastics Age*. 36:5.

Falade, J. A. 1984a. Variability in soils and cashew tree size. *J. Plantation Crops*. 12: 30-37.

Falade, J. A. 1984b. Effect of lime on the efficiency of two nitrogen fertilizers applied to cashew in acid sandy soils. *J. Plantation Crops*. 12(2):140-145.

Farooki, S. K. and Sreeramu, A. A. 1999. Cashew. In Bose, T. K., Mitra, S. K., Farooki, S. K. and Sadhu, M. K. (eds). *Tropical Horticulture Volume-1*, NAYA PROKASH, Bindan Sarani, Culcutta 700 006. pp. 563-575.

Fernandez, V. and Eichert, T. 2009. Uptake of hydrophilic solutes through plant leaves: Current state of knowledge and perspectives of foliar fertilization. *Critical Rev. in Pl. Sci*. 28:36-68.

Fernandez, V., I., Abadia, O. J. and Abadia, A. 2009. Foliar iron-fertilisation of fruit trees: Present knowledge and future perspectives - a review. *J. Hort. Sci. & Biotech*. 84: 1-6.

Fernandez, V., Sotiropoulos, T. and Brown, P. 2013. *Foliar Fertilization: Scientific Principles and Field Practices*, First edition, IFA, Paris, France. www.fertilizer.org

Fisher, E. G and Walker, D. R. 1955. The apparent absorption of phosphorous and magnesium from sprays applied to the lower surface of McIntosh apple leaves. *Proc. Amer. Soc. Hort. Sci*. 65:17-24.

Flore, J. A. and Bukovac, M. J. 1982. Factors influencing absorption of 14C (2-chloroethyl) phosphonic acid by leaves of cherry. *J. Amer. Soc. Hort. Sci*. 107: 965-968.

Flexas, J., Bota, J., Loreto, F., Cornic, C. and Sharkey, T. D. 2004. Diffusive and metabolic limitations to photosynthesis under drought and salinity in C_3 plants. *Plant Biol.* 6:269-279.

Fofifa 1981. Foibe Fiompiana sy Fambolena. Recherches sur les anomalies constatees dans la floraison et la fructification d'anacardiers. Rapport Final. Centre National de la Recherche Appliquee au Oeveloppement rural (Madagascar). pp. 98.

Fragoso, H.A. 1999. Macronutrient removal by both cashew nut and apple of two dwarf cashew clones. *Revista Brasileira De Ciencia Do Solo*. 23:603-608.

Franke, W. 1967. Mechanism of foliar penetration of solution. *Annu. Rev. Plant Physio*. 18:281-300.

Fred, R. R. and Melinda, M. R. 1998. The cashew nut, *Anacardium occidentale* (anacardiaceae), and its perennial association with ants: extra floral nectary location and the potential for ant defense. *American J. Botany*. 85(6):835-849.

Free, J. B. and Williams, I. H. 1976. Insect pollination of *Anacardium occidentale* L., *Mangifera indica* L., *Blighia sapida* Koenig and, *Persea americana* Mill. *Tropical Agriculture (Trinidad)*. 53(2):125-139.

Freire, E., Das, C. O. and Cardoso, J. E. 1995. Doencas do cajueiro. In: de Araujo, J.P.P. and da Silva, V.V. (eds) Cajucultura: Modernas Tecnicas de Producao. EMBRAPA/CNPAT, Fortaleza. pp. 249-267.

Gajbhiye, R. C., Sapkal, B. B., Patil, B. P. ans Deshpande, S.B. 2005. Vengurla-8: A New Bold Nut and Big Apple Cashew Variety. *The Cashew*. XIX(2):11-13.

Gajbhiye, R. C., Jalgaonkar, V.N., Bendale, V. W. and Gawankar, M. S. 2007. Effect of Growth Regulator to Improve Fruit Set and Yield in Cashew. *Paper presented in National seminar on Research, Development and Marketing of Cashew.* In: Souvenir and Extended summaries, ICAR Research Complex for Goa, 20-21st November, 2007. pp. 136-137.

Gajbhiye, R. C., Padhiar, B. V. and Shah, N. I. 2014. Effect of foliar spray of water soluble nutrients on growth and flowering in cashew (*Anacardium occidentale* L.) cv. Vengurla-4. *Global conference on technological challenges and human resources for climate smart horticulture-Issues and strategies.* 28-31 May, 2014 ACHF, NAU, Navsari (Gujarat), *Book of Abstracts.* p: 280.

Gajbhiye, R. C., Padhiar, B. V., and Shah, N. I., Patil, S. J. and Kolambe, B.N. 2013. Effect of foliar spray of water soluble nutrients on fruit set, fruit retention and yield of cashew (*Anacardium occidentale* L.) cv. Vengurla-4. *National Seminar on Tropical and Subtropical Fruits*. 9-11 January, 2013. ACHF, NAU, Navsari. *Book of Abstracts*. p. 95.

Gary, Y. and Grigg, M. G. 1999. Foliar nutrition of turf grasses: Superintendents can save money and reduce leaching by feeding grass through the leaves. *Growers' Association Yearbook.* 19(1-4):31-32.

Gawankar, M. S., Sawale, R.D., Pawar, S. N. and Chavan, S. A. 2010. Effect of Ethrel® on flowering, sex-expression and yield in cashew. *J. Hort. Sci.* 5 (1):68-70.

Ghosh, S. N. 1988. Studies on the flowering phases and the sex ratio in different cashew types at cashew research station, Jhargram. *The Cashew*. 2(1):14-16.

Ghosh, S. N. 1990. Studies on the NPK requirement of cashew (*Anacardium occidentale* L.) in lateritic tract of West Bengal. *The Cashew*. 4:6-9

Ghosh, S. N. and Bose, T. K. 1986. Nutritional requirement of cashew (*Anacardium occidentale* L.) in lateritic tract of West Bengal. *Indian Cashew J.* 18:11-16.

Ghosh, S. N., Chattopadhyay, N. and Hore, J.K. 2000. Effect of plant spacing on yield of cashew. *Recent Advances in Plantation Crops Research*, 13:157-159. (Proceedings of PLACROSYM - XIII, 1998)

Godse, S.K., Bhole S.R., Patil, R.P., Shivpuje, P.R. and Sapkal, B.B. 2004. *Status of Management of insect pests of cashew in Maharashtra*. Report presented in National Group Meeting of Scientists of AICRP on Cashew at NRC for cashew, Puttur, Karnataka, India.

Godia, F., Casas, C. and Sola, C. 1987. Asurvey of continuous ethanol fermentation systems using immobilized cells. *Process Biochem*. 22:43-48.

Gogate, S. and Nadgauda, R. 2000. Induction of somatic embryogenesis in cashewnut (*Anacardium occi-dentale* L.). *In Vitro Cellular & Dev. Bio*. 36:41-46.

Gogate, S. S. and Nadgauda, R. S. 2003. Direct induction of somatic embryogenesis from immature zygotic embryo of cashew (*Anacardium occidentale* L.). *Sci. Hort.* 92:75-82.

Goldblott, P. 1984. Index to Plant Chromosome Numbers, 1979 & 1981. Missouri Bot. Garden.

Gondins, A.GF. 1973. A agro-industria do caju no Nordeste. Situacao actual e perspectivas. Banco do Nordeste di Brazil SA, Departamento de Estudis Economicos de Nordeste (ETENE), Fortaleza (Brazil). pp. 222. In: OHLER, J.G. 1979. Cashew. Communication 71. Koniklijk Instituut Voor de Tropen. Amsterdan.

Gonzalej, C., Zheng, Y. and Lovatt, C. J. 2010. Properly timed foliar fertilization can and should result in a yield benefit and net increase in grower income. *Acta Hort.* 868: 272-283.

Griss, F. 1844. Nouvelles experiences sur Paction des composes ferrigineaux soluble appliques a la vegetation et special ment de la chlorose et la debilite des plantes. *Compt. Rend. Acad. Sci. (Paris).* 19:1118-1119.

Grundon, N. J. 1999. Cashew Nuts in North Queensland Respond to Phosphorus and Sulphur Fertilizers. *Better Crops International.* 13(2):22-24.

Grundon, N. J. 2001. A desktop study to predict the fertilizer requirements of cashew trees in Northern Australia. *CSIRO Land and Water, Atherton, Technical Report.* 32/01.

Guilherme, E. A. 2006. Desenvolvimento de plantas de cajueiro anão precoce (*Anacardium occidentale* L.) irrigadas com águas salinas. *M.Sc. Dissertation* Fortaleza, Universidade Federal do Ceará.

Gupta, C. R. 1999. Intercropping in cashew orchard under rainfed condition - a model for Bastar plateau zone of Madhya Pradesh. *The Cashew.* 13:18-22

Guruprasad, T.R., Khan, M. M., Ahamed, L. and Hanumanthappa, H. 1988. Success of cashew top working technologyat farmer½s doors. *The Cashew.* 2(1):10-11.

Guruprasad, T. R., Hanumanthappa, M., Sudhir, K. V. and Dhananjaya, B. 2009. Technologies for improvement of senile cashew gardens. *The Cashew and Cocoa J.* 1(3):18-22.

Haldankar, P. M., Deshpande, S.B., Chavan, V. G. and Rao, E.V.V.B. 2003. Weather associated yield variability in cashewnut. *Agrometeorology J.* 5:73-76.

Hamilton, J.M., Palmiter, D. H. and Anderson L. C. 1943. Preliminary tests with uramon in foliage sprays as a means of regulating the nitrogen supply of apple trees. *Proc. Amer. Soc. Hort. Sci.* 42:123-26.

Hammonds, T. W. 1977. The distribution of cashew shell liquid type compounds in the cashew plant. *Trop. Sci.* 19:155-59.

Hariharan, Molly, U. and Krishnan, K. 1985. Effect of gibberellic acid and total protein contents in developing kernel of cashew. *Acta Hort.* 108:25-30.

Harishu Kumar, P., Nagabushanam, S. 1981. Leaf nutrient content of cashew as influenced by different methods of fertilizer application. *Indian Cashew J.* 13(3):9-11.

Harishu Kumar, P. and Sreedharan, C. 1986. Nut characters as influenced by different levels of NPK in cashew (*Anacardium occidentale* L.). *Indian Cashew J.* 18:15-17.

Harishu Kumar, P., Nair, B. P., Ramkhiyappan, P., Nagabhushanan, S. and Mohan, E. 1982. Variation in mineral composition of leaves of cashew (*Anacardium occidentale* L.) as affected by season, position and age. *Indian Cashew* J. 14(1):7-10.

Havlin, J. L., Beaton, J. D., Tisdale, S. L. and Nelson, W. L. 2005. Soil Fertility and Fertilizers: An Introduction to Nutrient Management. 7th Edition. Pearson/Prentice Hal, Upper Saddle River, NJ. USA. pp. 515.

Haynes, R. J. and Goh, K. M. 1977. Review on physiological pathways of foliar absorption. *Sci. Hort.* 7:291-302.

Heard, T. A., Vithanage, V. and Chacke, E. X. 1990. Pollination Biology of cashew in the Northern Territory of Australia. *Aus. J. Agric. Res.* 41: 1101-14.

Hegde, M., Kulasekaran, M., Shunmugavelu, K.G and Jayasanker, S. 1990. *In vitro* culture of cashew seedlings and multiple plantlets from mature cotyledons. *Indian Cashew J.* 20:19-24.

Hegde, M., Kulasekaran, M., Jayasankar, S. and Shanmugavelu, K. G. 1994. *In vitro* embryogenesis in cashew (*Anacardium occidentale* L.). *Indian Cashew J.* 21(4):17-25.

Hull, H. M., Morton, H. L. and Wharrie, J. R. 1975. Environmental influences on cuticle development and resultant foliar penetration. *Bot. Rev.* 41:421-451.

Impey, R.L. and Jones, W. W. 1960. Rate of absorption of urea by intact leaves of Washington navel orange. *Proc. Amer. Soc. Hort. Sci.* 76:181-185.

Ingle, S.T., More, M.R. and Narvankar, D.S. 2005. *Management of irrigation to cashew plantation through drip irrigation.* Drainage and Irrigation Water Mgt, College of Agril Engineering and Technology, Dr. B.S. Konkan Krishi Vidyapeeth, Dapoli – 415 712. pp. 165-169.

Inglese, P., Gullo, G. and Pace, L. S. 2002. Fruit growth and olive quality in relation to foliar nutrition and time of application. *Acta Hort.* 586:507-509.

Inoue, T., Lefuji, H., Fujii, T., Soga, H. and Satoh, K. 2000. Cloning and characterization of a gene complementing the mutation of an ethanol- sensitive mutant of sake yeast. *Biosci. Biotechnol. Biochem.* 64:229-236.

Inoue, T., Lefuji, H., Fujii, T., Soga, H. and Satoh, K. 2010. Immobilized cells. *Process Biochem.* 22:43-48.

Jabeen, R. and Ahmad R. 2012. *Provision of Essential Minerals Through Foliar Sprays, Water Quality, Soil and Managing Irrigation of Crops.* Dr. Teang Shui Lee (Ed.), ISBN: 978-953-51-0426-1, In Tech.

Jeffree, C. E. 2006. The fine structure of the plant cuticle. In: Biology of the Plant Cuticle. M.M.C. Riederer, editor. 23:11-125.

Jha, T. B. 1988. In vitro morphogenesis in cashewnut Anacardium occidentale L. *Indian J. Exper. Bio.* 26:505-507.

Johnson, D. 1973. The botany, origin, and spread of the cashew *Anacardium occidentate* L. *J. Plantation Crops.* 1:1-7.

Johnson, D. 1977. Cashew apple products in Brazil. *Indian Cashew J.* 11(2):5-7.

Johnson, R. S., Rosecrance, R, Weinbaum, S., Andris, H. and Wang, J. 2001. Can we approach complete dependence on foliar-applied urea nitrogen in an early-maturing peach?. *J. Amer. Soc. Hort. Sci.* 126: 364-370.

Joseph, K. T. 1981. Seed-borne fungi on cashew (Anacardium *occidentale* L.). *Cashew Causerie.* 3(1):12.

Joshi, D. H. 1982. Cashew Industry in India- The initial stages and growth. *Indian Cashew J.* 15:5-9.

Kannan, S. 1980. Mechanisms of foliar uptake of plant nutrients: accomplishments and prospects. *J. Plant Nutrition.* 2:717-735.

Kannan, S. 2010. Foliar fertilization for sustainable crop production. *Sust. Agri. Rev.* 4:371-402.

Kannan, S. and Wittwer, S. H. 1965. Effects of chelation and urea on iron absorption by intact leaves and enzymically isolated leaf cells. *Plant Physiol.* 40:12.

Kang, M.S., Aulakh, M. S. and Dhiman, J. S. 2008. Agricultural development in Punjab: Problems, possible solution and new initiatives. *Agalumnus.* 38:26-34.

Kannaiyan, S., Parthasarathy, V. A. and Prasath, D. 2010. *Intellectual Property Rights- Initiatives in Horticultural Crops.* In: Recent Initiative in Horticulture, (Eds) K.L Chadha, A. K. Singh and V. B. Patel, Westville Publishing House, New Delhi. pp. 123-139.

Kerstiens, G. 2010. Plant cuticle. ELS.

Kesavachandran, R. and Khader, M.D. 1990. *Procs. Natl. Sem. Biotech. approaches for upgradation of Agri. and Hort. Crops*, TNAU, Coimbatore, India. pp. 57- 58.

Khader, K. B. A. 1986. Distribution of cashew roots in the laterite soil of West Coast of India. *Indian Cashew J.* 17(2):15-19.

Khalatkar, A.S. and Bhargava, Y.R. 1987. Effect of gamma radiations on the nuts (seeds) of *Anacardium occidentale. Acta Hort.* 215:45-50.

Khan, M. M., Hiremath, I.G. and Kumar, D. P. 1988. Topworking in cashew – A rapid method of producing scion material. *Cashew Bull.* 25:11-13.

Khan, M. M., Kumar, D. P. and Hiremath, I. G. 1987. Pruning studies in cashew. *The Cashew.* 1(1):11-12.

Khosla, P. K., Sareen, T. S. and Mehra, P. N. 1973. Cytological studies on Himalayan Anacardiaceae. *Nucleus.* 16:205-209.

Koch, K., Hartmann, K. D., Schreiber, L., Barthlott, W. and Neinhuis, C. 2006. Influence of air humidity on epicuticular wax chemical composition, morphology and wettability of leaf surfaces. *Environ. Exp. Bot.* 56:1-9.

Konhar, T. and Arun Mech 1988. Effect of growth regulators on flowering, fruit set and fruit retention in cashew. *Indian Cashew J.* 18:17-19.

Kozlowski, T. T. 1997. Responses of woody plants to flooding and salinity. *Tree Physiol. Monograph.* 1:1-29.

Kramer, P. J. and Boyer, J. S. 1995. Water Relations of Plants and Soils. Academic Press, San Diego.

Krishnakumar, R., Maheswari, P. and Dongre, T. K. 2003. Effect of apple extract of *Anacardium occidentale* in the management of red palm weevil. *Indian J. Coconut.* 34(1):14-18.

Krishnamurthy, K., Khan, M.M., Hiremath, I. G., Nagaraju, A.P., Mallik, B. and Hedge, M. 1985. Three decades of cashew research at the Agricultural Research Station, Ullal (1954-84) *Station. Tech. Bull. No. 3.* NARP, UAS Agricultural Research Station, Ullal, Karnataka. pp. 34.

Kubo, I., Ochi, M., Vieira, P. C. and Komatsu, S. 1993. Antitumour agents from cashew (*Anacardium occidentale* L.) apple juice. *J. Agri. food Chemistry.* 41:1012-15.

Kuepper, G. 2003. Foliar Fertilization. *ATTRA*, March 2003 Publication. pp. 1-10. www.attra.ncat.org

Kumar, P.H., Nair, B.P., Rakhiappan, P., Nagabhushanam, S. and Mohan, E. 1982. Variation in mineral composition of leaves of cashew (*Anacardium occidentale* L.) as affected by season, position and age. *Indian Cashew J.* 14:7-10.

Kumar, D. P. 1990. The growth and yield performance of top-worked cashew trees. *Cashew Bull.* 27:7-10.

Kumar, D. P., Subbarayappa, A. and Khan, M. M. 1990. Soft wood grafting- Successful method in clonal propagation in cashew. *The cashew.* 3:3-5.

Kumar, D. P. and Hegde, M. 1999. Recycling of plant nutrients through cashew leaf litter. *The Cashew.* 13(2):2-8.

Kumar, D. P. Khan, M. M, and Shridhar, P. H. 1997. Effect of NPK nutrients on vegetative growth and floral characters of cashew (*Anacardium occidentale* L.). *Cashew Bull.* 34(4): 3-9.

Kumar, D. P., Hegde, M. and Khan, M. M. 1993. Effect of nitrogen, phosphorus and potassium on growth and yield of cashew in the coastal soils of Karnataka. *Cashew Bull.* 30(12):9-12.

Kumar, D. P., Khan, M. M. and Subbarayappa, A. 1988. Effect of foliar application of fertilizers on the yield of cashew. *The Lal Bough.* 30(1):28-30.

Kumar, D. P. and Hedge, M. 1994. "Ullal-3" - A new cashew variety for Karnataka. *The Cashew.* 1(94):11-14.

Kumar, D.P., Hegde, M. and Guruprasad, T.R. 1998. Fertilisation for higher nut production in cashew (*Anacardium occidentale* L.). *Cashew Bull.*, 35(4):2-4.

Kumar, D.P., Khan, M. M. and Venkataramu, M. N. 1995. Effect of NPK and growth regulators on harvesting, nut yield, shelling per cent and kernel grade of cashew. *J. Plantation Crops.* 23:96-104.

Kumar, D.P., Khan, M.M. and Melanta, K.R. 1994. Effect of growth regulators on sex expression, fruit set, fruit retention and yield of cashew. *Proc. of the XI Symposium on Plantation Crops*. 30 November - 3 December 1994, Calicut, Kerala, India. pp. 610-627.

Kumar, G. K. and Aravindakshan, M. (1985). Studies on some qualitative aspects of cashew apple. *Indian Cashew J.* 17:19-21.

Kumar, G.N.M., Narayanaswamy, P., Mokashi, A.N. and Mohankumar, G.N. 1984. Fruit growth and development in cashew (*Anacardium occidentale* L.). *J. Plantation Crops*. 12(1):81-84.

Kumar, H. P. 1985. 'Effect of NPK on seed progenies and air layers of cashew'. *Ph. D. Thesis,* Kerala Agricultural University, Vellanikkara, India.

Kumar, N. S., Aggarwal, P. K., Rani, S., Jain, S., Saxena, R. and Chauhan, N. 2011. Impact of climate change on crop productivity in Western Ghats, coastal and northeastern regions of India. *Cur. Sci.* 101(3):332-341.

Kutty, N.M.C. 2000. Development pattern, storage behaviour and variability in processing character of cashew apple. *Ph.D. Thesis*. Kerala Agriculture University, Thrissur.

Lakshmi Sita, G. 1989. Differentiation of embryos and leafy shoots from callus cultures of cashew (*Anacardium occidentale* L.). In: XIII Plant Tissue Culture Conference, October 18-20, Shillong. *Plant Tissue Culture Association of India, Shillong.* Abstract 71.

Lakshmipathy, R. 2000. Symbiotic response of cashew root stocks to different VAM fungi. *The Cashew*. 14:20-24

Latha, A. 2000. Cashew varieties tolerant to N deficient soils. *The Cashew*. 14: 29-39.

Latha, A. and Salam, M. A. 2001. Cashew varieties tolerant to K-deficient soils. *J. Plantation Crops*. 29:16-21.

Latha, A. and Salam, M. A. 2003. Response of cashew seedlings to soil moisture regimes. *The Cashew*. 17:17-20.

Latha, A., John, P.S. and George, M. 1994. Productivity of cashew as influenced by chlorophyll and leaf nitrogen content. *J. Trop. Agril.* 32:86-88.

Lawlor, D. H. 2002. Limitation to photosynthesis in water stressed leaves: stomata vs. metabolism and the role of ATP. *Ann. Bot.* 89:275-294.

Leach, R. 1973. Insect injury stimulating fungal attack on plants. *Ann. Appl. Biol*. 22:525-537.

Leece, D. R. 1978. Foliar absorption in *Prunus domestica* L. Nature and development of surface wax barrier. *Australian J. Plant Physiol.*, **5**:749-766.

Lefebvre, A. 1969. L'anacardier, une richesse de Madagascar. *Paris Bull. De Madagascar.* 19(272):3-50.

Lefebvre, A. 1973. Mineral nutrition of the cashew. *Fruits*. 28(10):691-697.

Lenka, P. C. and Pattnaik, A. K. 2010. *Cashew: Research and Development.* (Eds) K. L. Chadha, A. K. Singh and V. B. Patel, Recent Initiative in Horticulture, Westville Publishing House, New Delhi. pp. 266-274.

Lenka, P. C., Mohapatra, K. C., Mishra, N. K. and Dash, S. 1998. Studies on apple characters of cashew types and its physico-chemical characters. Orissa *J. Hort.* 10(1):20-24.

Lester, G. E., Jifon, J. L. and Makus, D. J. 2006. Supplemental foliar potassium applications with or without a surfactant can enhance netted muskmelon quality. *Hort. Sci.* 4:741-744.

Lester, G. E., Jifon, J. L. and Makus, D. J. 2010. Impact of potassium nutrition on food quality of fruits and vegetables: A condensed and concise review of the literature. *Better Crops.* 94(1):18-21.

Leva, A. R. and Falcone, A. M. 1990. Propagation and organogenesis in vitro of *Anacardium occidentale* L. *Acta Hort*. 280:143-145.

Levitt, J. 1972 Responses of Plants to Environmental Stresses, Academic Press, Bostan.

Lidster, P. D., Tung, M. A. and Yada, R. G. 1979. Effects of pre-harvest and post-harvest calcium treatments on fruit calcium content and the susceptibility of "Van' cherry to impact damage. *J. Amer. Soc. Hort. Sci.* 104:790-793.

Lievens, C., Pylyser, M. and Boxus, P. H. 1989. First results about micropropagation of *Anacardium occidentale* by tissue culture. *Fruits*. 44:553-557.

Lima, R. de L.S. de., Fernandes, V.L.B. and Oliveira, V. H. de. 2001. Growth of seedling dwarf-precocious cashew influenced by organic and mineral fertilizations. *Revista Brasileira De Fruticultura*. 23:391-395.

Lin, M. L. and Staba, E. J. 1961. Peppermint and spearmint tissue cultures I. Callus formation and sub-merged culture. *Lioydia*. 24:139-145.

Machado, 1944. Estudos novos sobre uma planta velha cajueiro (*Anacardium occidentale* L.). *Rodriguesia*. 8:19-48.

Mahajan, S. S. and Patil, P. J. 2012. Exploring Asymmetries in Production, Export and Import in Post-Globalization Era: A Case of Indian Cashew-Nut Industry. *Excel Inter. J. Multidiscipl. Manag. Study*. 2(3):58-77.

Majule, A. E., Topper, C. P. and Nortcliff, S. 1998. The environmental effects of sulphur dusting cashew trees in southern Tanzania. In: Topper, C.P., Caligari P.D.S., Kullaya, A.K., Shomari S.R, Kasuga, L.J., Masawe, P.A.L. and Mpunami, A.A. (eds.) *Proc. International Cashew and Coconut Conference*. BioHybrids International, Reading. pp. 84-93.

Mandal, R. C. 1992. *Cashew Production and Processing Technology.* Agro Botanical Publishers (India), Bikaner. pp. 195.

Mandal, R. C. 2007. *Cashew Production and Processing Technology*, pub. Agrobios (India), Jodhpur 342 002. pp.212.

Mandal, R. C. and Rethinam, P. 1994. Nutrition of cashew. In: *Advances in Horticulture Vol. 9 - Plantation and Spice Crops Part 1*, Eds. K. L. Chadha and P. Rethinam, Malhotra Publishing House, New Delhi, India. pp. 457-466.

Mandal, R. C., Yadukumar, N. and Mohan, E. 1991. Cashew based farming system. *J. Plantation Crops*. 18(Supplement):56-150.

Mane, M. S., Mahadkar, U. V., Ayare, B. L. and Thorat, T. N. 2009. Performance of mechanical soil conservation measures in cashew plantation grown on steep slopes of Konkan. *Indian J. Soil Conser*. 37:181-184

Manjunatha, D. 2001. Effect of nutrients (NPK) supplied through irrigation water on growth of rootstocks and grafts of cashew. *The Cashew*. 15:13-18.

Manoj P.S. and George T.E. 1993. Evaluation of F_1 hybrids. *The Cashew*. 6(2):3-4.

Manoj, P. S. and George T. E. 1993. Heterosis in cashew. *The Cashew*. 6(3):7-9.

Mantell S.H., Boggetti, B., Bessa, A.M.S., Lemos, E.E.P., Abdelhadi, A. and Mneney, E.E. 1998. Micropropagation and micrograiting methods suitable for safe international transfers of cashew. In: Topper, CP., Caligari, P.D.S., Kullaya, AJC, Shomari, S.H., Kasuga, L.J., Masawe, PAX. and Mpunami A.A. (eds). *Proceedings of the International Cashew and Coconut Conference*. BioHybrids International, Reading. pp. 95-107.

Maranan, F. S. and Mendioro, M. S. 2008. Isozyme analysis of Philippine cashew (*Anacardium occidentale* L.) germplasm. *The Philippine Agri. Scient*. 91:329-333.

Marriappan, S., Prabhakaran, J. and Sambandamoorthy, S. 1995. Effect of growth regulators on sex expression and fruit set in cashew (*Anacardium occidentale* L.). *The Cashew*. 9:11-13.

Masawe, P. A. L. 1994. Aspects of breeding and selecting improved cashew genotypes. *Ph. D. Thesis*. The University of Reading.

Marschner, H. 1995. *Mineral Nutrition of Higher Plants*. Academic Press, London. pp. 887.

Marschner, H. 2003. *Mineral Nutrition of Higher Plants*. 2nd edition (6th printing), Academic Press, 525 B Street, Suite 1900, San Diego, California, USA. http:/www. academicpress. com.

Massari, F. 1994. Introduction. In: The World Cashew Economy, (Eds.) A.M. Delogu and G Haeuster. NOMISMA, Tnchiostroblu, Bologna, Italy. pp. 3-4.

Mathew, J., Mini, C. and Sobhana, A. 2010. Cashew Apple - Economic Utilization through Value Addition. In: Recent Initiative in Horticulture, (Eds) K.L Chadha, A. K. Singh and V. B. Patel, Westville Publishing House, New Delhi. pp. 612-623.

Mathew, R. 1990. Yield prediction in cashew based on foliar nutrient levels. *M.Sc. (Ag) Thesis*, Kerala Agricultural University, Vellanikkara, India.

McLellan, T. 2005. Correlated evolution of leaf shape and trichomes of Begonia dregei (Begoniaceae). *American J. Bot.* 92:1616-1623.

Menon, M. A., Chandra Mohanan, R., Harishu Kumar, P., Rakhiyappan, P. and Nair, B. P. 1978. Yellow leaf pot of cashew. *Indian Cashew J.* XII(3): 15-17

Mika, A. 1986. Physiological responses of fruit trees to pruning. *Hort. Rev.* 8:339-378.

Mikkelsen, R. L. 2007. Biuret in urea fertilizers. *Better Crops*. 91(3):6-7.

Millanzi, K.J.K. 1998b. In situ bud vigour method: a field technique for screening of cashew for resistance to powdery mildew disease, Topper, C.P., Caligari P.D.S., Kullaya, A.K., Shomari S.H., Kasuga, L.J., Masawe, R.A.X. and Mpunami A.A. (eds) *Proceedings of the International Cashew and Coconut Conference*. BioHybrids International, Reading, UK. pp. 138-140.

Miller, R. H. 1982. Apple fruit cuticles and the occurrence of pores and transcuticular canals. *Ann. Bot.* 50:355-371.

Mini, A., Salam, M. A. and Kamalam, N.V. 1995. Foliar application of 32P by cashew (*Anacardium occidentale* L.). *J. Plantation Crops*. 23(2):72-77.

Mini, C., Thomas, J. K. and Mathew, J. 2006. Economic potential of cashew apple processing. *National Symposium on Underutilized Horticultural Crops*, IIHR, Bangalore, 8-9 June, 2006.

Mishra, J. N., Paul, J. C. and Pradhan, P. C. 2008. Response of cashew to drip irrigation and mulching in coastal Orissa. *J. Soil and Water Conservation*. 7(3):36-40.

Mitchell, J. D. and Mori, S. A. 1987. The Cashew and its relatives (*Anacardium*: Anacardiaceae). *Memories of the New York Botanical Garden*. 42:1-76.

Mneney, E. E., Mantell, S. H. and Bennett, M. 2001. Use of random amplified polymorphic DNA (RAPD) markers to reveal genetic diversity within and between populations of cashew (*Anacardium occidentale* L.). *J. Hort. Sci. Biotech.* 76:375-383.

Mneney, E. E., Mantell, S.J.H.L., Tsoktouridis, G, Amin, S., Bessa, A.M.S. and Thangavelu, M. 1998. RAPD-profiling of Tanzanian cashew (*Anacardium occidentale* L.). In: Topper, C.P., Caligari, P.D.S., Kullaya, A.K., Shomarj, SJL, Kasuga, L.J., Masawe, RAX. and Mpunami, A.A. (eds) *Proceedings of the International Cashew and Coconut Conference*. BioHybrids International, Reading, UK. pp. 108-115.

Mneney, E.E. and Mantell, S.H. 2001. *In vitro* micrografting of cashew. *Plant Cell, Tissue and Organ Culture*. 66:49-58.

Mobini-Dehkordi, M., Nahvi, I., Ghaedi, K. and Tavassoli, M. 2007. Isolation of high resistant species of *Saccharomyces cerevisiae*. *Res. Pharmacol. Sci.* 2:10-28.

Mohan, E. and Rao, M. M. 1995. Effect of growth regulators and pruning on the growth and yield of cashew *Environment & Ecology*. 13:675-679.

Mohan, E. and Singh R. 1989. Effect of time and severity of pruning in cashew. *Proc. of VIII Symposium on Plantation Crops*. 28-30 December 1988, Kochi, Kerala, India. pp 7-10.

Mohankumar, G. N., Swamy N. P., Mukashi, A. N. 1984. Fruit growth and development of cashew. *J. Plantation Crops*. 12(1):81-85.

Mohapatra, A. R., Kumar, K. V. and Bhat, N. T. 1973. A study on nutrient removal by the cashew tree. *Indian Cashew J.* 9(2):19-20.

Morada, E. K. 1941. Cashew culture. *Philipp. J. Agric.* 12:89-106.

Murthy, M.M.K., Sayi, I.V., Rao, S.V. and Rao, B.H.K. 1975. Note on the biology of shoot and blossom webber *Macalla moncusalis* (Lepidoptera: Pyralidae) of cashew. *Indian J. Ent.* 36(1):76-77.

Murtic, S., Civic, H., Duric, M., Sekularac, G., Kojovic, R., Kulina, M. and Krsmanovic, M. 2012. Foliar nutrition in apple production *African J. Biotech.* 11(46):10462-10468.

Muroi, H., Kubo, A. and Kubo, I. 1993. Antimicrobial activity of cashew apple flavour compounds. *J. Agri. Food Chem.* 41:1106-1109.

Mutter, N. E. S. and Bigger, M. 1962. Cashew. *Bulletin of the Ministry of Agriculture, Tanzania.* 11:1-5. In: (OHLER, J. G. 1979. Cashew. Communication 71, Koniklijk Instituut Voor de Tropen, Amsterdan).

Nagaraja K. V. 1996. Biochemical changes during storage of cashew. *Plant Physiol. and Biochem.* 23(2):163-168.

Nagaraja, K. V. 2000. Harvest and post harvest handling of cashew. Singh, H. P., Balsubramanian, P. P. and Huballi, V. N. (Eds.) *Indian Cashew Industry*, Directorate of Cashew and Cocoa Development. Kochi-11. pp. 118-123.

Nagaraja, K. V. and Giridhar Prabhu, G. 1996. Investigations into the factors influencing the kernel rejects during storage - A preliminary study. *Paper presented at National Seminar on Development of Cashew Industry in India.* 14-15 December 1996, Bhubaneswar.

Nair, M. K., Bhaskara Rao, E.V.V., Nambiar, K.K.N. and Nambiar, M.C. 1979. *Monograph of plantation crops- I Cashew (Anacardium occidentale L.)* Central Plantation Crops Research Institute, Kasargod, Kerala, India. pp. 153.

Nair, P., Sundaresan., George, C. M. and Tajuddin, E. 1972. Studies on foliar and soil application of fertilisers on cashew (*Anacardium occidentale*). *Kerala Agric. Res. J.* 10:10-13.

Nair, S. R. and Mohankumar, N. 1993. Standardization of tissue/apical meristem culture techniques on Horticultural crops of Kerala. *Final Technical Report*, USDA, IN-AES-34 Thiruvananthapuram, Kerala. pp. 3-15.

Nallni, P. V. and Santhakerman, S. 1994. Study on performance of selected types of cashew at cashew research station, Anakkayam, Kerala. *The Cashew.* 5(3):3-6.

Nakasone, H. Y. and Paull, R. E. 1998. *Tropical fruits.* CAB International, U.K. pp. 432.

Nambiar, K. K. N., Sarma, Y. R. and Pillai, G. B. 1973. Inflorescence blight of cashew. *J. Plantation Crops.* 1:44-46.

Nambiar, M. C. 1983. *Annual Report of All India Co-ordinated Spices and Cashew Improvement Project*, Submitted to Central Plantation Crops Research Institute, Kasargod, India.

Nambiar, M. C., Rao, E.V.V.B. and Thankarnma Pillai, P. K. 1990. Cashew. In: Bose, T.K. and Mitra, S.K (eds) *Fruits: Tropical and Subtropical.* Naya Prokash, Calcutta. pp. 386-419.

Narayanan, P. V. and Dordi, M. C. 2000. Modern Trends in Packaging of Cashew Kernels. Singh, H. P., Balsubramanian, P. P. and Huballi, V. N. (Eds.) *Indian Cashew Industry*, Directorate of Cashew and Cocoa Development, Kochi, Kerala. pp. 156-160.

Nawale, R. N., Sawke, D. P. and Salvi, M. J. 1985. Effect of black polyethylene mulch and supplemental irrigation on fruit retention in cashewnut. *Cashew Causerie.* 7:8-9.

Nayak, M. G. 1996. Training and pruning practices for cashew. *The Cashew.* 10(2):5-9.

Nayak, M.G. 2011. Pruning and training practices for cashew. *DCR Tech. Bull. No. 13.* pp. 19.

Nayak, M. G. and Bhat, M. G. 2004b. Registration of cashew germplasm. Compiled information on registered germplasm has been communicated for publication in Indian Journal of Plant Genetic Resources. NBPGR, New Delhi.

Nayak, M. G. and Bhat, M. G. 2005. *Cashew.* In: Biodiversity of Horticulture Crops. Vol. 1, (Ed.) K. V. Peter and Abraham, Z. Daya Publishing House. pp. 330-339.

Nayak, M. G. and Bhat, M. G. 2010. Recent approaches in canopy management and rejuvenation in Cashew. In: *Proc. & recommendations. State level Seminar on 'New Challenges in Cashew Production under the Current Trends of Climate Change'.* 21st August, 2010, ICAR Research Complex for Goa, Ela, Old Goa. pp. 26-33.

Nayak M. G. and Swamy, K. R. M. 2010. Studies on the effect of Paclobutrazol on growth and yield of cashew (*Anacardium occidentale* L.). *J. Plantation Crops.* 38(1):73-74.

Nayak M. G., Swamy, K. R. M., Yadukumar, N. and Raviprasad, T. N. 2008. Effect of limb pruning on growth and yield of cashew (*Anacardium occidentale* L.) varieties. *J. Plantation Crops*. 36(3):300-303.

Nayak, M.G., Saroj, P.L. and Meena, R. K. 2013. High Density Planting and Canopy Management in Cashew. *Article presented in National Seminar on Canopy Management*. 24-25 March, 2013, Horticulture College, TNAU, Coimbatore.

Neelakandan, T. and Usharani, G. 2009. Optimization and Production of Bioethanol from Cashew Apple Juice Using Immobilized Yeast Cells by *Saccharomyces cerevisiae*. *American-Eurasian J. Sci. Res.* 4(2):85-88.

Neto, P. S., Maruta, I., Takaiwa, F., Oono, K., Matsumoto, K. and Gomes, J. A. 1995. Identification of cashew (*Anacardium occidentale* L.) seedlings with RAPD markers. *Acta Hort.* 370: 21-26.

Nijjar, G. S. 1985. *Nutrition of Fruit Trees*. Kalayani Publishers, New Delhi India. pp. 10-50.

Norris, R. F. and Bukovac, M. J. 1968. Structure of the pear leaf cuticle with special reference to cuticle penetration. *Amer. J. Bot.* 61:74-79.

Northwood, P. J. 1966. Some observations on flowering and fruit-setting in cashew (*Anacardium occidentale* L.). *Trop. Agri.* (*Trinidad*), 43(1):35-42.

O'Farrell, P. J., Armour, J. D. and Reid, D. J. 2010. Nitrogen use for high productivity and sustainability in cashew. *Sci. Hort.* 124:19-28.

O'Farrell, P. J., John Armour and David Reid. 2002. The effect of nitrogen on cashew in north Queensland 1995-99. RIRDC Publication No W02/001. RIRDC Project No. DAQ-257A.

Oblisami, G., Santhanakrishnan, P., Pappaiah, C.M. and Shanmugavelu, K.G. 1985. Effect of Azotobacter inoculant and growth regulators on the growth of cashew. *Acta Hort.* 108:40-45.

Ohler, J.G. 1966. Cashew. Department of Agricultural Research, *Royal Tropical Institute*, Amsterdam. p. 26

Ohler, J.G. 1979. Cashew processing. *Tropical Abstracts*. 21:1792-2007.

Ohler, J. G. 1979a. *Cashew.* Communication 71, Department of Agricultural Research, *Koninklijk Instituut voor de Tropen,* Amsterdam, Netherlands. pp. 260.

Ohler, J. G. 1979b. *Cashew.* Kohinklijk Instuut voor de Tropeu, *Amsterdam*. pp. 50-78.

Ohmstedt, U. 1991. Protection des champs d'anacardier par les moyens naturels, renegerable et diaponible au niveau local. *Report Projet PASA*, Sokone, Senegal.

Okwelogu, T. N. and Mackay, P. J. 1969. Cashew moisture relations. *J. Sci. Food and Agric* (UK). 20:697-702.

Oliveira, V.H. de., Barros, L. de M. and Lima, R. N. 2003. Influence of irrigation and genotype on the nut production of precocious dwarf cashew. *Pesquisa Agropecuaria Brasileira.* 38:61-66.

Oliveira, V.H., Miranda, F.R., Lima, R. N. and Cavalcante, R.R.R. 2006. Effect of irrigation frequency on cashew nut yield in Northeast Brazil. *Sci. Hort.* 108:403–407.

Olunloyo, O. A. 1975. A leaf blight disease of *Anacardium occidentale* caused by *Pentalotia paeoniae*. *Plant Disease Rep.* 59(10):829-830.

Panda, J. M. 1990. Effect of pruning severity on yield of cashew nut. *Cashew Bull.* 27:15-17.

Pappaiah, C. M., Hameed, A. S. and Mustaffa, M. M. 1978. Effect of growth regulators on fruit set and yield in cashew (*Anacardium occidentale* L.). *Proceedings of Placrosym-I* (ed. E.V. Nelliat), pp. 360-368. (cited by Subbaiah, 1983).

Parameswaran, N. K. 1979. Factors effecting yield in cashew. *M.Sc. thesis* submitted to Kerala Agri. University, College of Horticulture. Vellanikkara, Trichur, India.

Parente, J. I. G. 1972. Cajueiro. Aspectos economicos e agronomicos. Circulara, institute de Pesquisa A gropecuaria do Nordeste (Brazil) No. 19.

Patnaik, H. P., Das, M. S. and Panda, J. M. 1985. Studies on the fruit set and fruit drop in cashew (*Anacardium occidentale* L.) under Orissa Conditions. *Cashew Causerie*. 7(4):7-8.

Paul, W. R. C. 1936. The cashew nut industry of South India. *Trop. Agri.* 87:166-173.

Pelleschi, S., Rocher, J. P. and Prioul, J. L. 1997. Effect of water restriction on carbohydrate metabolism and photosynthesis in mature maize leaves. *Plant Cell Environ.* 20:493-503.

Peixoto, A. 1960. Caju. Servvico de In formacao Agaricola, Ministerio de Agricultura Rio de Joneiro (Brazil), pp.61. Pursegle, J.W. (ed), 1988. Tropical Crops: Dicotyledons. English Language Book Society, Longman, London. pp. 18-32.

Pfeiffer B., Eis, B., Zimmer, J. and Fieger-Metag, N. 2008. Optimizing crop loading of apples and pears - results (foliar fertilizers, thinning). In: Boos, Markus (Ed.) Ecofruit - 13th International Conference on Cultivation Technique and Phytopathological Problems in Organic Fruit-Growing: Proceedings to the Conference. pp. 324-329.

Philip, V. J. 1984. *In vitro* organogenesis and plant formation in cashew (*Anacardium occidentale* L). *Ann. Bot.* 54:149-152.

Picchioni, G A., Weinbau, S. A. and Brown, P. H. 1995. Retention and the kinetics of uptake and export of foliage applied, labelled boron by apple, pear, prune and sweet cherry leaves. *J. Amer. Soc. Hort. Sci.* 120:28-35.

Pillai, G B. 1975. Pests of cashew nut and how to combat them. *Cashew News Letter* (Oct.-Dec):31-33.

Pillai, P.K.T. and Pillai, GB. 1975. *Indian J. Agric. Sci.* 45:233-234.

Pillai, G B. 1979. *In: Cashew Monograph* (*Anacardium occidentale* L.). Eds. Nair, M. K., Rao, E. V. V. B., Nambiar, K.K.N. and Nambiar, M. C. CPCRI, Kasaragod, India. pp. 55-72.

Pillai, G B., Dubey, O. P. and Singh, V. 1976. Pests of cashew and their control in India. *J. Plantation Crops.* 4(2):37-50.

Pillai. G B., Singh, Vijay, Dubey, O. P. and Abraham, V. A. 1979. Seasonal abundance of tea mosquito elopeltis antonii. On cashew tree in relation to meteorological factors. Int. cashew symp. Cochin.

Prabhu, G 2001. Quality aspects of raw cashew nuts imported in to India. *Cashew Bulletin.* 39(2):22-24.

Prasad, R. 2012. Fertilizers and manures. *Cur. Sci.* 102(6):894-898.

Prasanna Kumar, B., Reddy, M.L.N. and Radhakrishna, Y. 2006. Studies on the NPK requirement of clonally multiplied cashew in sandy soils of Bapatla, Andhra Pradesh. *The Cashew.* 20:23-29.

Pratt, R. K. and Mendoza, D. B. Jr. 1980. Influence of nut removal on growth and ripening of cashew apple. *J. Amer. Soc. Hort. Sci.* 105(4):545-562.

Press, M. C. 1999. The functional significance of leaf structure: a search for generalizations. *New Phytol.* 143:213-219.

Purseglove, J. W. 1988. *Tropical Crops - Dicotyledons.* English Language Book Society, Longman, London. pp. 18-32.

Quinalan, J. D. and. Tobutt, K.R. 1990. Manipulating fruit tree structures chemically and genetically for improved performance. *Hort. Sci.* 25(1):60-63.

Rai, R. and Vatsala, P. 1985. Epidermal morphology and stomatal development in Anacardiaceae. *Acta Hort.* 108: International Cashew Symposium, DCCD, 2011. *The Cashew and Cocoa J.* 3(1):23.

Rejani, Nadgauda, Subramanian, J. and Litz, R. E. 2005. *Anacardium occidentale* Cashew. In *Biotechnology of Fruit and nut crops,* (Eds.) Richard E. Litz, CABI Publishing, CAB International Wallingford, Oxfordshire OX10 8DE, UK. pp. 30-39.

Rejani, R. and Yadukumar, N. 2010. Soil and water conservation techniques in cashew grown along steep hill slopes. *Sci. Hort.* 126:371-378.

Rajapakse, R. 1998. The integrated management of important cashew insect pests of Sri Lanka: a case study of the tea mosquito, *Helopeltis antonii* and the root and stem borer, *Plocaederus ferrugineus*. Ire Topper, CP., Caligari P.D.S., Kullaya, A.K., Shomari, S.H.,

Kasuga, L.J., Masawe, P.A.L. and Mpunami A.A. (eds) *Proceedings of the International Cashew and Coconut Conference*. BioHybrids International, Reading, UK. pp. 165-169.

Raju, V. K. 1979. Studies on fruit set and fruit drop in Cashew (*Anacardium occidentale* L.). *M. Sc. Thesis*, submitted to Kerala Agric. University, College of Horticulture, Vellanikkara. Trichur, Kerala.

Rakotovao, Z. 1999. L'amelioration de la productivite des plantations d'anacardier dans la region de Mahajanga. Table ronde. Doc. FOFIFA and LDI. pp.25.

Ramanayake, S.M.S.D. and Kovoor, A. 1999. *In vitro* micrografting of cashew (*Anacardium occidentale* L). *J. Hort. Sci. & Biotech.* 74:265-268.

Ramaiah, M. S. 1976. Progress of Research in cashew industry. *Fette Seifen Anstrichmittel.* 78(12):472-477.

Ramesh, N., Lingaiah, H. B. and Khan, N. M. 1999. Economics of bio-fertilizer application in cashew graft production under eastern dry zone of Karnataka. *Crop Res.*, (*Hisar*). 17:336-341.

Ramsey, R. J. L., Stephenson, G. R., and Hall, J. C. 2005. A review of the effects of humidity, humectants, and surfactants composition on the absorption and efficacy of highly water-soluble herbicides. Pesticide Biochem. *Physiol.* 82:162-175.

Ravishankar, H., Kumar, K., Singha, A. and Adak, T. 2010. Integrated nutrient management in mango and guava. *Indian J. Fert.* 6(11):12-21.

Roe, D. J. 1994. Some ecophysiological aspects of cashew (*Anacardium occidentale* L.) with emphasis on possible flower manipulation in Maputaland. *Thesis*. Dept. Hortic. Sci. Univ. Natal, Pieterrnaritzburg. South Africa. pp. 164.

Rao, C. B., Dasarathi, T. B. and Rao. Y. Y. 1962. Studies on fruit development in cashew (*Anacardium occidentale* L.). *South Indian Hort.* 10:18-21.

Rao, E. V. V. B. 1989. Cashew Research-Programmes, Achievements and Thrust. *The Cashew.* 2(4):15-20.

Rao, E. V. V. B. 1989. Released Cashew varieties. *The cashew*. 3(1):6-17.

Rao, E. V. V. B. 1998. Cashew crop improvement programmes in India. In: Topper, C. P., Caligari, P. D. S., Kullaya, A. K., Shomari, S. H., Kasuga, L. J., Masawe, P. A. L. and Mpunami, A. A. (eds.) Proceedings of the International Cashew and Coconut Conference. BioHybrids International, Reading, UK. pp. 141-145.

Rao, E. V. V. B. 1999. Cashew development - a futuristic approach. In: Ratnambal, M.J., Kumaran, P.M., Muralidharan, K., Niral, V. and Arunachalam, V. (eds). Improvement of Plantation Crops. CPCRI Publications, Kasargod. pp. 86-90.

Rao, E. V. V. B. and Bhat, M. G. 1996. Cashew breeding - achievements and priorities. *In: Proceedings of the Seminar on Crop Breeding in Kerala.* University of Kerala, Jan. 25, 1996. pp. 47-53.

Rao, E.V.V. B. and Swamy, K.R.M. 1994. Genetic Resources of Cashew. In: *Advances in Horticulture Vol. 9: Plantation and Spice Crops, Part 1*, (Eds.) K.L. Chadha and P. Rethinam. Malhotra Publishing House, New Delhi. pp. 79-97.

Rao, E.V.V.B. and Swamy, K.R.M. 2000. Cashew Research Scenario in India. Singh, H. P., Balsubramanian, P. P. and Huballi, V. N. (Eds.) Indian Cashew Industry, Directorate of Cashew and Cocoa Development. Kochi-11. pp. 30-41.

Rao, G.S.L.H.V.P. and Gopikumar, C. S. 2010. Climate variability and cashew production across the West Coast of India. In: *Proc. & recommendations. State level Seminar on 'New Challenges in Cashew Production under the Current Trends of Climate Change'*. 21st August, 2010, ICAR Research Complex for Goa, Ela, Old Goa-403 402. pp. 18-25.

Rao, G.S.L.H.V.P., Giridharan, M.P. and Jayaprakash Naik, B. 2001. Influence of weather factors on bud-break, flowering and nut quality of cashew. *Indian J. Agril. Sci.* 71:399-402.

Rao, M. R., Reddy, S.K and Rao, R. R. 1984. Effect of NPK on nut yield of cashew. *Cashew Causerie*. 6(1):3.

Rao, V.N.M. 1956. Multiply better yielding cashewnut. *Indian Farming*. 6:15-17.

Rao, V.N.M. 1974. Establishment and management of cashew plantation. *Mimeo. Report of the all India Summer Institute on improvement and management of plantation crops*. Central plantation Crop Research Institute, Kasargod. pp.123-127.

Rao, V.N.M. 1985. Top working in cashew- commercial feasibility. *Cashew Cousserie*. 7(4):5-6.

Rao, V.N.M. and Hassan, M.V. 1957. Preliminary study on the floral biology of cashew (*Anacardium occidentale* L.). *Indian J. Agric. Sci.* 27(3):277-288.

Raviprasad, T. N., Bhat, P. S. and Sundararaju, D. 2009. Integrated pest management approaches to minimize incidence of cashew stem and root borers (*Plocaederus spp.*). *J. Plantation Crops*. 37:185-189.

Reddy Narayana, M. A., Shiranandam, Y.N., Godwa, C.M. and Raju Sidda, M. 1988. A note on the occurrence of floral abnormalities in cashew (*Anacardium occidentale* L.). *The Cashew*. 2(1):12-13.

Reddy Narayana, M.A., Krishnappa, V.S., Gowda, C.M. and Thirumala Raju, G.T. 1989. Studies on sex ratio in cashew (*Anacardium occidentale* L.) selections. *The Cashew*. 2(3):6-8.

Reddy, A. V., Rao, P. V. N., Ankaiah, S. and Rao, I. V. S. 1982. Cashew NPK nutrition in relation to growth under graded doses of nitrogen fertilisation. *Indian Cashew J.* 14(4):15-19.

Reddy, K.S., Rao, R.R., Raju, M.R., Rao, N.H.P. and Rao, A.M. 1985. Promising cashew clones from cashew Research Station, Bapatla. A. P. *Cashew Causerie*. 4(83):2-4.

Reddy Satyanarayana, K. and Rao, R. R. 1985. Cultivation of cashew in Andhra Pradesh. India. *The Cashew*. 7(2):9-11.

Richard, N. K. 1992. Cashew tree nutrition related to biomass accumulation, nutrient composition and nutrient cycling in sandy red earths of Northern territory, Australia. *Sci. Hort.* 52:125-142.

Richards, N. K. 1993. Report on cashew research in northern territory of Australia. *(Tech.Bulletin No. 202)*. CSIRO, Coastal Plains Research Station, Northern Territory, Australia.

Rocchetti, G 1970. *Comportamento alla salsedine dell'anacardio nei primi stadi vita. Rivista di agricoZtura Subtropicale* (Italy). 64:10-12.

Rocha, M., Souza, V. P. and Machiel, R. 2007. *Appl. Biochem. Biotechnol.* 136: 140:185-194.

Rosecrance, R. C., Johnson, R. S. and Weinbaum, S. A. 1998. The effect of timing of post-harvest foliar urea sprays on nitrogen absorption and partitioning in peach and nectarine trees. *Hort. Sci. & Biotech. J.* 73:856-861.

Rudeco 1989. Potential analyse und Vermartungsstudie fUr Chashewusse. Unpublished Report. GTZ, Eschbom. In: BEHRENS, R. 1996. Cashew as an agroforestry crop: prospects and potentials. *Trop. Agri.* 9: 83.

Rupa, T. R. and Bhat, M. G. 2010. Integrated nutrient management in cashew. *Indian J. Fert.* 6(11):62-66.

Sack, L. and Holbrook, N. M. 2006. Leaf hydraulics. *Annual Plant Biol. Rev*. 57: 361-381.

Salam, M. A. 1997. Cashew - a nitrogen lover. *Cashew Bull.* 34(8):5.

Salam, M. A. 1997a. High density planting in cashew- principles and practices. *The Cashew.*11(4):12-13.

Salam, M. A. 1999. Cashew and pepper living in harmony. *The Cashew.* 13(3): 54 -55.

Salam, M. A. and Kamalan, N. V. 1993. Molecular absorption of urea by cashew (*Anacardium occidentale* L.). *J. Plantation Crops.* 21(2):107-109.

Salam, M. A., Pushpalatha, P. B. and Suma, A. 1995. Root distribution patterns of seedlings raised cashew tree. *J. Plantation Crops*. 23(1):59-61.

Salam, M. A. and Peter, K. V. 2010. *Cashew- A Monograph.* Stadium Press (India) Pvt. Ltd., New Delhi. pp. 256.

Salam, M.A. 1999. Cashew varieties suitable for high density planting. *Cashew Bull.* 37:3-5.

Salam, M.A., John, P.J., Joseph, M., Poduwal, M., Kumar, P., Yadukumar, N. and Bhat, M.G. 2008. Quantitative estimation of soil fertility and fertilizer recommendations (QUEFC) for Cashew (*Anacardium occidentale* L.). *Plantation Crops J.* 36:86-94.

Salunkhe, D. K. and Kadam, S. S. 1995. Handbook of Fruit Science and Technology, Pub. Marcel Dekker. pp. 509.

Samuel, J. C. and Singh, H. P. 2000. Strategies for Improved Productivity Through Efficient Water Management for Cashew. Singh, H. P., Balsubramanian, P. P. and Huballi, V. N. (Eds.) *Indian Cashew Industry*, Directorate of Cashew and Cocoa Development. Kochi-11. pp. 77-91.

Sapkal, B. B. 2000. Integration of foliar application of urea and insecticidal spray for yield maximization in cashew. *The Cashew.* 14:41-44

Saroj, P. L. 2013. The Cashew: Issues, Challenges and Strategies. *Cashew News*, Pub. Directorate of Cashew research, Puttur. p. 1-2.

Saroj, P. L. and Nayak, M. G. 2013. *Potential of high density plantation for enhancing quality production of cashew.* Canopy Management and High Density Planting in Subtropical Fruit Crops, 2013, Eds. Singh, V. K. and Ravishankar, H., CISH, Lucknow. pp. 55-63.

Satapathy, C. R. 1993. 'Bioecology of major insect pests of cashew (*Anacardium occidentale* L.) and evaluation of certain pest management practices'. *Ph.D. Thesis,* University of Agricultural Sciences, Bangalore, India.

Satpathy, B. 1988. Thinning, training and pruning of cashew to increase production and productivity of cashewnut. *The Cashew.* 2:6-8.

Sawke, D. P., Gunjate, R. T. and Limaye, V. P. 1985. Effect of nitrogen, phosphorus and potassium fertilization on growth and production of cashew nut. *Acta Hort.* 108:95-99.

Schaper, H., Chacko, E. K. and Blaikie, S. J. 1996. Effect of irrigation on leaf gas exchange and yield of cashew in northern Australia. *Aust. J. Exp. Agric.* 36:861-868.

Schenk, R. U. and Hildebrandt, A. 1972. Medium techniques for induction and growth of monocotyledonous and dicotyledonous plant cell cultures. *Canadian J. Botany.* 50:199-204.

Schonherr, J. 2000. Calcium chloride penetrates plant cuticles via aqueous pores. *Planta.* 212:112-118.

Schonherr, J. 2001. Cuticular penetration of calcium salts: effects of humidity, anions, and adjuvants. *J. Plant Nutrition & Soil Sci.* 164:225–231.

Schonherr, J. 2002. Foliar nutrition using inorganic salts: laws of cuticular penetration. *Acta Hort.* 594:77–84.

Schonherr, J. 2006. Characterization of aqueous pores in plant cuticles and permeation of ionic solutes. *J. Experimental Bot.* 57:2471-2491.

Schonherr, J. and Bukovac, M. J. 1972. Penetration of stomata by liquids- dependence on surface tension, wettability, and stomatal morphology. *Plant Physio.* 49:813-819.

Shanmugavelu, K. G. 1985. Studies on the effect of plant growth regulator in cashew. Cashew Research and Development: *Proceedings of the International Cashew Symposium,* Cochin, Kerala, India. 12-15 March, 1979. pp. 32-9.

Shingre, D.V., Gawankar, M .S. and Jamadagni, B. M. 2003. Effect of irrigation and nitrogen on scion yield in cashew. *The Cashew.* 17:19-22.

Shirly, R.A. and Thimmappaiah 2005. Somatic embryogenesis in nucellar callus of cashew (*Anacardium occidentale* L.). *J. Hort. Sci. & Biotech.* 80:327-331.

Shi-WenGe., Qiao-GuanMing., Wang-YongGang and Zhu-XinTian. 2005. Experiment of applying nitrogen and phosphate fertilizers for cashew nut. *South-China Fruits.* 1: 30 (Research Institute of Forestry, Honghe Harni and Yi Nationality Autonomous Prefecture, Yunnan, China)

Simmond, N. W. 1954. Chromosome behaviour in some tropical plants. *Heredity.* 8:139.

Singh, Gorakh 2010. *Practical Manual on Canopy Management in Fruit Crops.* pp. 20-26.

Singh, H. P. 1998. Keynote address. *National Seminar on Cashew Development in India: Challenges and Opportunities.* Held at Goa on 1st and 2nd March, 1998.

Singh, S., Sehgal, H. S., Pandey, P. C. and Bakshi, B. K. 1967. Anthracnose disease of cashew, its increase epidemiology and control. *Indian Forester.* 93(6):374-376.

Singh, S., Singh, G. and Shyam Singh 1989. A simple method of measuring leaf area in different crop plants. *J. Andaman Science Association.* 5:146-148.

Singh, S.K., Syamal, M.M. and Maurya, A.N. 1992. Effect of NAA and Ethrel on vegetative growth, flowering, fruiting and yield of cashewnut. *The Cashew.* 6:11-12.

Singh, S.N. 1997. Effect of bio-fertilizer (*Azotobacter chroococcum*) in cashew plantation. *Environ. Ecol.* 15:482-484.

Singh, V. and Pant, J. C. 1986. Influence of moisture content of cashew kernel and relative humidity to storage pests. *J. Plantation Crops.* 14(1):29-35.

Sinha, N. 1999. Leaf development in angiosperms. *Annual Plant Physiol. & Plant Mol. Biol. Rev.* 50:419-426.

Sivaprasad, P., Sulochana, K.K., George, B. and Salam, M. A. 1992. Growth and phosphorous uptake of cashew as influenced by inoculation with mycorrhizae. *The Cashew.* 6:16-18.

Smith, F. G. (1958). *Bee World.* 29:29-36.

Souza, I.H., Andrade, E.M. and Bezerra, F.M.L. 2004. Initial growth of early dwarf cashew plants under different soil water tensions. *Proc. of the Interamerican Soc. for Tropical Hort.* 47:157-160. (Escola Agrotecnica Federal de Iguatu, CEP: 63.500-000, Iguatu, CE, Brazil).

Sreenath, K. H. and Jeffries, T. W. 1996. A variable tilt fermentation rack for screening organisms in micro tube. *Biotechnol. Tech.* 24: 239-242.

Sridar, R., Shah, H.A. and Manivannan, K. 1990. The effect of VA mycorrhizae inoculation on certain growth characters of cashew. *Cashew Bull.* 27:15-16.

Sriram T. A. 1970. Cashew Nut. *Indian Farming.* 20: pp 33-35. In: Ashok, T. H. *et al.*, 1983. Effects of certain growth regulators on the sex expression and its relation to fruit set in cashew (*Anacardium occidentale* L.). *Cashew Causerie.* 2(3):8-10.

Srihari Babu. 1981. Identifying superior mother trees for propagation of cashew in Andhra Pradesh. *Cashew Causerie.* 3(2):15-18.

Srivastava, A. K. and Singh, S. 2003. Foliar fertilization in citrus – A Review. *Agri. Rev.* 24:250-264.

Srurtz, J. D. 1985. Cashew-*Anacardium occidentale* L. In: Page, P.E. (ed.) *Tropical Tree Fruits for Australia.* Queensland Department of Primary Industries, Brisbane. pp. 20-24.

Stassen, P.J.C., Grove, H.G. and Davie, S.J. 1999. Tree shaping strategies for higher density mango orchards. *J. Appl. Hort.* 1(1):1-4.

Subbaiah, C. C. 1984. A polyethylene glycol based medium for in vitro germination of cashew pollen. *Canadian J. Botany.* 62:2473-2475.

Subbaiah, C. C., Manikandan, P. and Joshi, Y. 1986. Yellow leaf spot of cashew: a case of molybdenum deficiency. *Pl. and Soil.* 94:35-42.

Sumrit, Feungehan, Taweekiat Yimsawat and Sorot Chindaprasert 1989. Physiological development of cashew nut on canopy flower and fruit. *Khon Kaen Agricultural J.* 17(4):239-253.

Sundararaju, D. 1993b. Compilation of recently recorded and some new pests of cashew in India. *The Cashew.* 7:15-19.

Sundararaju, D. 1998. 'Screening of cashew accessions to tea mosquito bug (*Helopeltis antonii* L.) Sign. '*The Cashew*'.

Sundararaju, D. and Bakthavatsalam, N. 1990. Cashew pest management for coastal Karnataka. *The Cashew*. 4:3-6.

Swamy, K.R.M. 1995. Top working of cashew in Goa and Maharashtra States - a case study. *The Cashew*. 9:12-17.

Swamy, K.R.M., Bhaskara Rao, E.V.V. and Bhat, M. G 1997. *Catalogue of Minimum Descriptors of Cashew* (*Anacardium occidentale* L.) *Germplasm Accessions-I*, National Research Centre for Cashew, Puttur, Karnataka, pp. 41.

Swamy, K.R.M., Bhaskara Rao, E.V.V. and Bhat, M. G., 1998. *Catalogue of Minimum Descriptors of Casheto (Anacardium occidentale L.) Germplasm Accession-II*, National Research Centre for Cashew, Puttur, Karnataka, India. pp. 54.

Swamy, K.R.M., Bhaskara Rao, E. V. V. and Bhat, M. G 2000. *Catalogue of Minimum Descriptors of Cashew* (*Anacardium occidentate* L.) *Germplasm Accessions-III*, National Research Centre for Cashew, Puttur, Karnataka, India. pp. 54.

Swietlik, D. and Faust, M. 1984. Foliar nutrition of fruit crops. *Hort. Rev*. 9:319-355.

Sy, M. O., Martinelli, L. and Scienza, A. 1991. In *vitro* organogenesis and regeneration in cashew (*Anacardium occidentale* L.). *Acta Hort*. 289: 267-268.

Tagliavini, M. and Marangoni, B. 2002. Major nutritional issues in deciduous fruit orchards of north Italy. *Hort. Tech*. 12:26-31.

Tariq, M., Sharif M., Shah Z. and R. Khan. 2007. Effect of foliar application of micronutrients on the yield and quality of sweet orange (Citrus sinensis L.). *Pak. J. Biol. Sci*. 10(11): 1823-1828.

Thamburaj, S., Jaypal, R. and Pillai, O. A. A. 1980. Studies on growth and development of fruit in cashew (*Anacardium occidentale* L.). *South Indian Hort*. 28(1):26-27.

Taiz, L. and Zeiger, E. 2002. Plant Physiology. 2nd ed. Sinauer Associates, Massachusetts.

Thimmappaiah and Samuel S. R. 1999. *In vitro* regeneration of cashew (*Anacardium occidentale* L.). *Ind. J. Exptl. Biol*. 37(4):384-390.

Thimmappaiah and Shirly, R. A. 2000. Micropropagation in mature tree of cashew (*Anacardium occidentale* L.) In: *Recent Advances in Plantation Crops* (Eds: N. Muralidharan and R. Raj Kumar), Allied publishers Ltd., New Delhi. pp. 45-48.

Thimmappaiah. 1997. *'In vitro* studies in cashew (*Anacardium occidentale* L.)' *Ph.D. Thesis*, Mangalore University, India. pp. 264.

Thimmappaiah., Melwyn, G S., Shobha, D. and Shirly, R. A. 2009b. Assessment of cashew species for molecular diversity. *J. Plant. Crops*. 37:146-151.

Thimmappaiah, Puthra, G.T. and Shirly, Raichal Anil 2002c. *In vitro* grafting of cashew (*Anacardium occidentale* L). *Sci. Hort*. 92:177-182.

Thimmappaiah, Santhosh, W. G., Shobha, D. and Melwyn, G S. 2009a. Assessment of genetic diversity in cashew germplasm using RAPD and ISSR markers. *Sci. Hort*. 120:411-417.

Thimmappaiah, Shirly, R.A. and Sadhana P.H. 2002b. *In vitro* propagation of cashew (*Anacardium occidentale* L.) from shoot explants of young trees. *In Vitro Cell. Dev. Biol.Plant*. 38:152-156.

Thomas, R. J. 1995. Studies on *in vitro* propagation of cashew (*Anacardium occidentale* L.). *Ph. D. dissertation*, Tamil Nadu Agricultural University, Coimbatore.

Totten, F. W., Liu, H., McCarty, L. B., Baldwin, C. M., Bielenberg, D. G and Toler, J. E. 2008. Efficiency of foliar versus granular fertilization: a field study of creeping bentgrass performance. *J. Plant Nutrition*. 31(5):972-982.

Tsakiris, A. and Northwood, P. J. 1967. Cashew nut production in Southern Tanzania - IV: The root system of cashew nut tree. *East Afr. Agric. Forestry J*. 33(1):83-87.

Turkey, H. B. and Marczynski, S. 1984. Foliar nutrition old ideas rediscovered. *Acta Hort*. 145: 205-212.

Tyman, J. H. P. 1978. Long-chain phenols. XIII. Quantitative analysis of the phenolic omposition of natural cashew nut-shell liquid (*Anacardium occidentale*) by thin-layer chromatography, densitometry and ultraviolet spectrophotometry. *J. Chromatography.* 166:159-72.

Tyman, J. H. P. 1980. Cultivation, processing and utilisation of the cashew. *Chemistry and Industry.* 59-62.

Vaidehi, M. P. 1994. By-product utilization of cashew with specific reference to economic exploitation of cashew apple for the benefits of rural woman folk. *Proc. 5th National Seminar on "Cashew Development in 8th Plan".* 15-16 April 1994, Panaji, Goa. Directorate of cashew and cocoa development, Cochin. pp. 47-6.

Vaidehi, M. P. 2000. Commercial Exploitation of Cashew Apple. Singh, H. P., Balsubramanian, P. P. and Huballi, V. N. (Eds.) *Indian Cashew Industry*, Directorate of Cashew and Cocoa Development. Kochi-11. pp. 166-173.

Valeriano, C. 1972. Cajuoiro. *Boletim do Instituto Biologi CO de Bahia (Brazil)* 11(1):19-58.

Van Eijnatten, C. L. M. 1991. *Anacardium occidentale* L. *In: Plant Resources of South –East Asia, No.2. Edible Fruits and Nuts (Eds)* Verheij, E.W.M. and Coronel, R.E. PudocDLO, Wageningen, Netherlands. pp. 446.

Van Eijnatten, C. L. M. 1992. *Anacardium occidentale* L. In: Verheij, E.W.M. and Coronel, R.E. (eds*) Plant Resources of South-East Asia No. 2*: Edible Fruits and Nuts. Pudoc-DLO, Wageningen, Netherlands. pp. 60-64.

Varma S. K. 2000. Nutritional Values of Cashewnuts. Singh, H. P., Balsubramanian, P. P. and Huballi, V. N. (Eds.) *Indian Cashew Industry*, Directorate of Cashew and Cocoa Development. Kochi-11, pp. 124-129.

Veeraraghavan, P. G., Celin, V. A. and Balakrishnan, S. 1985. Studies on the fertilizer requirement of cashew (*Anacardium occidentale* L.). *Cashew Causerie.* 7(2):6-8.

Venkatesh, M. 2003. Fruit waste utilization for pectinase production through solid state fermentation. *M.Sc. Thesis*. Kerala Agriculture University, Vellanikkara, India. pp. 99.

Vidyadharan, K. K. and Peethambaran, C. K. 1979. Root studies in cashew air layers and seedlings, *In: Cashew Research and Deuelopment, Proceedings of the International Symposium on Cashew.* (Eds.) Bhaskara Rao, E.V.V and Hameedkhan, H., Cochin, Indian Society for Plantation Crops, Kasaragod, India. pp. 296.

Vieira, M.R., Lacerda, C. F., Cândido, M.J.D., Carvalho. P.L., Costa, R.N.T. and Tabosa, J. N. 2005. Produtividade e qualidade da forragem de sorgo irrigado com águas salinas. *Rev. Bras. Eng. Agric. Amb.* 9:42-46.

Vishnuvardhana, G. T., Thirumalanju, Y., Raghuramulu, N., Janakiraman, V., Shankaranarayma and Shivappa. 2002. Influence of different levels of nitrogen, phosphorus and potassium on growth and yield of cashew under eastern dry zone of Karnataka. *The Cashew.* 16:39-42.

Wahid, P. A., Kamalam, N. V., Ashokan, P. K and Vidyadharan, K. K. 1989. Root activity pattern of cashew (*Anacardium occidentale* L.) in laterite soil. *J. Plantation Crops.* 17:85-89.

Wait, A. J. and Jamieson, G. I. 1986. The Cashew: its botany and cultivation. *Queensland Agr. J.* 112:253-257.

Wang, F., Robotham, J.M., Teuber, S.S., Tawde, P., Sathe, S. K. and Roux, K.H. 2002. A cashew (*Anacardium occidentale*) allergen of the vicilin seed storage protein family. *J. Aller. Clinic. Immunol.* 110:160-166.

Weinbaum, S. A. 1988. *Foliar nutrition of fruit trees*. In: P.M. Neumann (ed.), Plant growth and leaf applied chemicals, CRC Press, Boca Raton, Florida. pp. 81-100.

Weinbaum, S. A., Brown, P. H. and Johnson, R. S. 2002. Application of selected macronutrients (N, K) in deciduous orchards: physiological and agro technical perspectives. *Acta Hort.* 594:59-64.

Weinbaum, S. A., Merwin, M. L. and Muroaka, T. T. 1978. Seasonal variation in nitrate uptake efficiency and distribution of absorbed nitrogen in non-bearing prune trees. *J. Amer. Soc. Hort. Sci.* 103:516-519.

Wilson, R. J. 1975. Marker for cashew kernel and cashewnut shell liquid. *G(VK) No. 91*. TPI (London). pp. 128.

Wunnchit, W. and Sedgley, M. 1992. Floral Structure and Phenology of Cashew in Relation to Yield. *J. Hort. Sci. & Biotech.* 67(6):769-778.

Wittwer, S. H. 1963. Foliar application of fertilizer. Paper presented at fertilizer symposium for farm edition, Sept. 30, 1963, Musle Shoals, Alabana in: "New Horizons in the Fertilizer World". pp. 36-50.

Wittwer, S. H. and Teubner, F. G. 1959. Foliar absorption of mineral nutrients. *Annual Pl. Physiol. Rev.* 10:13-32.

Wojcik, P. 2004. Uptake of mineral nutrients from foliar fertilization. *Fruit Ornam. J. Plant Res.* 12:201-218.

Wolcott, G. N. 1944. How to make wood unpalatable to the West Indian drywood termite *Cryptotermes brevis* Walker with organic compounds. *Caribean Forester*, 5:171-80. In: Behrens, R. 1996. Cashew as an agroforestry crop: prospects and potentials. *Trop. Agri.* 9:83.

Wolfgang, F. 1986. The basis of foliar absorption of fertilizers with special regard to the mechanisms. In: *Foliar Fertilization*, Edited by A. Alexander, Martinus Nijhoff Publisher Dordrecht. pp. 17-25.

Ximenes, C.H.M. 1995. Adubação mineral de mudas de cajueiro anão precoce cultivadas em diferentes substratos. Fortaleza, Universidade Federal do Ceará. *M.Sc. dissertation.*

Yadukumar, N. 2007. Organic farming in cashew. *Proceedings of the National Seminar on Research, Development and Marketing of Cashew*, 20-21st November, Goa. pp. 36-46.

Yadukumar, N. and Mandal, R. C. 1994. Effect of supplementary irrigation on cashewnut yield. Water Management for Plantation Crops - Problems and Prospects. Centre for Water Resource Development and Management, Calicut, Kerala. pp. 79-84

Yadukumar, N., Nagaraja, B. and Dixit, S. 1994. Pineapple as an intercrop with cashew. *SAIC News Letter*. 4(2):6.

Yadukumar, N. and Nandan, S.L. 2005. Recycling organic wastes of cashew plantations by aerobic composting. *J. Plantation Crops*. 33:99-102.

Yadukumar, N. and Nayak, M. G. 2000. Canopy Management and Utilization of Interspace in Cashew. (Eds.) Singh, H. P., Balsubramanian, P. P. and Huballi, V. N., *Indian Cashew Industry*, Directorate of Cashew and Cocoa Development. Kochi-11. pp. 101-103.

Yadukumar, N. and Rejani, R. 2008. Soil and water management in cashew plantations. *Technical Bulletin No. 10.* NRC-Cashew, Puttur, Darbe (PO), D.K., Karnataka. pp 20.

Yadukumar, N., Rao, E.V.V.B. and Mohan, E. 2001. High density planting of cashew. *Trop. Agri.* 78:19-28.

Yadukumar, N., Rao, E.V.V.B. and Nayak, M. G. 2002. High density planting of cashew. *Technical Bulletin No. 8*, NRC-Cashew, Puttur, Darbe (PO), D.K., Karnataka.

Yadukumar, N., Rejani, R., Prabhakar, B., 2009. Fertigation for efficient water and nutrient management in high density cashew plantation. *J. Plantation Crops*. 37(2):102-110.

Yadukumar, N., Raviprasad, T.N., Nagaraja, K.V., Haldankar, P.M., Godase, S.K., Susanamma, K., Gajendran, G., Mahalingam, T., Lenka, P.C., Mohapatra, R.N. and Bandyopadhyay, B. 2003. National Agricultural Technology Project, *Final Report on developing integrated production packages for enhancing productivity of cashew*. National Research Centre for Cashew, Puttur, D.K., Karnataka. pp 95.

Yadukumar, N., Rejani, R. and Nandan, S.L. 2008. Studies on green manuring in high density cashew orchards. *J. Plantation Crops*. 36:265-269.

Yamada, Y., Jyung, W, H., Wittwer, S. H. and Bukovac, M. J. 1965. Effects of urea on ion penetration through isolated cuticular membranes and ion uptake by leaf cells. *Proc. Amer. Soc. Hort. Sci.* 87:429-432.

Yaseen, M., Nadeem, M. and Hussain, S. 2004. Investigating the effectiveness of micro-power foliar spray on growth and yield of different crops. *Pak. J. Life Soc. Sci.* 2(2):156-158.

Yu, J., Zhang, X. and Tan, T. 2007. A novel immobilization method of *Saccharomyces cerevisiae* to sorghum baggase for ethanol production. *J. Biotechnol.*129: 415-420.

Zhang, Q. and Brown, P. 1999. The mechanism of foliar zinc absorption in pistachio and walnut. *J. Amer. Soc. Hort. Sci.* 124: 312-317.

Zhu, J. K. 2001. Plant salt tolerance. *Trends Plant Sci.* 6:66-71.

□□□□